AVIONIC SYSTEMS

OPERATION & MAINTENANCE

JAMES W. WASSON

JEPPESEN.
Sanderson Training Products

JS312667B

About the Author

James W. Wasson was born December 9, 1951 in Pittsburgh, Pennsylvania. He is the son of Dolores H. Wuerl and George F. Wasson, a Mechanical Engineer by trade and retired executive from U.S. Steel Corporation, and has two brothers, Daniel, a Computer Programmer, and Gerard, an Attorney.

Mr. Wasson first became interested in electronics while in grade school. It was then that he learned on his own how to design and build power supplies and audio systems. It was during high school that he took electronic courses at the University of Pittsburgh, worked as a retail salesman and technician at Olson Electronics and Radio Shack, and earned his Federal Communications Commission (FCC) Commercial Radio-Telephone and Amateur Radio licenses. As a Ham Radio Operator, he organized the South Hills Catholic High School Amateur Radio Club and often designed and built his own telecommunications equipment.

Following high school, Jim Wasson attended Pittsburgh Institute of Aeronautics and earned an Associate in Specialized Technology (AST) degree and a Federal Aviation Administration (FAA) Airframe and Powerplant (A&P) Mechanics license. Following his graduation in November 1972, Mr. Wasson landed a job with Antilles Air Boats, a seaplane airline in the U.S. Virgin Islands, as an A&P mechanic. Shortly after hiring in, Jim realized that there was an acute shortage of qualified aircraft electronics (avionics) technicians. In fact, based on his previous experience as a Ham Radio Operator, he was the only employee at Antilles that knew how to troubleshoot and repair aircraft radio systems. It was at this point that he decided to pursue a career in avionics.

Naturally, Jim's first thought was to have a formal education in avionic systems operation and maintenance. Hence, he enrolled in the twelve-week avionics curriculum at Northrop Institute of Technology in Los Angeles, California. It was at this time, he also took private pilot ground school and flight school training at Long Beach Community College and at Hawthorne Airport, respectively.

Jim stayed at Northrop University and graduated Magna Cum Laude in June 1981 with a Bachelor of Science degree in Engineering Technology (BSET). While attending Northrop, now Rice Aviation School, he was employed full-time at Catalina Airlines, K.C. Aircraft, Great Western Aircraft Radio, and finally Garrett AiResearch Aviation as an avionics technician. In 1978, he was promoted to the position of Avionics Design Engineer at Garrett, where he was responsible for the design, development, integration, and flight test of avionic systems on the Lockheed Jetstar and Hawker-Siddley HS-125 corporate jet aircraft retrofit programs. He also designed and developed production avionics equipment, such as the Garrett AiResearch Model 149 Annunciator Coupler Unit, which provided signal conversion and lamp driver dimming and test functions for the navigation/flight director annunciator panel.

During 1980, Jim was President of the Xi Beta Chapter of Tau Alpha Pi, the National Engineering Technology Honor Society, and was named "Student Engineer of the Year " at Northrop University. In 1981, he was honored at Pittsburgh Institute of Aeronautics (PIA) with the "Distinguished Alumnus Award " and was named advisor to the President of PIA. Also in 1981, he was asked to join the Northrop University Industry Advisory Committee to develop a one-year avionics technician curriculum, which is the genesis of this book.

Jim Wasson Displaying His Annunciator Coupler Unit Design. *(Courtesy Allied Signal Aerospace Company, 1980)*

In April of 1981, Mr. Wasson accepted a position with Northrop Aircraft Division as Senior Technical Specialist, in which he was responsible for leading a team of engineers and scientists in the design and development of military avionic systems for the U.S. Air Force's Advanced Tactical Fighter YF-23, the U.S. Navy's F/A-18 fighter/attack aircraft, and the export F-20 fighter. In January 1984, Jim accepted a position as Senior Program Manager for Advanced Avionic Systems at McDonnell Douglas Corporation, the world's largest military aerospace contractor, where he is currently employed managing multi-million dollar avionics research and development programs.

Jim attended the University of Phoenix in Arizona and earned a Master of Business Administration (MBA) degree in March of 1989. He was the first to receive a Business Research Award for his Master's Thesis entitled "Business Opportunities in Artificial Intelligence. " (Artificial Intelligence is an advanced method of computer programming whereby a computer can "reason " based on a knowledge base built from subject matter expertise and a network of rules.)

Mr. Wasson is currently an Adjunct Professor of Business Management at the University of Phoenix and serves on the Board of Directors of two corporations. He was co-founder of Avionics Engineering Services, Inc. and Leading Edge Technologies, Inc., a producer of innovative consumer electronic systems. He served as National Chairman of the American Helicopter Society Avionics Committee, and was Senior Vice President of the Arizona Chapter of the Army Aviation Association of America. Jim is also a member of the Association for Avionics Education, the National Society of Professional Engineers, the American Defense Preparedness Association, and the Institute of Electrical and Electronic Engineers. He has authored numerous technical papers and appears in Who's Who in America and the International Who's Who. In May of 1991, he was inducted into the Career Colleges Association Hall of Fame.

Jim is married to Evelyn Fay Gonzales from Kauai, Hawaii. The Wassons reside in Mesa, Arizona with their sons, Robert and Brian, and spend their summer weekends at their ranch resort in the beautiful White Mountains of Northeastern Arizona.

[illegible]

[illegible]

Preface

In the past decade, new developments in electronics and computer technology have revolutionized the aircraft electronics (avionics) industry. Applications of high-speed microprocessors, large memories, and data distribution networks have caused a transformation from conventional analog communication, navigation, radar, and flight control systems to digital avionic systems that are less costly, smaller in size and weight, consume less power, and are more reliable and easier to maintain than their predecessors.

For example, the highly advanced, avionic system architectures currently being employed on our modern military fighter aircraft, such as the F/A-18 and F-16, use high-speed, digital computers that form a distributed data processing network. Within this network, the various avionic subsystems communicate with the central computer via messages sent on a multiplex digital data bus. In place of electromechanical instrumentation, cathode-ray tube (CRT) displays are used to present the pilot with a multitude of navigational, flight management, and weapon system information.

Civil aircraft are also integrating their avionic systems into a distributed processing network as seen in the new generation of commercial airliners, such as the McDonnell Douglas MD-11 and the Boeing 757/767. One of the significant achievements in digital avionic system integration is the flight management system, which stores flight plans, indicates flight progress, issues control steering commands to the flight director and autopilot, and monitors the aircraft performance for optimum safety and efficiency throughout the entire regime of the flight.

The flight deck of the McDonnell Douglas MD-11, shown on the front cover (courtesy Honeywell Commercial Flight Systems), has two Control Display Units (CDUs), located on the center console, which are actually computer terminals used to enter and receive data to and from the flight management computer. Video CRT screens on the CDUs display flight performance alphanumerically, while the Electronic Flight Instrumentation System (EFIS) projects a color graphic presentation of flight information as shown in Figure P-1. Even the conventional electromechanical engine instruments have been replaced with color CRT displays to bring a new level of cockpit sophistication to help ease the crew workload.

Figure P-1. Rockwell Collins Proline 4 Primary Flight Display.
(Courtesy Rockwell Collins Avionics)

With these advances in electronics technology, a resultant reduction in ownership cost has allowed modern private jet aircraft to also use integrated digital avionic systems. Most of these aircraft are installing EFIS, while some have elected to use the digital color weather radar indicator as a multifunction display (MFD) to present a variety of navigation operational/emergency checklist, and diagnostic formats, in addition to weather/turbulence avoidance and ground mapping functions.

The Beechcraft Starship is a typical example of a long-range corporate aircraft incorporating a modern avionics suite. As shown in Figure P-2, the crewstation of the Starship has one MFD located on the top center instrument panel for the display of the Engine Indication and Crew Alerting System (EICAS), which centralizes and prioritizes engine and caution status.

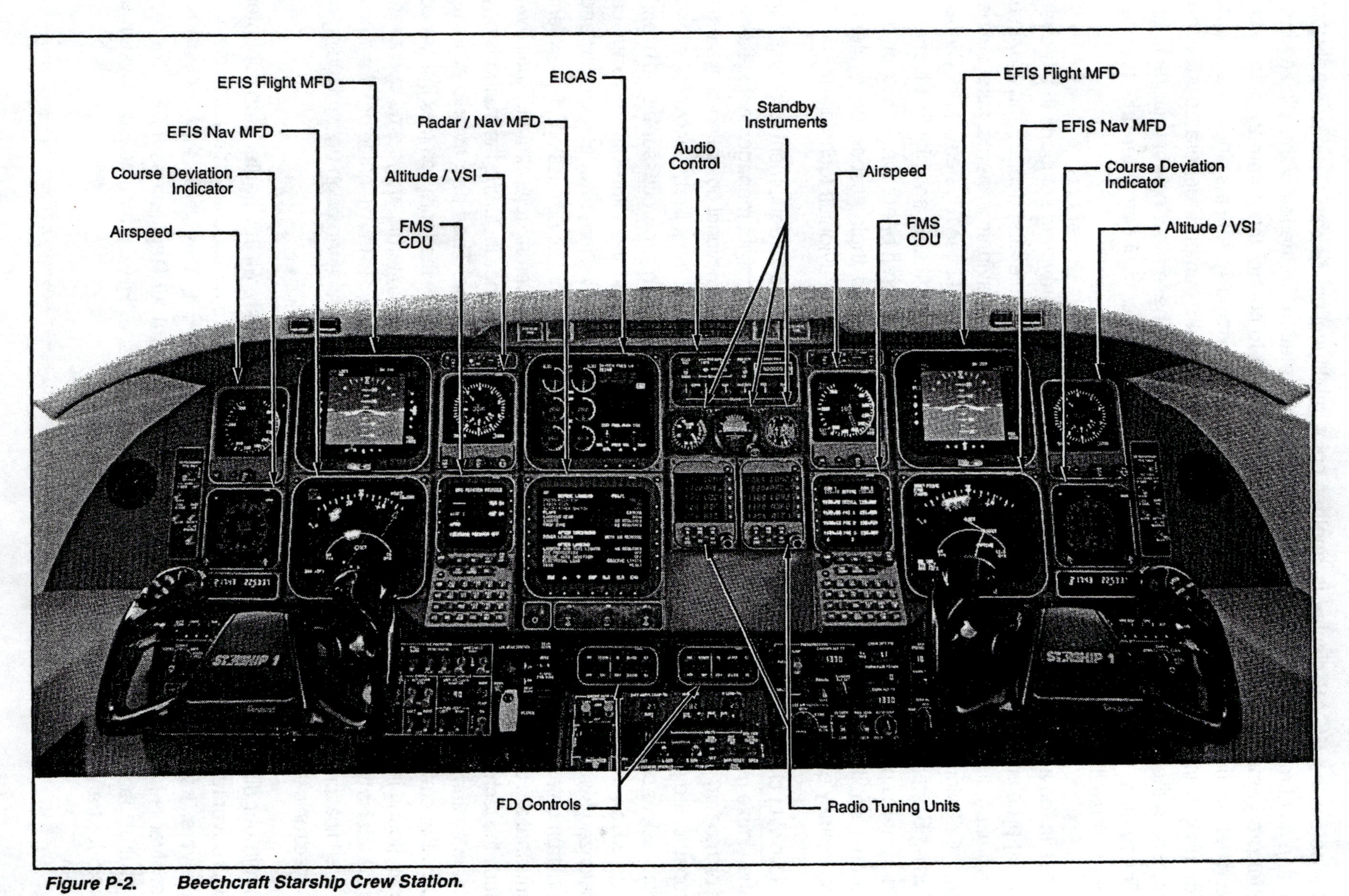

Figure P-2. Beechcraft Starship Crew Station.

Directly below the EICAS display is another color multifunction display used primarily for presenting flight management, navigation and radar information. To the left of this display is a CDU, for programming the flight management system and to control the communication and navigational functions of the avionic system. The Starship's avionic system also includes a flight director (FD) system to present the pilot and/or copilot with computergenerated steering commands on the EFIS MFDs. The pilot may elect to couple these steering commands to an autopilot system, whereby, a command signal from the flight control computer drives servo motors to cause the proper control surface deflection on the aircraft.

Advances in the application of solid-state digital electronics are providing solutions to the ever increasing problems of navigation and air traffic control as available airspace around terminal areas is limited due to the increase in air traffic. Newly developed avionic systems, such as the Terminal Collision Avoidance System (TCAS), Windshear Detection System, and Global Positioning System (GPS) are just being introduced. These sophisticated and highly advanced avionic systems provide much greater efficiency and safety in aircraft operation. However, the continued reliability and performance of avionic equipment and their associated systems depend entirely on the skills of the highly-trained professional, the avionics technician.

In the following chapters, we will explore the entire spectrum of avionic equipment and systems from the simple magnetic compass to the most advanced integrated flight management systems found on modern jet aircraft. Each chapter attempts to illustrate the functional operation of systems developed in the past 10 to 15 years that the avionics technician will most likely encounter, along with discussions concerning the operation of more recently developed microprocessor-based hybrid systems which the student will encounter in increasing numbers in the future as they replace transistorized systems of the past. The basic concepts of communication, navigation, radar and flight instrumentation will be presented, followed by a detailed technical discussion of prevalent systems that are currently being employed in private, corporate and commercial aircraft. Finally, we will study the techniques of avionic systems operation and maintenance including the methodology of avionic systems operational analysis and performance validation. The basic avionic systems found in all aircraft are discussed in the context of circuit theory, while the more complex integrated systems (found in larger jet transport aircraft) are discussed more in terms of systems operation, since the later is typically a matter of knowing which black box to return to the manufacturer for repair and recertification. The contents of this text assume the student has a thorough understanding of basic electronic theory.

We begin our discussion with a review of aircraft electrical systems in Chapter One. Chapter Two will cover radio communication theory as a primer for more detailed technical discussions of receivers, transmitters, and audio systems used in aircraft. The VHF Communications Transceiver is heavily relied upon, and is required by Federal Air Regulations (FARs) for any aircraft that passes through an airport's controlled airspace. HF Communication Systems, required for transoceanic flights, will also be discussed.

The next four chapters deal with various types of navigation equipment, which is used for determining the aircraft's location and for plotting a course to a desired destination. Automatic Direction Finders (ADF) are installed in aircraft to aid in flying in and out of small airports not equipped with VHF Omnirange Navigation (VOR) stations. VOR Navigation Receivers allow aircraft to fly any desired course radial to or from a VOR ground station. ADF and VOR systems are discussed in Chapters Three and Four, respectively.

Instrument Landing Systems (ILS), presented in Chapter Five, are used for take-offs and landings from larger airports, even in near zero visibility. The ILS consists of a Localizer (LOC), which provides an indication of whether the aircraft is left or right of the runway; a Glideslope (GS), which provides an indication of the desired angle of descent; and a Marker Beacon (MB), which marks the distance from the runway along the glidepath at two or three distinct intervals.

Chapter Six contains a discussion of Distance Measuring Equipment (DME), which provides the pilot with an indication of the distance to or from a ground station. DMEs are required in aircraft with gross weight of 10,000 pounds or more. Also in compliance with the FARs, aircraft flying through Terminal Control Areas (TCA) must be equipped with a Radio Beacon Transponder. The Transponder, discussed in Chapter Seven, assists the Air Traffic Controller in identifying aircraft within the TCA on his/her radar screen.

Weather is a threat to any aircraft flying any appreciable distance. Smaller planes rely on tuning in the Air Terminal Information Service (ATIS) on the VHF Communications Receiver to hear local weather advisories. Larger aircraft have onboard Weather Radar Systems to allow the pilot to reroute his or her flight path around storm activity. Weather Radar Systems are discussed in Chapter Eight.

Chapter Nine presents a different kind of onboard radar, the radio altimeter, which is used to measure the actual distance the aircraft is from the terrain when flying at low altitudes. The radio altimeter is much more accurate than the barometric altimeter, which measures height above sea level at a given barometric pressure.

This leads into a discussion on Flight Instrumentation. Chapter Ten describes pitot-static systems, which drive basic pneumatic instruments, such as the airspeed indicator and the altimeter. It also covers air data computers, which convert air pressure from the pitot-static system to electrical signals to drive a variety of instruments in larger aircraft. Altimeters, airspeed indicators, machmeters, vertical speed indicators, and other instruments will be examined in detail in this chapter.

This book concludes with an introduction to some of the more sophisticated avionic systems found primarily in larger transport and business jet aircraft. Chapter Twelve, titled Flight Management and Display/Control Systems, also discusses EFIS. However, to fully understand Flight Management Systems (FMS), one must be familiar with the principles of operation of Inertial Navigation Systems (INS) and other long-range navigation systems. These systems will be discussed in the context of FMS operation in Chapter Eleven.

The author wishes to thank Allied-Signal Aerospace Company, Baker Electronics, Boeing Commercial Aircraft Company, the Federal Aviation Administration, Honeywell Commercial Flight Systems Group, ITT Avionics, Kollsman Division of Sequa Corporation, Litton Aero Products, Magnavox, McDonnell Douglas Aerospace Corporation, Rockwell Collins Avionics, and S-TEC Corporation for their contributions toward making possible the publication of this book. We sincerely hope that those who study the following pages will become proficient in the techniques of avionic systems operation and maintenance and will join us in promoting safer and more efficient air travel.

Contents

About the Author iii

Preface vii

Chapter 1 Aircraft Electrical System 1

Introduction 1
Light Aircraft Electrical System Operation 2
Heavy Aircraft AC Electrical Circuit Theory 4
Variable-Speed, Constant-Frequency (VSCF) Power Systems 6
Boeing 747 Electrical System Operation 7
Conclusion 10
Review Questions 11

Chapter 2 Audio and Communication Systems 13

Introduction 13
Radio Frequency Bands 14
Radio Communication Theory 16
Aircraft Audio Systems 17
Radio Transmitter Principles 21
Radio Receiver Principles 26
VHF Communication Systems 29
Rockwell Collins VHF-20A VHF Communications Transceiver Operation 31
Bendix/King KX-170A/KX-175 VHF Communications Transceiver Operation 34
HF Communication Systems 36
SELCAL Decoder 40
Satellite Communications (SATCOM) 41
Emergency Locator Transmitters (ELT) 48
Conclusion 49
VHF Transceiver Laboratory Project 50
Review Questions 53

Chapter 3 Automatic Direction Finders 57

Introduction 57
Principles of ADF System Operation 57
ADF Antenna Theory 58
ADF Circuit Theory 59

Contents (Continued)

Bendix/King KR-86 ADF System Operation 61
S-TEC ADF-650 System Operation 63
UHF ADF Systems 66
Installation Techniques 67
Performance Validation 69
Operating Procedures 70
ADF Receiver Laboratory Project 76
Review Questions 78

Chapter 4 VHF OmniRange Navigation 79

Introduction 79
VOR Navigation Concepts 81
Principles of VOR Operation 81
VOR Circuit Theory 84
S-TEC VIR-351 Navigation Receiver Operation 84
Bendix/King KX-170A/175 Navigation Receiver Operation 91
Performance Validation 91
Operating Procedures 93
Conclusion 97
Review Questions 98

Chapter 5 Instrument Landing Systems 99

Introduction 99
Principles of Localizer (LOC) Operation 99
Localizer Circuit Theory 100
Principles of Glideslope (GS) Operation 102
Principles of Marker Beacon (MB) Operation 104
Glideslope Circuit Theory 104
Marker Beacon Circuit Theory 107
Operation of the Localizer Portion of the S-TEC VIR-351 Navigation Receiver 108
S-TEC GLS-350 Glideslope Receiver Operation 109
S-TEC MKR-350 Marker Beacon Receiver Operation 112
Rockwell Collins VIR-30 VOR/ILS Navigation Receiver Operation 113
Rockwell Collins VIR-432 Navigation Receiver 115
Automated Test Equipment 118
Microwave Landing Systems 118
Conclusion 119

Contents (Continued)

VOR/ILS Navigation Receiver Ramp Test Project....120
VOR/ILS Navigation Receiver Laboratory Project....122
Review Questions....124

Chapter 6 Distance Measuring Equipment....127

Introduction....127
DME Navigation Concepts....128
Principles of DME System Operation....130
S-TEC DME-451 System Description....132
Rockwell Collins 860E-5 DME Transceiver Operation....136
DME Navigation Procedures....140
Conclusion....142
Review Questions....143

Chapter 7 Radio Beacon Transponders....145

Introduction....145
Principles of ATC Radar Surveillance System Operation....145
Principles of Radio Beacon Transponder Operation....148
S-TEC TDR-950 Transponder Operation....151
S-TEC TDR-950 Transponder Circuit Theory....152
Bendix/King KT-76A/78A Transponder Operation....157
Traffic Alert and Collision Avoidance System (TCAS)....159
Rockwell Collins TCAS-II With Mode-S Transponder System....160
Conclusion....161
Review Questions....163

Chapter 8 Weather Radar Systems....165

Introduction....165
Weather Radar (WX) System Description....165
Analog Versus Digital Radar Systems....167
Weather Radar Theory....168
Principles of Weather Radar System Operation....170
Rockwell Collins WXR-300 Weather Radar System Operation....173
Rockwell Collins TWR-850 Turbulence Weather Radar System Operation....180
Installation Procedures....184
Passive Weather Detection Systems....186
Conclusion....187
Review Questions....188

Contents (Continued)

Chapter 9 Radar Altimeter Systems **189**

Introduction 189
Radio Altimeter (RAD ALT) System Description 189
Radar Altimeter Circuit Theory 192
Rockwell Collins 860-F4 Radar Altimeter RT Operation 193
Radar Altimeter Installation and Test Procedures 197
Conclusion 199
Review Questions 200

Chapter 10 Flight Instrumentation **203**

Introduction 203
Turn-and-Bank Indicator Operation 205
Angle-of-Attack System Operation 206
Introduction to Pitot-Static Systems 207
Altimeter Principles 210
RADBAR Encoding Altimeter System Operation 211
Altitude Alerter Operation 215
Airspeed Indicator Principles 216
Maximum-Allowable Airspeed/Mach Indicator Operation 218
TAS/SAT Indicator Operation 220
Vertical Speed (VS) Indicator Principles 223
Electric Vertical Speed Indicator Operation 225
TCAS Resolution Advisory/Vertical Speed Indicator Operation 226
Central Air Data Computer (CADC) Operation 226
Conclusion 229
Review Questions 230

Chapter 11 Long-Range Navigation Systems **231**

Introduction 231
The Gyrosyn Compass System 231
Inertial Navigation Systems (INS) 233
Strapdown Inertial Navigation Systems 236
Operation of the Honeywell AHZ-600 Attitude and Heading Reference System (AHRS) 238
Operation of the Rockwell Collins AHS-86 Attitude and Heading Reference System 242
Laser Inertial Navigation Systems 244
Long-Range Radio Navigation (LORAN) 248
Very Low Frequency (VLF)/Omega Radio Navigation 249

Contents (Continued)

Operation of the Litton LTN-311 VLF/Omega Navigation System 252
Global Positioning System (GPS) Navigation 256
Operation of the Litton LTN-2001 Global Positioning System Sensor Unit (GPSSU) 259
Integrated GPS/INS 260
Operation of the Litton GPS Integration Module 261
Conclusion 263
Review Questions 264

Chapter 12 Flight Management and Display/Control Systems **265**

Introduction 265
Automatic Flight Control System (AFCS) 265
Rockwell Collins APS-85/86 AFCS System Description 266
Flight Director Systems (FDS) 267
Electronic Flight Instrumentation Systems (EFIS) 268
Rockwell Collins EFIS-85 Functional Operation 279
Principles of Area Navigation (RNAV) 284
Bendix/King KLN-90 GPS RNAV Functional Operation 286
Flight Management Systems (FMS) 290
Honeywell FMZ-800 Flight Management System Operation 292
Conclusion 297
Review Questions 298

Appendix A Glossary of Terms **299**

Appendix B DME Channel/Frequency/Spacing Correlation **311**

Figures

Figure — Page

1-1 Typical Light Aircraft Electrical System Schematic ... 2
1-2 Combined Sine Waves of a 3-Phase System ... 4
1-3 Wye-Connected Brushless AC Generator ... 5
1-4 Typical Variable-Speed, Constant-Frequency Power System ... 6
1-5 Boeing 747 Electrical Power Distribution System Block Diagram ... 8
1-6 Aircraft Electrical Control Panel ... 10
2-1 Radio Wave Propagation Characteristics ... 15
2-2 Radio Communication System ... 16
2-3 M1035 Audio Control Unit ... 18
2-4 Schematic Diagram of a Typical Audio Control Unit ... 19
2-5 Radio Transmitter Block Diagram ... 22
2-6 AM and FM Waveforms ... 23
2-7 Phase-Lock-Loop Block Diagram ... 24
2-8 AM Sideband Generation ... 25
2-9 SSB Transceiver Block Diagram ... 26
2-10 Crystal Receiver Circuit Diagram ... 27
2-11 Superheterodyne Receiver Block Diagram ... 28
2-12 ARINC 429 Data Bus Message Formats ... 31
2-13 Collins VHF-20/21/22 Transceiver ... 32
2-14 Collins VHF-20A Block Diagram ... 32
2-15 Collins VHF-20A Synthesizer Block Diagram ... 33
2-16 Bendix/King KX-170/175 VHF Communications Transceiver Block Diagram ... 35
2-17 Rockwell Collins CTL-200 HF Control Head ... 37
2-18 Rockwell Collins HF-220 Signal Flow Diagram ... 38
2-19 Typical Single-Channel SELCAL System Block Diagram ... 42
2-20 SATCOM System Segments ... 43
2-21 Typical SATCOM (AES Segment) Components ... 44
2-22 Low Data Rate System ... 45
2-23 SATCOM Network Configuration ... 47
2-24 Emergency Locator Transmitter (ELT) ... 51
2-25 VHF Transceiver Bench Test Diagram ... 52
3-1 Loop Antenna Operation ... 59
3-2 Typical ADF System Block Diagram ... 59
3-3 Simplified ADF Block Diagram Using a Stationary Loop Antenna ... 60
3-4 Bendix/King KR-86 ADF System Operation ... 61

Figures (Continued)

Figure | Page

3-5 KR-86 ADF Block Diagram 62
3-6 ADF-650 System Components 64
3-7 S-TEC ADF-650 System Operation 65
3-8 Rockwell Collins DF-301 System Operation 67
3-9 S-TEC ADF-650 Interconnect Wiring Diagram 68
3-10 Interception of an Inbound Bearing 71
3-11 Tracking 72
3-12 Determining a Fix 74
3-13 ADF Receiver Bench Test Diagram 76
4-1 VOR Station Shown on Aeronautical Chart 79
4-2 VOR Radials and Magnetic Courses 80
4-3 VOR Phase Relationships 82
4-4 Phase Relationships of a 260 Degree VOR Radial 83
4-5 Flying a VOR Radial 84
4-6 S-TEC VIR-351 VOR Converter Block Diagram 85
4-7 IND-350 Course Deviation Indicator 86
4-8 VIR-351 Navigation Receiver 87
4-9 S-TEC VIR-351 Navigation Receiver Block Diagram 87
4-10 VIR-351 Interconnect Wiring Diagram 90
4-11 Bendix/King KX-170A/175 Navigation Receiver Block Diagram 92
4-12 VOR Orientation 94
4-13 VOR Interception and Tracking 95
4-14 VOR Cross-Check 96
5-1 Flying a Localizer Course 101
5-2 Typical Localizer Instrumentation Block Diagram 102
5-3 Flying the Glidepath 102
5-4 Localizer and Glideslope Operation 103
5-5 Marker Beacon Station Locations 106
5-6 Typical Marker Beacon Lamp Circuit 108
5-7 S-TEC VIR-351 Localizer Converter Block Diagram 108
5-8 S-TEC GLS-350 Receiver Block Diagram 109
5-9 GLS-350 Glideslope Instrumentation Block Diagram 110
5-10 GLS-350 Interconnect Wiring Diagram 111
5-11 S-TEC MKR-350 Marker Beacon Block Diagram 112
5-12 S-TEC MKR-350 MB Receiver 113
5-13 Rockwell Collins VIR-30 Navigation Receiver Signal Interface Diagram 114
5-14 Typical Radio Magnetic Indicator 115

Figures (Continued)

Figure		Page
5-15	VIR-432 Navigation Receiver Block Diagram	116
5-16	VIR-432 ATE Display Screen	118
5-17	VIR-432 Test Setup Diagram	119
6-1	DME Slant-Range Distance	128
6-2	DME System Operation	129
6-3	Typical DME System Block Diagram	130
6-4	Pulse Sequence Process	131
6-5	S-TEC IND-151 DME Indicator	133
6-6	S-TEC DME-451 Transceiver Block Diagram	134
6-7	S-TEC DME-451 Receiver Block Diagram	134
6-8	S-TEC DME-451 Interconnect Wiring Diagram	135
6-9	Rockwell Collins 860E-5 DME Transceiver Block Diagram	137
6-10	DME Arc Procedure	141
7-1	ATC PSR/SSR System	146
7-2	Typical PPI Display	147
7-3	SSR Interrogation Modes	148
7-4	Propagation Pattern of SSR Interrogation Signal	149
7-5	Side-Lobe Detection and Reply Suppression	150
7-6	Transponder Reply Code Pulses	151
7-7	S-TEC TDR-950 Transponder	151
7-8	S-TEC TDR-950 Functional Block Diagram	153
7-9	Ditch Digger and Comparator Pulses	154
7-10	S-TEC TDR-950 Interconnect Wiring Diagram	156
7-11	Bendix/King KT-76A/-78A Block Diagram	158
7-12	Rockwell Collins TCAS-II System Block Diagram	160
8-1	Precipitation Presentation on Weather Radar CRT	166
8-2	Typical Digital Weather Radar System Simplified Block Diagram	168
8-3	Bendix/King RDR-1400C Color Radar Indicator	169
8-4	Radar Antenna Reflector Size Versus Beam Width	171
8-5	Typical Weather Radar System Expanded Block Diagram	172
8-6	Collins WXT-250A RT Mixer/Duplexer Block Diagram	174
8-7	Collins IND-300 Indicator	175
8-8	Collins IND-300 Scan Control Circuit Block Diagram	177
8-9	Collins ANT-310/312/318 Weather Radar Antenna System Block Diagram	179
8-10	Rockwell Collins TWR-850 Turbulence Weather Radar System	180
8-11	RTA-85X Receiver/Transmitter/Antenna/Major Subassemblies	181

Figures (Continued)

Figure		Page
8-12	WXP-850A and WXP-850B Weather Radar Panels, Front Panel View	182
8-13	Collins TWR-850 Turbulence Weather Radar System Block Diagram	183
8-14	Radar Waveguide Installation Drawing	184
9-1	Typical Radio Altimeter System	190
9-2	Radar Altimeter Indicators	191
9-3	Frequency Shifting Techiques for Determining Altitude	192
9-4	Radar Altimeter System Block Diagram	193
9-5	Rockwell Collins 860F-4 Radio Altimeter RT Block Diagram	194
9-6	AID Bias Chart	198
10-1	DC-10 First Officer's Instrument Panel	204
10-2	Turn-and-Bank Indicator	205
10-3	Angle-of-Attack System (A - Indicator, B - Transmitter)	206
10-4	Airstream Direction Detector	207
10-5	Pitot/Static System Head	208
10-6	Typical Pitot-Static System Diagram	209
10-7	Typical Pneumatic Altimeter Mechanism	211
10-8	RADBAR Altimeter System	212
10-9	RADBAR Altimeter	212
10-10	Altitude Alerter/Preselector	215
10-11	Typical Altimeter/Alerter Wiring Diagram	216
10-12	Pneumatic Airspeed Indicator Mechanism	218
10-13	Maximum-Allowable Airspeed/Mach Indicator	219
10-14	TAS/SAT Indicator	221
10-15	Kollsman TAS/SAT Indicator Functional Block Diagram	222
10-16	Pneumatic Vertical Speed Indicator Mechanism	224
10-17	Electric Vertical Speed Indicator	225
10-18	TCAS Resolution Advisory/Vertical Speed Indicator	226
10-19	Central Air Data Computer	228
11-1	Vertical and Directional Gyros	232
11-2	Accelerometer Operating Principles	233
11-3	The Integrators Convert Acceleration into Distance as a Function of Time	234
11-4	Accelerometer Error Due to Pitch Angle	235
11-5	Stabilized Platform	235
11-6	The Gyro Controls the Level of the Platform	236
11-7	Strapdown Mechanization	237

Figures (Continued)

Figure		Page
11-8	AHZ-600 System Components	238
11-9	AC-800 AHRS Controller	239
11-10	AHZ-600 AHRU Simplified Block Diagram	239
11-11	Rockwell Collins AHS-86 System Block Diagram	243
11-12	Typical Laser INS Interface Diagram	245
11-13	Laser Gyro Diagram	245
11-14	Readout Optics	248
11-15	Omega Transmission Format	249
11-16	Omega Frequency Relationships	251
11-17	LTN-311 Omega Navigation System	252
11-18	LTN-311 System Block Diagram	254
11-19	Navstar GPS Segment Operation	256
11-20	LTN-2001 Sensor Block Diagram	259
11-21	LTN-92/Data Base/GPS Integration Module Interface	262
12-1	Typical Autopilot System	266
12-2	Collins APS-85/86 AFCS System Block Diagram	267
12-3	Typical Flight Director System	268
12-4	Attitude Director Indicator	269
12-5	Horizontal Situation Indicator	269
12-6	Electronic Flight Instrumentation System Block Diagram	270
12-7	EFIS Display Controller	271
12-8	EFIS Source Controller	272
12-9	Electronic ADI Display	273
12-10	Electronic HSI Display	275
12-11	EHSI Composite Display Formats	278
12-12	Collins EFIS-85/86 System	280
12-13	Collins EFIS-85 ADI	281
12-14	Collins EFIS-85 Navigation Display	282
12-15	RNAV Waypoint	284
12-16	RNAV Distance and Crosstrack Deviation	285
12-17	Bendix/King KLN-90 System Components	286
12-18	KLN-90 Block Diagram	289
12-19	RNAV Mode Hierarchy	291
12-20	Honeywell FMZ-800 Flight Management System	293
12-21	Honeywell CD-800 (Mono)/810 (Color) Flight Management System Control Display Unit	294

Tables

Table		Page
2-1	RADIO FREQUENCY BANDS	14
2-2	2-OUT-OF-5 FREQUENCY SELECTION	30
5-1	CORRESPONDING LOCALIZER AND GLIDESLOPE FREQUENCIES	105
8-1	IND-300 COLOR COMBINATIONS	176
11-1	OMEGA LANE WIDTHS	252
12-1	DPU-85 DISPLAY PROCESSOR INPUTS	283
12-2	DPU-85 DISPLAY PROCESSOR OUTPUTS	284

Chapter 1
Aircraft Electrical Systems

INTRODUCTION

The aircraft electrical system provides electrical power to the onboard aircraft electrical and avionic equipment. The configuration of the electrical system and the type of power it provides is dependent on the size and category of the aircraft. The electrical power source for a light, single-engine aircraft typically consists of an engine-driven direct current (DC) generator or alternator. For purposes of discussion, a DC generator derives a DC voltage from the rotating armature through the use of a commutator and brushes; an alternator converts the alternating current (AC) output from the stator to a DC output through the use of rectifiers. The generator or alternator, whose output is controlled by a voltage regulator to approximately 14 volts, supplies power to the aircraft electrical distribution bus and, in addition, provides a charging current for the 12-volt storage battery. The storage battery also acts as a filter capacitor to smooth the ripple voltage from the generator.

The output of the generator must be in excess of the voltage supplied by the battery in order for the charging current to flow. If the generator output falls below the battery voltage, a reverse current relay (RCR) will disconnect the generator from the electrical bus to prevent the battery from discharging through the generator windings. The lead-acid or nickel-cadmium storage battery is intended only for starting the engine or for providing emergency power when the generator is inoperative. Normally, all load currents are supplied by the generator during flight.

Since larger aircraft have a much greater electrical load than smaller aircraft, they employ 28-volt instead of 14-volt systems. This higher voltage results in a lower current requirement for a given load. With less current required, the wire sizes used in the power distribution circuits can be appreciably smaller, which results in less weight, and thus greater payload capacity. In larger, multi-engine aircraft, DC generators are replaced with AC generators, which are more efficient and lighter in weight. These generators supply 115-volt power at 400 Hz (cycles per second) to the AC electrical buses. This high-frequency AC results in the requirement for less iron, and thus less weight, in transformer cores and armatures. This is due to the fact that inductive reactance increases directly with frequency. If the AC generator is driven directly from the engine shaft, the frequency will vary with the engine speed and the generator will output what is termed "wild" AC. Wild AC can be used for purely resistive loads, such as heating elements; however this frequency variation induces power factor problems when reactive (capacitive or inductive) loads are introduced. For this reason, a constant speed drive (CSD) is normally used to hydraulically drive the AC generator at a constant speed to supply a stable 400-Hz output. If a CSD is not used, the frequency of the variable-speed generator is controlled by a solid-state frequency converter.

Transformer-rectifier units (TRU) convert the 115-volt AC input to a 28-volt DC output which is fed to the DC electrical buses and to the battery charging circuits for the 24-volt storage batteries. The 115-volt AC power from the generator is also applied to step-down transformers to supply a 26-volt AC output for the reference voltage required by the various synchro circuits used in the flight instruments. In the event there is a loss of AC power, the batteries supply emergency DC power to the essential DC bus, and also to a standby inverter which supplies emergency AC power. Only equipment that is considered essential for safety of flight is provided power during an emergency power condition.

LIGHT AIRCRAFT ELECTRICAL SYSTEM OPERATION

A typical electrical system for a light, single-engine aircraft as shown in Figure 1-1, consists of a battery circuit, generator circuit, engine-start circuit, ammeter, control switches, and a bus bar

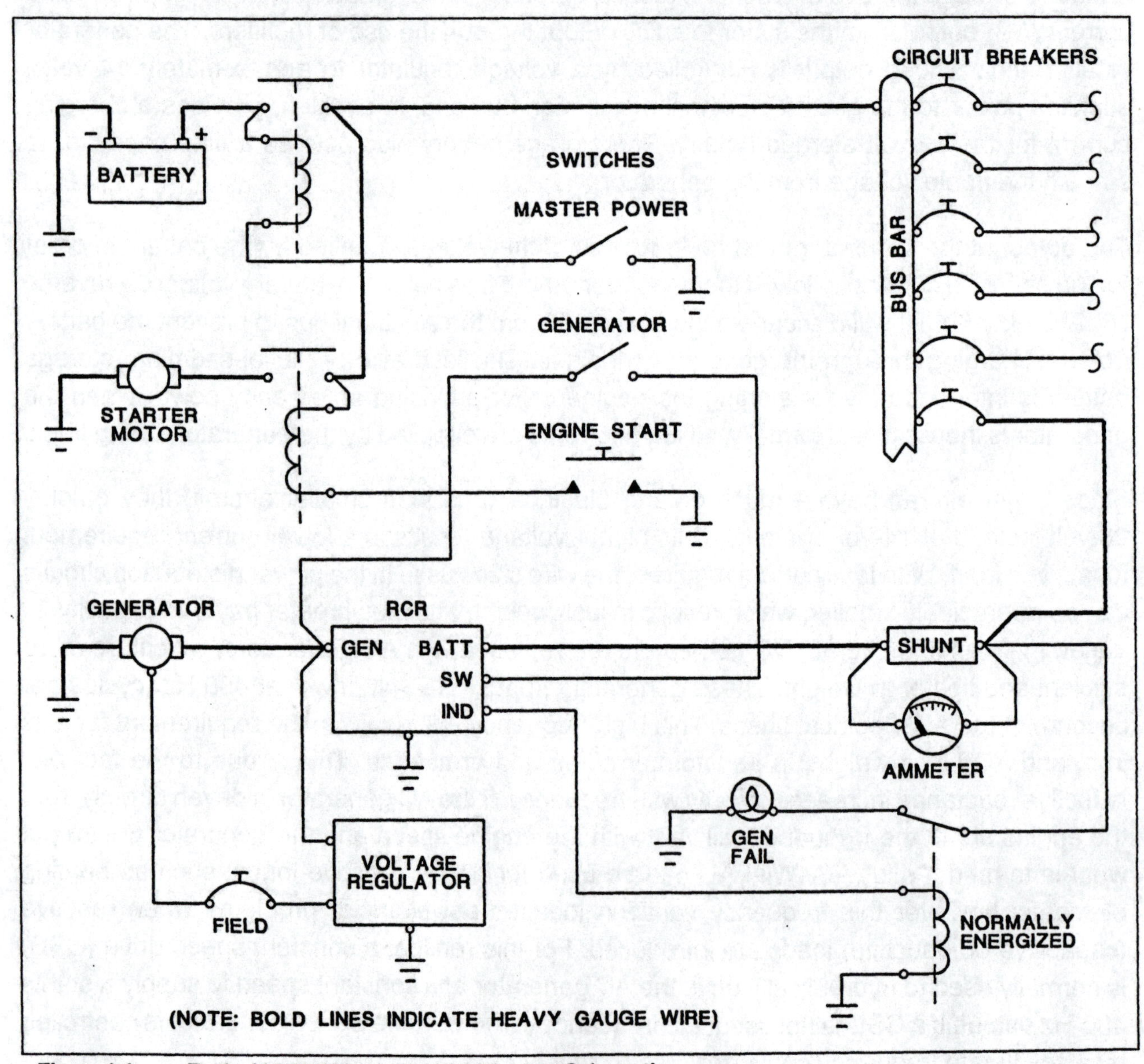

Figure 1-1. Typical Light Aircraft Electrical System Schematic.

with circuit breakers to distribute the electrical power to the various loads. The voltage supplied by the battery is applied through the battery relay, which is activated by the master switch, to the bus bar and starter relay. Depressing the engine-start button, activates the starter relay and applies the battery voltage to the starter motor mounted on the engine. The aircraft structure serves as the ground return for all electrical signals.

Once the engine is started, the generator develops an output voltage which is applied to the voltage regulator and reverse current relay. Since the generator output voltage will vary with the engine speed and the applied load, a voltage regulator is used to control the voltage output of the generator by regulating the amount of current flow through the generator field circuit.

The RCR opens whenever the generator output falls below the battery voltage, thereby preventing the battery from discharging through the generator windings. Should the RCR open, a "generator fail" warning light will illuminate in the cockpit. A switch or overvoltage relay is also provided which either opens the generator field circuit or prevents the RCR from closing should an overvoltage condition occur.

The generator output from the RCR is routed through the ammeter shunt and on to the DC bus bar. The current flowing through the very small resistance of the shunt develops a voltage drop which is applied to the ammeter in the cockpit to indicate the amount of current being drawn from the generator. The ammeter shunt is sometimes inserted in series with the battery to indicate the amount of charging current demanded by the battery.

The DC bus distributes the power to the thermal circuit breakers, which will automatically open if the current from the connected load exceeds the circuit breaker rating. If AC power is required, a circuit breaker from the DC bus will feed the input of either a rotary or static inverter, which will convert the DC input to an AC output. Rotary inverters consists of a constant-speed DC motor, which is mechanically coupled to drive an AC generator.

Rotary inverters have since been replaced with solid-state static inverters which have an internal multi-vibrator oscillator that produces an AC voltage at the desired frequency (60 or 400 Hz). The output of the oscillator is passed through a transformer and filter circuits to produce a sine-wave output at the proper voltage (26- or 115-volt AC). The AC from the inverter is then applied to an AC bus bar for load distribution.

Power distribution for the battery and engine start circuits requires the use of heavy gauge wiring due to the high current demand of the starter motor. The power distribution wiring for the generator circuit is also of a heavy gauge, but not quite as heavy as the battery/starter circuit since the generator does not power the starter motor. The control wiring from the switches to the relays, from the shunt to the ammeter, and from the voltage regulator to the generator is of a much lighter gauge. The circuit breaker ratings and wiring sizes from the bus bar are determined by the individual equipment loads. Refer to the FAA Airframe and Powerplant Handbook AC 65-15 for determining circuit breaker ratings and wire sizes.

HEAVY AIRCRAFT AC ELECTRICAL CIRCUIT THEORY

Light aircraft electrical systems typically only use 14-volt or 28-volt DC as the primary power source since the power requirements are moderate (perhaps no more than 10 amps) for the limited amount of onboard electrical and avionic equipments. However, since large commercial aircraft require many times more power (typically 1,500 to 3,000 amps per generator), 115-volt or 200-volt AC, three-phase, 400-Hz power is used as the primary source.

Phase Relationships

The term phase is used to relate to the number of separate voltage waves in an alternating current supply. The phase angle is the difference in degrees of rotation between two alternating currents or voltages. In a three-phase system, each phase differs from the other by 120°. The phase sequence, as shown in Figure 1-2, begins when phase one increases in amplitude from zero volts at time zero. When phase one has completed 120° of its 360° cycle, phase two begins to increase from zero volts. When phase one is at 240° and phase two is at 120°, phase three begins its sine-wave cycle.

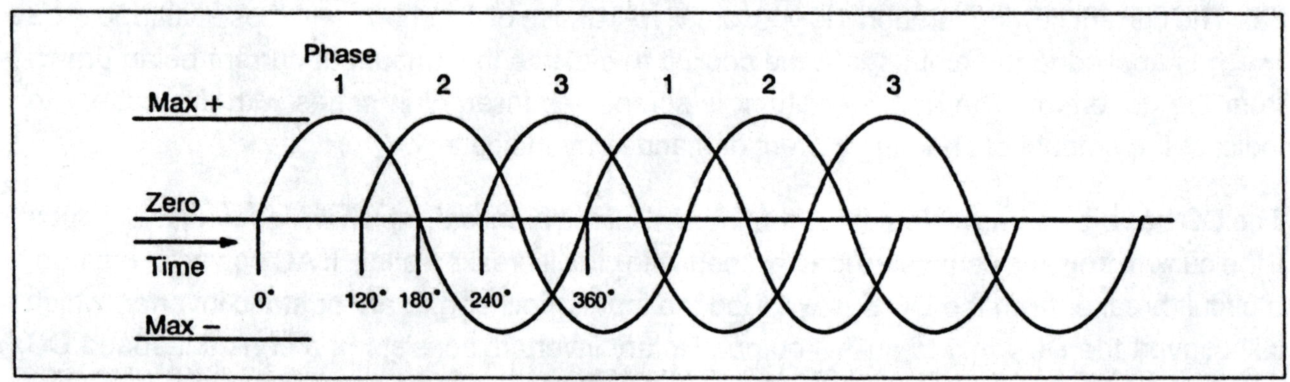

Figure 1-2. ***Combined Sine Waves of a 3-Phase System.***

AC Generators

As previously mentioned, the AC generators used on heavy aircraft provide 115-volt or 200-volt AC, three-phase power at a frequency of 400 Hz. The three-phase generator may be considered as three single-phase generators in one machine. It requires a minimum of three wires to deliver three-phase service. The output terminals of a three-phase generator are marked to indicate the phase sequence. These terminals are connected to individual buses labeled A, B, or C, which indicate either phase 1, 2, or 3. Equipment that requires 115-volt AC three-phase power will have one wire connected to each of the three buses and one to ground, while single-phase equipment will have one wire connected to a single bus and one to ground. For maximum efficiency, the loads should be distributed as evenly as possible across all three buses on three-phase systems.

Three-phase generators may have either delta (D) or wye (Y) connected windings; however, those used for aircraft are typically wye-connected with the neutral lead attached to ground. The phase-to-phase voltage of a wye-connected generator is the vector sum of the voltage of two of the windings, which can be calculated by multiplying the square root of three by the phase-to-neutral voltage. For example, a typical wye-connected generator with a 115-volt AC phase-to-neutral voltage will have an approximate 200-volt AC phase-to-phase voltage. For this reason, 200-volt AC power can be obtained by connecting the load between two different phase buses, instead of between one bus and ground. As shown in Figure 1-3, each wye-connected generator has a brushless exciter which supplies a DC excitation voltage to the generator during start-up. The exciter field and rotor assembly is actually a DC generator without a commutator to convert the resultant AC to DC. Instead, a rectifier assembly, which rotates with the armature, provides the DC excitation for the main field of the AC generator. When the generator output has reached approximately 50% of its operating voltage, a start relay opens which causes the voltage regulator to take control of the field excitation in order to for the generator to maintain a constant voltage output.

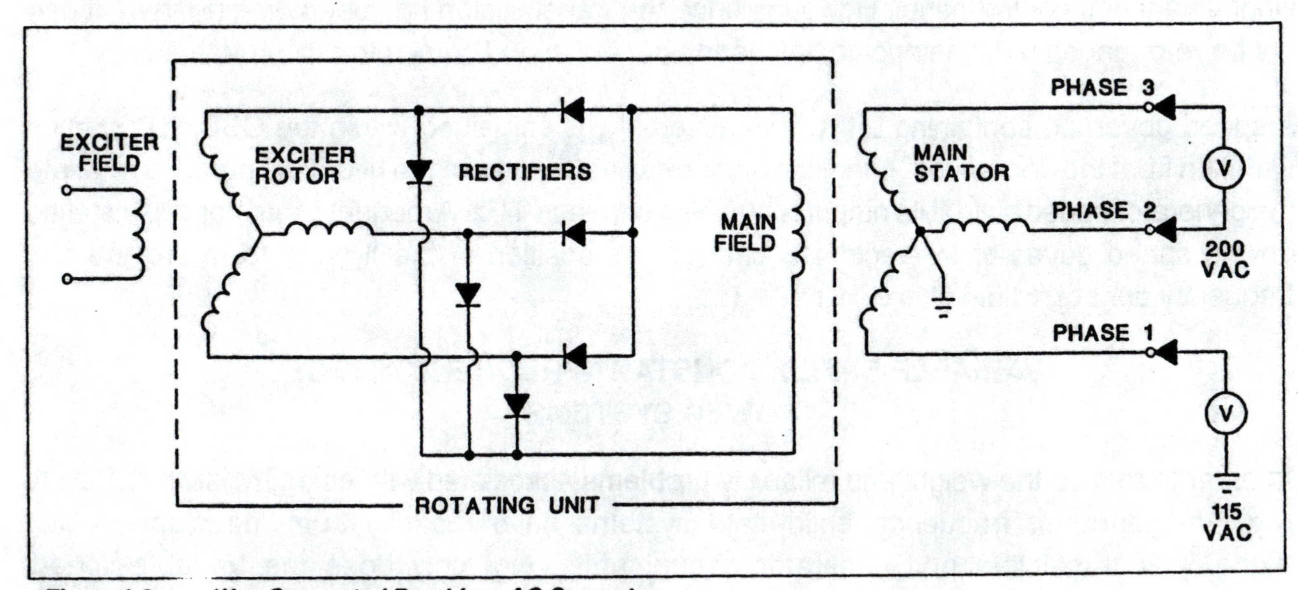

Figure 1-3. Wye-Connected Brushless AC Generator.

Aircraft that have more than one generator use a paralleling circuit, whereby all operating generators are connected to the AC bus through bus-tie relays. When the generators are paralleled, the equalizing circuits in the voltage regulators monitor the outputs of the other generators on the bus and adjust the voltage of the generators so that each shares the same load.

Constant Speed Drives (CSD)

The frequency of the AC generator is dependent on the number of poles it has and its speed of rotation, as given in the following formula:

Frequency (cycles per second) = (Revolutions per minute × Pairs of Poles)/60

It can be seen from the preceding formula that in order for the AC generator to have a steady 400-Hz frequency output, its speed of rotation must be kept constant regardless of the engine RPM. This is usually accomplished by means of a generator drive transmission, known as a CSD. A typical CSD will turn the generator rotor at a constant speed of 8,000 plus or minus 80 RPM.

The CSD assembly is basically an engine-driven, variable-ratio transmission which drives the input shaft of a variable-displacement hydraulic pump. The fluid pressure of the hydraulic pump is then applied to a hydraulic motor which drives the output spline connected to the AC generator. A means is usually provided to disengage the transmission from the generator at the flight engineer's control panel. However, once the transmission has been dis-engaged, it can not be re-engaged until the engine has been shut down and comes to a complete stop.

A speed governor, containing a flywheel assembly, is contained within the CSD that senses variation from the "on-speed" condition and ports fluid to or from the hydraulic pump to regulate the generator speed so that its output is 400 plus or minus 4 Hz. A magnetic trim head is installed on the speed governor to electrically change the position of the flyweights to provide fine frequency control, within plus or minus 1 Hz.

VARIABLE-SPEED, CONSTANT-FREQUENCY (VSCF) POWER SYSTEMS

In order to reduce the weight and reliability problems associated with using hydraulic CSDs to regulate generator frequency, solid-state systems have recently been developed using high-power transistors and generator control units. Not only does the Variable-Speed, Constant-Frequency power system eliminate all moving parts, except for the generator rotor

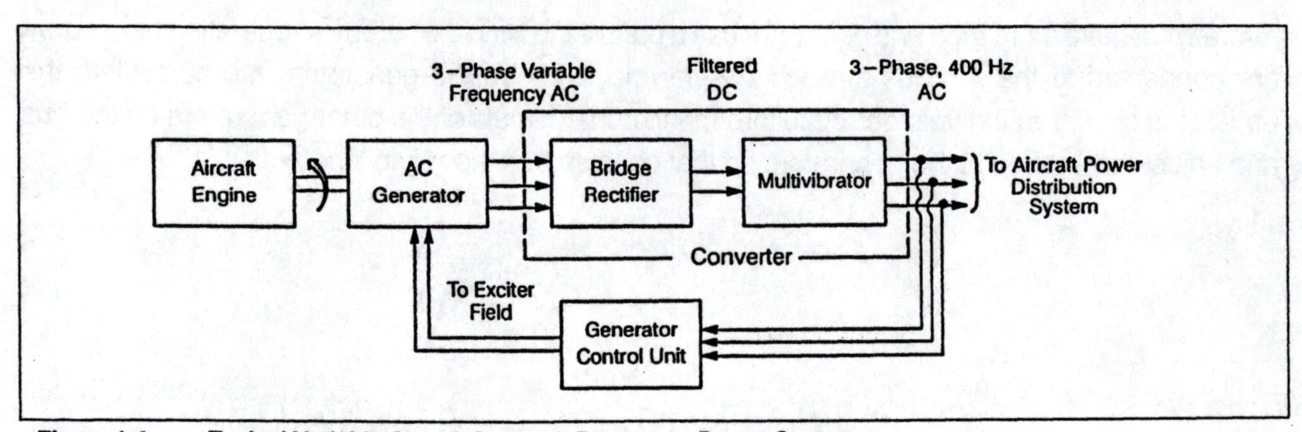

Figure 1-4. Typical Variable-Speed, Constant-Frequency Power System.

and possibly an oil pump for cooling and lubrication, but the electronic subassemblies can be remotely mounted in the fuselage, whereas the CSD must be mounted on the generator in the engine nacelle.

A typical VSCF system, as shown in Figure 1-4, consists of three units: an AC generator, a converter, and a Generator Control Unit (GCU). The three-phase, variable-frequency AC from the generator is first rectified and filtered into pure DC, which is fed to a multi-vibrator oscillator (such as in a static inverter) to produce three-phase 400-Hz AC. The GCU monitors the 400-Hz output of the converter and adjusts the current flow to the exciter field of the generator to maintain the voltage output within specified limits.

Modern aircraft also use computerized Bus Power Control Units (BPCU) for power distribution and control. Current transformers installed on the various electrical buses monitor the load continuously and send this information to the BPCU. The BPCU also receives information from the GCU on available power output from the generators. By knowing the power available and the power required, the BPCU automatically distributes the power from the various generators as evenly as possible to the various buses demanding power from the system.

BOEING 747 ELECTRICAL SYSTEM OPERATION

A simplified schematic diagram of the electrical system for a Boeing 747 commercial transport is shown in Figure 1-5. During ground handling and servicing, electrical power may be obtained from either the onboard 90 KiloVolt-Ampere (KVA) Auxiliary Power Unit (APU), which is an AC generator driven from a separate small gas-turbine engine, or from an external ground power unit (GPU). This power may be connected to any aircraft AC bus through use of the APU circuit breaker (APB) or external power contactor (XPC), when used in conjunction with the bus-tie breakers (BTB). In flight, electrical power is obtained from four engine-driven, three-phase, 60-KVA, 400-Hz generators. Each of the generators are connected to their respective AC bus through the generator circuit breakers (GCB). These AC buses may be paralleled or isolated through the BTBs.

All 115-volt AC loads, except essential, standby, ground service, ground handling, and galley loads, are connected to one of the four main AC buses. Essential power, which is that considered necessary for safety of flight, is normally supplied from the No. 4 main AC bus. However, the essential bus can be connected to the generator side of GCBs 1, 2, or 3 by means of the essential power select relay. The standby AC bus supplies 115-volt AC power to the most important equipment needed for safe flight. During normal operation, the standby bus is supplied power from the essential AC bus; however, in the event that all AC power is lost, the standby AC can be provided by a battery-powered inverter.

A 28-volt DC system is provided to supply those loads requiring DC power. This power is supplied by four 75-amp TRUs, which are energized from the No. 1, 2, 3, and essential 115-volt AC buses, respectively. Each TRU powers an associated 28-volt DC bus. The TRUs are

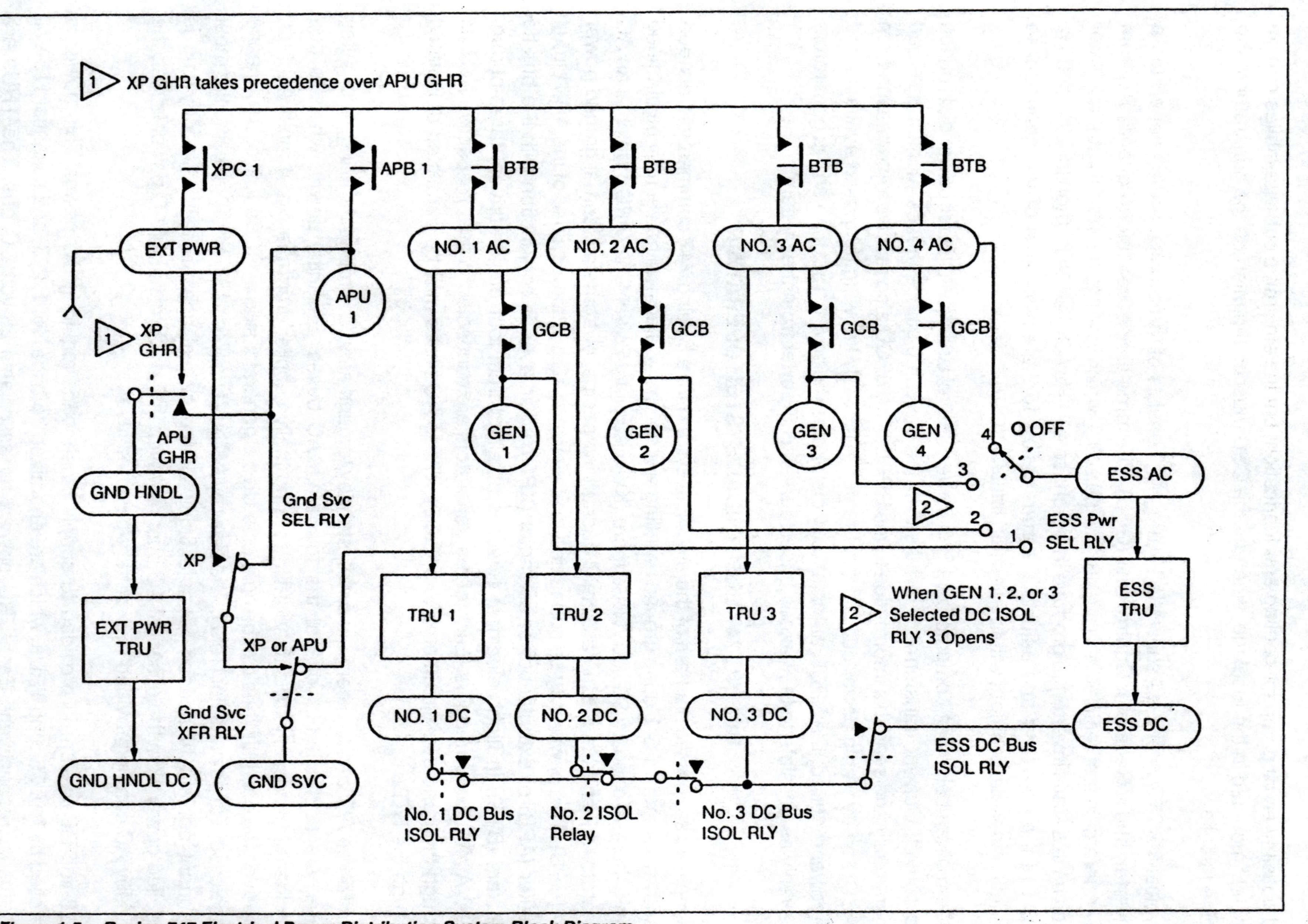

Figure 1-5. Boeing 747 Electrical Power Distribution System Block Diagram.

normally operated in parallel; however, a blocking diode prevents the essential TRU from feeding any bus other than the essential DC bus. A 20-amp TRU energized by the AC ground-handling bus supplies DC power required during ground operation.

In addition to the TRUs, a 24-volt nickel-cadmium battery permits starting the electrical system without an external source, supplies back-up power, and can be used as a standby source for minimum communication, navigation, and instrumentation systems should all other sources fail. A battery charger, energized by the ground service bus, maintains full capacity in the battery. The "hot" battery bus, for those loads requiring uninterrupted power, is always connected to the battery. The battery bus is normally energized by the essential DC bus. However, in the event of loss of essential DC bus power, it is automatically transferred to the hot battery bus. Similarly, the standby bus is normally energized by the essential bus, but can be transferred to the battery bus.

Control and Indication

The electrical control panel consists of the necessary gauges, annunciator lights, and switches to provide the flight engineer a visual indication and manual control of the electrical generation and power distribution system. A typical control panel for a four-engine turbine jet aircraft, such as the one for the Boeing 747 illustrated in Figure 1-6, has four CSD oil temperature gauges; four generator drive low oil pressure lights; four AC loadmeters calibrated in percentage of maximum generator rating; one generator over-load indicator light, which is annunciated when any one generator exceeds its maximum load; four generator on/off switches with a reset position for "flashing" the field; four generator unparalleled lights, which annunciate if a bus-tie relay opens; and four emergency electrical control switches for load-shedding a particular bus.

The control panel also has included four bus power failure indicator lights, four DC load meters calibrated in percentage of maximum TRU rating, AC and DC bus isolation switches, a generator paralleling switch, a battery power or external power selector switch, and a external power available light. In addition, a voltage and frequency selector switch is provided for measuring various AC and DC voltages and frequencies present throughout the electrical system. These indications are presented on the three gauges directly above the selector switch.

CONCLUSION

Modern day transports have replaced the flight engineer's station and his electrical control panel with a computerized system that automatically monitors and controls the aircraft's electrical system. In addition, constant speed drives are being replaced with solid-state variable-speed, constant-frequency power systems. However, the preceeding discussion on CSDs and electrical control panels is valid since the majority of airliners are still flying with this equipment.

The following chapters discuss the operation and maintenance of avionic systems that are dependent on the supply of aircraft electrical power in which to operate. We begin the next chapter with a description of aircraft audio and communication systems and proceed from there to aircraft navigation systems.

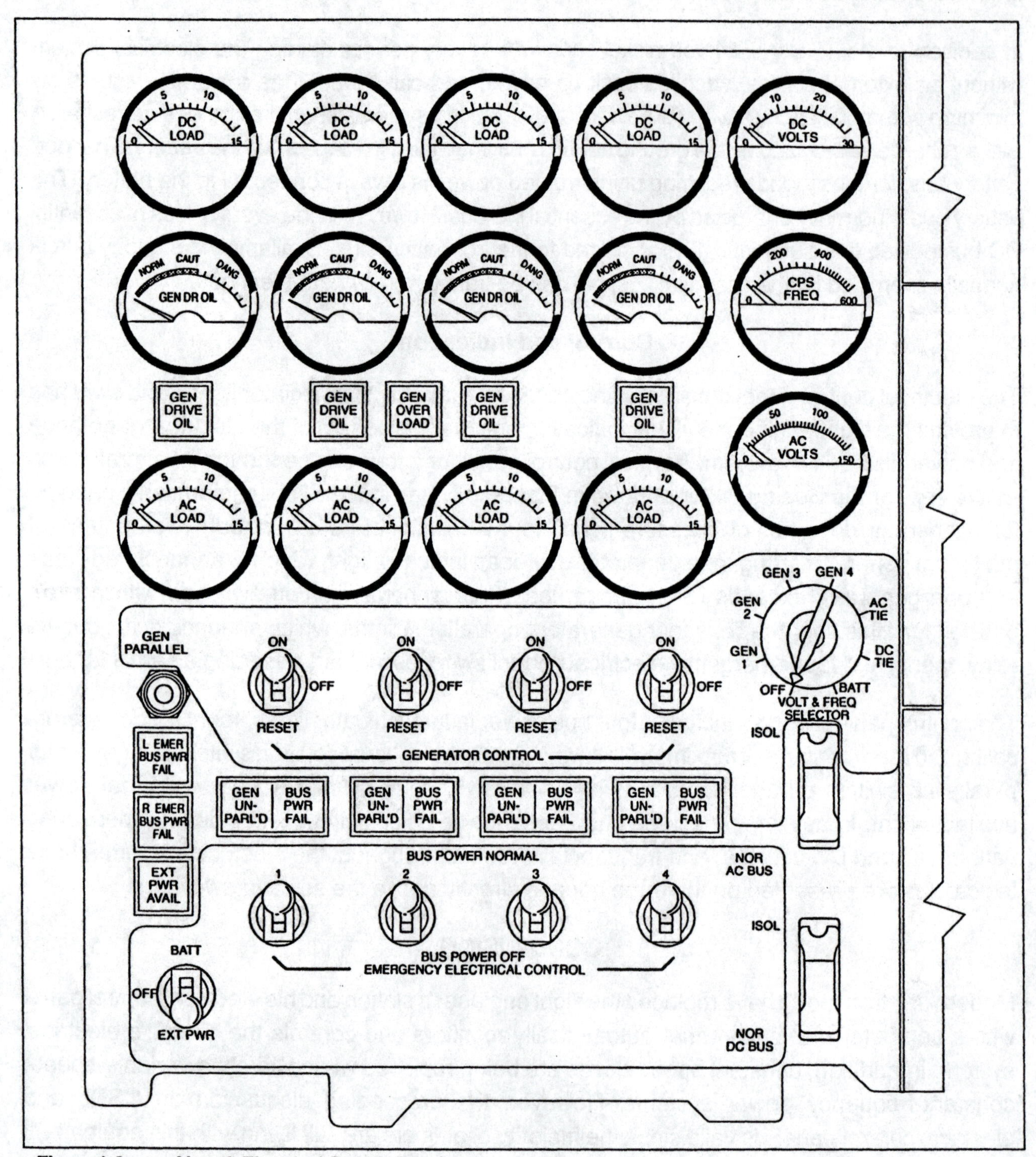

Figure 1-6. ***Aircraft Electrical Control Panel.***

REVIEW QUESTIONS

1. What three purposes does the storage battery serve in the operation of the aircraft electrical system?
2. Describe the difference in operation between a DC generator and a DC alternator.
3. What is the function of the voltage regulator and RCR in an aircraft electrical system?
4. Why do larger aircraft use 28-volt DC power as opposed to 14-volt DC systems used in smaller aircraft?
5. Explain why commercial electrical systems for a home or factory use 60-Hz AC power while electrical systems for heavy aircraft use 400-Hz AC power.
6. What is meant by "wild AC" and what two methods can be used to correct for this condition?
7. How does the voltage regulator maintain a constant voltage output from the generator?
8. What two methods are commonly used to interrupt the power output from a generator should an overvoltage condition occur?
9. What is the purpose of the shunt and where may it be connected in the electrical system?
10. Compare the function of a TRU versus an inverter. How does the operation of a static inverter differ from that of a rotary inverter?
11. Compare the primary and secondary power sources used in a heavy aircraft electrical system as opposed to the power sources used in a light aircraft electrical system.
12. Define phase and phase angle. What is the phase angle in a threephase system?
13. Describe the operation of a brushless AC generator.
14. What is meant by paralleling generators and what is the function of an equalizer circuit?
15. Calculate the phase-to-phase voltage of a wye-connected generator supplying 115-volt AC as measured from one phase output to ground.
16. If an AC generator has three pairs of poles and is rotating at a speed of 8,000 RPM, what is its output frequency?
17. Explain how a CSD controls the generator rotational speed as the engine speed varies. What is the purpose of the speed governor and magnetic trim head on a CSD?

18. What are the advantages of the VSCF power system compared to a CSD system? Explain how the VSCF system produces a constant frequency output while the speed of the generator varies.

19. What power sources are available to the essential AC bus, standby AC bus, essential DC bus, standby DC bus, battery bus, and hot battery bus during all modes of operation for the Boeing 747?

20. Briefly describe the control functions and indicators provided on a typical electrical control panel. Explain how BPCUs automatically provide these functions.

Chapter 2

Audio and Communication Systems

INTRODUCTION

Besides basic flight instrumentation that informs the pilot of the aircraft's attitude, altitude, airspeed, and heading, the communication system is the most essential part of the avionics suite for safety of flight. In fact, the FAA requires that all aircraft operating in high-traffic areas be equipped with two-way radio for communication with air traffic controllers and tower operators. Air-to-ground communication makes it possible for the pilot to keep informed of traffic conditions in the area in order to minimize or eliminate the possibility of a mid-air collision. While flying enroute, the pilot can communicate with ground facilities to obtain weather information, verify his/her position, report the progress of his/her flight, and report any emergency conditions which may develop.

In the 1920's, airborne communications was accomplished with the use of spark-gap transmitters and crystal receivers. These early systems operated in the low-frequency range of the RF spectrum, whereby signals where encoded and decoded with continuous-wave (CW) radio-telegraphy using Morse Code. At that time, the typical communication system was unreliable, consumed vast amounts of electrical power, weighed a great deal, and was highly susceptible to atmospheric noise. However, all of these constraints have since been overcome by the development and application of solid-state electronics. Modern day communication equipment is highly sophisticated and complex, incorporating microprocessors and digital displays.

The majority of air-to-ground communication in modern aircraft is accomplished with a Very-High Frequency (VHF) transmitter and receiver, or transceiver. The typical VHF communication transceiver operates in the frequency range between 118 and 135.975 MegaHertz (MHz) with 720 channels spaced at 25-kiloHertz (kHz) intervals. Modern day transceivers extend this frequency range to 151.975 MHz. At this portion of the radio frequency (RF) spectrum, the nature of radio-wave propagation is such that communication is limited only to line-of-sight distances. However, an airplane flying at an altitude of 30,000 feet has an effective VHF range of 200 miles, and these signals are seldom distorted by atmospheric interference.

Aircraft that require long-distance communication employ High-Frequency (HF) communication systems. These HF systems are usually found onboard commercial and corporate jet aircraft that operate outside the range of VHF communication ground stations, and therefore are used mainly for transoceanic flight. The HF communication system operates between the range of 2 MHz and 30 MHz. In this portion of the RF spectrum, radio-wave propagation is reflected back and forth between the earth and the ionosphere for several thousands of miles.

In order to achieve such long distant communication range, typical airborne HF transmitters usually provide a nominal RF power output to the antenna of 100 watts, compared with 20 watts RF output from a typical VHF transmitter. (More power output is required since the energy field radiated from the antenna decreases with the square of distance traveled.) In order for the HF system to operate effectively over such a wide frequency range, an antenna coupler unit is required to tune the resonant frequency of the aircraft antenna to the operating frequency of the HF transceiver.

RADIO FREQUENCY BANDS

Communication systems and most navigation systems are dependent on the use of radio waves to operate. Radio waves make it possible for a pilot to determine his or her location, to maintain an accurate course, and to communicate with ground personnel or other aircraft. The reflection characteristics of radio waves allows the Radio Detecting And Ranging (RADAR) system to warn the pilot of impending storm conditions in order to make possible all-weather flight.

Radio waves are produced by an alternating current that is fed to an antenna to produce radio-frequency waves. Radio waves travel as an energy field from the antenna at the speed of light, approximately 300,000,000 meters per second. An RF wave is composed of an electromagnetic field and a 90° displaced electrostatic field. The polarization of an RF wave is determined by the direction of the lines of force in the electromagnetic field. The direction of these lines of force are dependent on the polarization of the radiating element. Thus, a vertical antenna produces a vertically polarized wave, and a horizontal antenna emits a horizontally polarized wave. Simple vertical antennas have omnidirectional characteristics in that they transmit and receive in a 360° pattern, and therefore, are used for communication. Horizontally polarized antennas are basically directional, and hence, are primarily used for navigation.

The RF portion of the electromagnetic wave spectrum extends from approximately 3 kHz to 300 GHz. This portion of the spectrum is divided into frequency bands, as shown in Table 2-1.

Frequency Range	Band
Very Low Frequency (VLF)	3 to 30 kHz
Low Frequency (LF)	30 to 300 kHz
Medium Frequency (MF)	300 to 3000 kHz
High Frequency (HF)	3 to 30 MHz
Very High Frequency (VHF)	30 to 300 MHz
Ultra High Frequency (UHF)	300 to 3000 MHz
Super High Frequency (SHF)	3 to 30 GHz (Giga Hertz)
Extremely High Frequency (EHF)	30 to 300 GHz

Table 2-1. Radio frequency bands.

RF propagation characteristics are complex because the earth appears different to radio waves at various frequencies. The ground acts as a dielectric at frequencies above 5 MHz and as a conductor below 5 MHz. At low and medium frequencies, the wave formation follows the curvature of the earth, and therefore, is called a ground wave. This is illustrated in Figure 2-1.

Between 100 to 250 miles above the earth, there exists a vast region of ionized gas. Ultraviolet energy radiated from the sun dislodges electrons from the atoms of various gases forming positively charged atoms known as ions. These ions form layers that surround the earth making up the ionosphere. Since the velocity of radio waves traveling through an ionized layer is less than those traveling through ordinary air, refraction occurs, which causes the radio waves to be reflected back to earth. As the RF energy is reflected, the wave may "skip", or bounce back and forth between the ground and the ionosphere, for many thousands of miles, thereby making possible long-range HF communications.

VHF and above radio waves allow only line-of-sight communication. At these higher frequencies, the radio wave is not reflected by the ionosphere, but passes right through it. VHF communications are ideal in that they minimize interference with distant unrelated stations operating on the same frequency. Also, a VHF communication system requires a much smaller antenna than an HF system. This is due to the fact that as the frequency increases, the wavelength decreases, and the antenna length is usually sized to an even fraction of the operating wavelength. The relationship between frequency and wavelength is given in following formula:

$$\text{Wavelength (meters)} = 300{,}000{,}000/\text{Frequency (Hz)}$$

Where 300,000,000 meters per second is the speed of electromagnetic wave propagation.

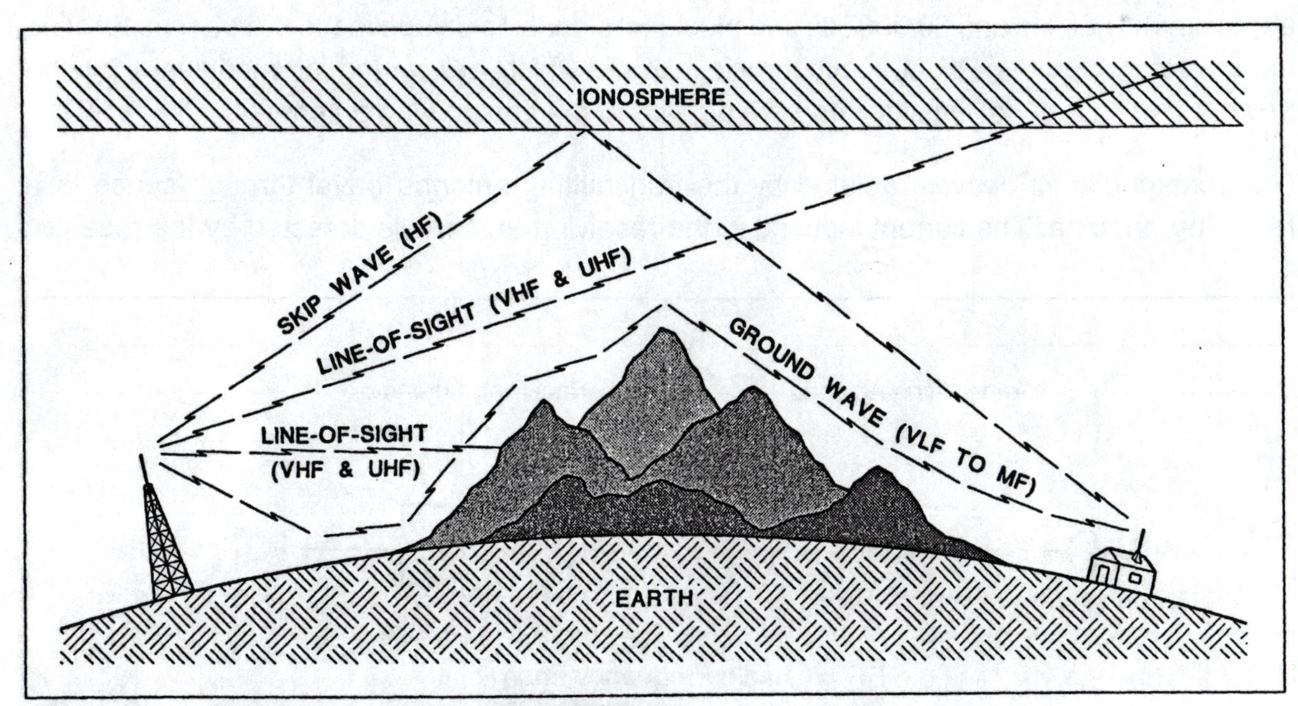

Figure 2-1. Radio wave propagation charcteristics.

Above 3,000 MHz, coaxial cable is replaced with waveguides and tuned circuits take the form of resonant cavities. As radio frequencies approach the frequency of light, they are able to be focused into narrow beams of energy using parabolic reflectors behind the antenna. This characteristic allows a radar system to transmit a burst of energy and receive its return echo using the same parabolic reflector.

RADIO COMMUNICATION THEORY

Communication systems primarily involve voice transmission and reception between aircraft and/or between aircraft and ground stations. The basic concept of radio communication is based on sending and receiving electromagnetic waves traveling through space. Alternating current passing through a conductor creates electromagnetic lines of flux or fields around a conductor. Energy is alternately stored in these fields and returned to the conductor. As the frequency of this AC is increased, less of the energy stored in the field returns to the conductor. Instead, the energy is radiated into space in the form of electromagnetic waves.

Electromagnetic waves traveling through space strike conductors which set electrons in the conductor in motion due to the energy induced. This electron flow constitutes a current that varies with changes in the electromagnetic field. Thus, a variation of the current in a radiating antenna from a transmitter causes a similar varying current in the conductor of the receiving antenna.

As shown in Figure 2-2, a basic communication system consists of a microphone, transmitter, transmitting antenna, receiving antenna, receiver and speaker. A microphone is an audio transducer which changes sound energy, or audio-frequency (AF) signals between approximately 20 Hz and 20,000 Hz, into electrical signals to be applied to the transmitter. The transmitter generates RF electromagnetic waves, which are varied or modulated by the microphone's electrical impulses and are applied to the transmitting antenna.

Electromagnetic RF waves radiated by the transmitting antenna travel through space to a receiving antenna. The current induced in the receiving antenna is detected by the receiver,

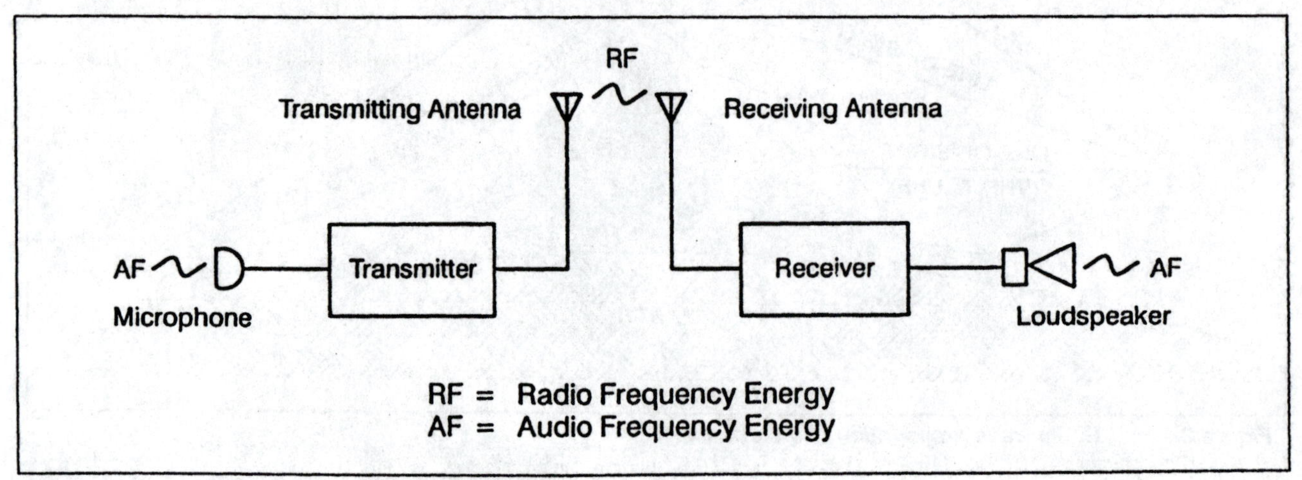

Figure 2-2. Radio Communication System.

which extracts the audio intelligence by changing the RF signal back to electrical impulses. Finally, the speaker converts these electrical signals into AF sound energy that is nearly identical to that input into the microphone at the transmitting station.

AIRCRAFT AUDIO SYSTEMS

As mentioned previously, a radio communication system requires the use of two audio transducers: a microphone to convert audio signals to electrical impulses for use by the transmitter, and a speaker to convert the electrical impulses from the receiver back into audio signals. The transmitter portion of an airborne VHF or HF communication transceiver requires a microphone audio input and a means to activate the transmitter. The transmitter may be activated or keyed by a push-to-talk (PTT) switch located directly on the microphone housing.

The receiver portion of the VHF or HF transceiver provides a 500-ohm (typical) low-level audio output to be used by the headphones, and sometimes an eight ohm (typical) high-level audio output for loudspeaker operation in cases where an external isolation amplifier is not used. In addition, the airborne navigation receivers, such as the ADF, VOR/ILS, Marker Beacon and DME, provide audio output signals to be reproduced by the headphone and/or loudspeaker.

Integrating the audio signals from these systems into an audio control unit allows the pilot to control the desired audio inputs and outputs to and from the respective avionic equipment. In larger aircraft, one each audio control unit is usually found positioned at the pilot and copilot side consoles and one at the flight engineer's console if the aircraft is equipped with a third crewmember station.

A typical audio control unit, such as the Baker Model M1035 shown in Figure 2-3, has provisions for up to eleven low-level receiver audio inputs, four transmitter key and microphone audio outputs, plus selection for crew interphone or cabin public address (PA). Also, an option is available to provide automatic selection of the corresponding receiver audio when the transmitter key and microphone audio is selected for a particular transceiver.

Inside the M1035, all communication, VOR/ILS navigation and DME inputs from 10 to 300 milliwatts are leveled to less than a 3-dB change in output. ADF and Marker Beacon inputs are not leveled so that the output is proportional to the input. The selected receiver inputs are processed through an internal audio mixing circuit and isolation amplifier before being output to the cockpit speaker and headphones. A 1,020-Hz filter (FILT) may be selected for coded ADF or VOR station identification signals.

Audio Control Unit Operation

The audio control unit is an integral component in the operation of aircraft communication and navigation systems. It not only provides audio selection, but also isolates and mixes the outputs from the various receivers to the speaker or headphones. The audio outputs from the various receivers must be isolated from each other so that one receiver's audio output will dissipate

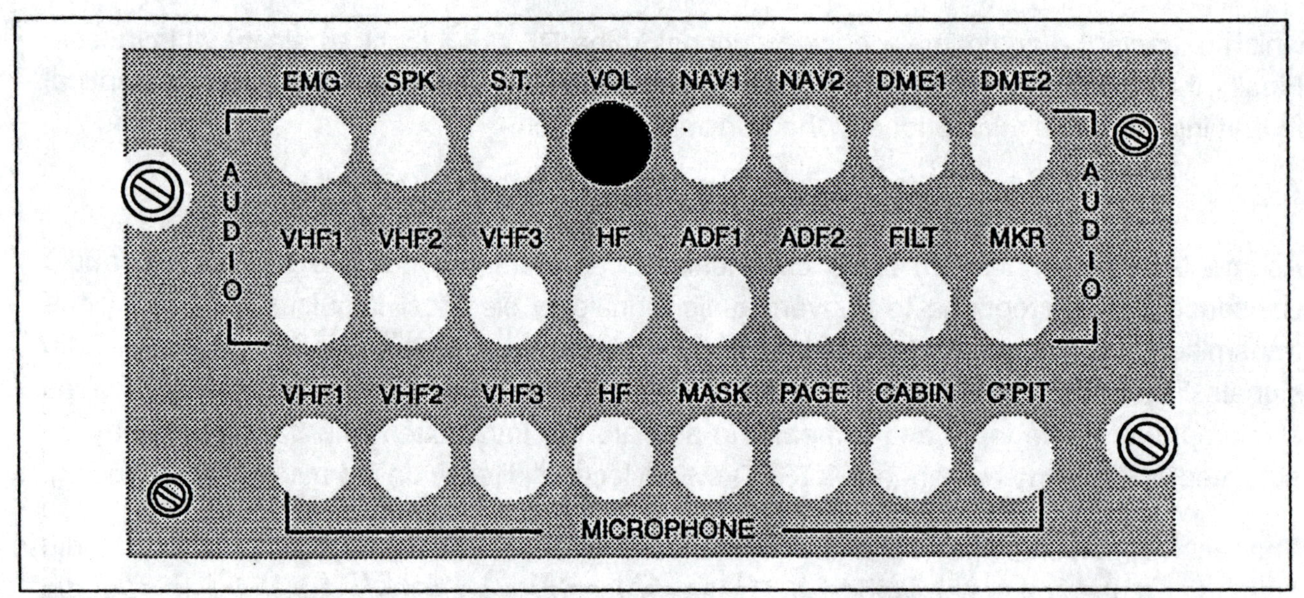

Figure 2-3. M1035 Audio Control Unit. *(Courtesy Baker Electronics, Inc.)*

power only to the speaker and not to the output stages of the other receivers. If the various audio outputs were simply connected in parallel to the speaker, the resultant load impedance would be less than the receiver's termination impedance. This condition could result in audio distortion caused by driving the receiver's audio output amplifier into saturation.

As illustrated in Figure 2-4, which is a schematic diagram of a typical audio control unit, isolation is obtained through the use of a mixing circuit in the audio control unit, which provides a high resistance isolation between the various receiver audio outputs when several receivers are connected to the same speaker. In addition, the audio mixer matches the output impedance to the receiver's termination impedance for maximum power transfer, and sends the mixed audio signals to an isolation amplifier to increase the audio gain before being output to the cockpit speaker. In the event the isolation amplifier fails, or there is a power loss to the audio control unit, selection of the emergency function allows continued use of transmit and receive functions. In the emergency mode, all inputs bypass the internal isolation amplifier and are connected directly to the headphones on a priority basis so that only one receiver's audio may be listened to at any one time.

A speaker volume control is provided which adjusts the gain of the isolation amplifier. In addition, the navigation and communication control heads have volume controls to adjust the individual receiver's audio output level to the audio control unit. Once the individual volume controls are set to approximately the same level, the speaker volume control can be used as a master level control. In all cases, the receiver's internal audio level adjustments should be adjusted to the manufacturer's specifications prior to installation.

The receiver audio output provided to the speaker should be disabled or muted when a microphone is keyed to prevent retransmission of received audio and to prevent possible audio feedback. Although the receiver in the respective transceiver is internally muted when its

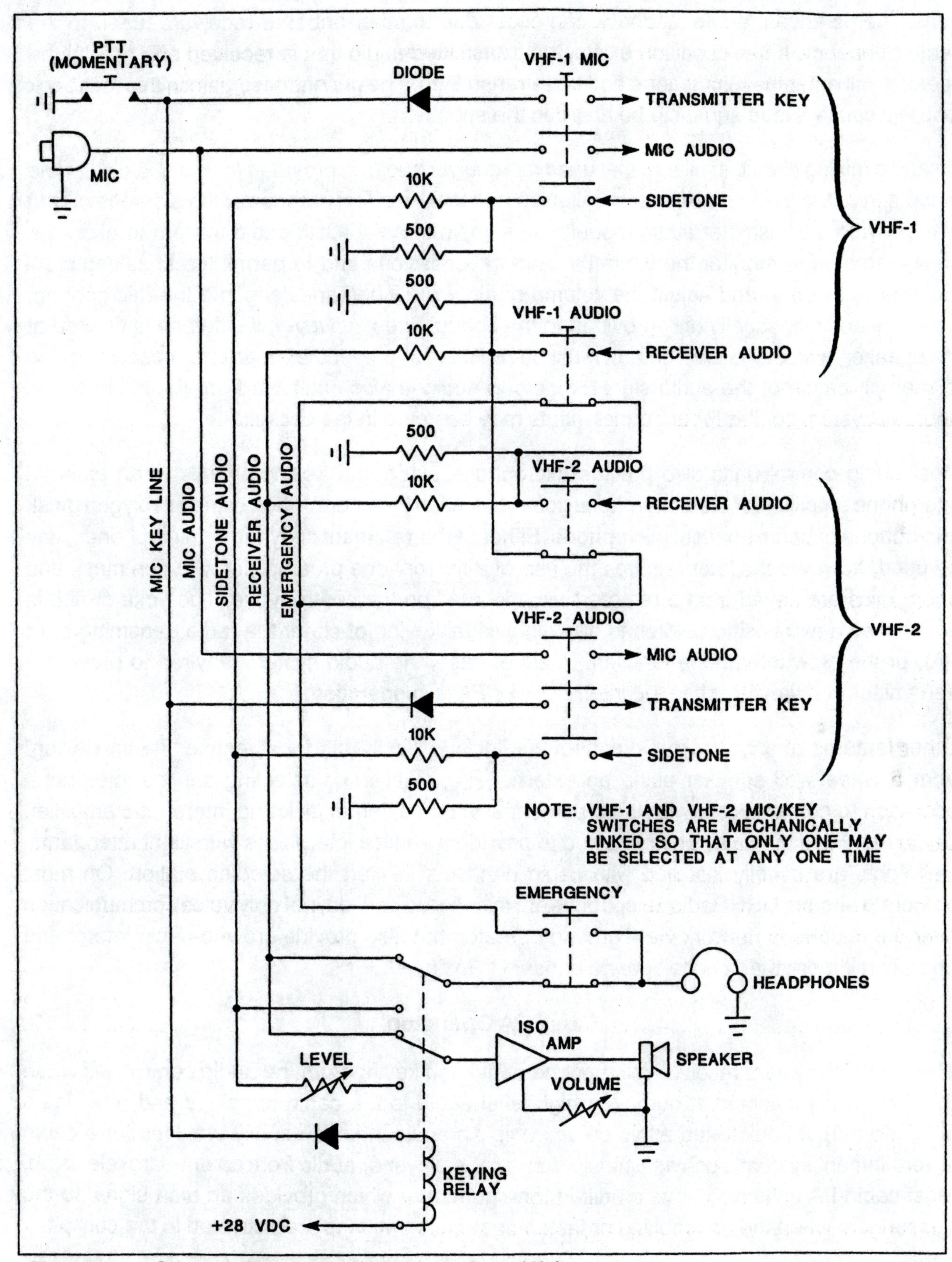

Figure 2-4. Schematic Diagram of a Typical Audio Control Unit.

transmitter is keyed, audio feedback can occur due to other onboard receivers tuned to the same frequency. If this condition exists, the transmitted audio that is received and sent to the speaker will enter the microphone and be retransmitted. The uncontrolled gain in this feedback loop will cause a loud squeal to be heard in the speaker.

An audio mixing circuit, similar to that used for receiver audio, is provided for the audio sidetone outputs from the VHF and HF communication transceivers. The purpose of the sidetone output is to provide a transmitter audio modulation sample to the aircraft audio system to allow the crew members to monitor the aircraft's radio transmissions and to permit the speaking crew member to listen to and adjust the volume of his voice when speaking into the microphone. Sidetone audio is usually only provided to the headphones; however, if sidetone is desired at the speaker, a sidetone level control must be provided to prevent feedback from occurring due to reamplification of the audio signal. Sidetone audio is also available from the cabin public address system so that PA announcements may be heard in the cockpit.

Most audio control units also provide microphone audio and keying signals to an external interphone amplifier for crew interphone communication (intercom) when using an oxygen mask microphone or boom headset microphone. Either carbon element or dynamic microphones may be used, however the later requires the use of a microphone preamplifier. Oxygen mask and boom mics are keyed from a remote switch located on the control yoke. The yoke switch is sometimes a two-position switch to allow individual keying of either the radio transmitters (or PA), or the crew interphone function. In either case, the audio system is wired to prevent a transmitter from keying when the interphone or PA is in operation.

Other features of a typical audio control unit include provisions for selecting "Ramp Hailing" from a wheel-well speaker using an external PA amplifier, or selecting service interphone operation to communicate with ramp personnel via an external jack and interphone amplifier. Cabin handsets may also be connected to provide a service interphone with flight attendants. Call lights are usually installed with cabin handsets to alert the selected station. On most corporate aircraft, UHF Radio Telephones are installed to provide, not only voice communication over the telephone network via a ground operator, but also provide crew-to-cabin interphone operation if more than one telephone handset is installed.

Cabin PA Operation

The cabin PA system receives microphone audio and keying from the audio control unit when this function is selected. It outputs a high-level audio to the cabin speakers and provides a sidetone output back to the audio control unit. An audio input is usually provided for a cabin entertainment system, such as a music from a tape player or audio from an onboard television. Most cabin PA units also have a built-in tone generator which provides an alert signal to the passengers when the no-smoking or fasten-seat-belts switches are activated in the cockpit.

The inputs to the PA amplifier are usually arranged on a priority basis with the pilots' control given first priority, the stewards' control given second priority, and cabin entertainment system as the last priority. The input relays in the PA amplifier are interconnected so that the pilot has full control in overriding inputs provided to the PA from the steward stations and cabin entertainment system. The steward's push-to-talk switch has second priority control to disable only the cabin entertainment input to the PA amplifier.

When any one of the input control wires are grounded by selecting PA (or Cabin) on an audio control unit, or by selecting a cabin entertainment switch, a corresponding relay is activated in the PA unit which applies operating power and the input signal to the amplifier. The amplified audio output is then applied to the cabin loudspeaker system.

Usually, the PA volume level can be automatically changed to overcome variations in noise level between ground and flight operations. This is accomplished by connecting the PA audio gain control wire to either a flight/ground switch in the cockpit or to a separate set of contacts on the landing gear switch. When the aircraft is on the ground, the flight/ground switch is closed, resulting in the addition of a gain control signal to the amplifier for reducing the PA audio output to the cabin speakers. When the aircraft is airborne, the gain control signal is removed to restore the cabin PA audio output to a sufficient level to compensate for the higher ambient noise level.

All audio leads between the audio control units, interphone amplifiers, PA amplifier, communication transceivers, navigation receivers, and other audio equipment, must be shielded to prevent the occurrence of audio interference due to inductive cross-coupling with other wiring. It is recommended that the audio wires be 22-gauge twisted-pair and shielded, rather than coaxial cable, so that the return current is not dependent on the conductivity of the shield. The copper-braided shield covering is intended only to reduce inductive coupling of unwanted signals, such as 400 Hz AC from the aircraft electrical system inverters. Shields should be grounded at one point to prevent loop currents, and this common ground point is usually provided at the audio junction box or terminal strip.

RADIO TRANSMITTER PRINCIPLES

A transmitter is basically a device which changes electrical power into electromagnetic radio waves. Its primary function is to generate a radio frequency signal, amplify the RF signal, and provide a means of placing intelligence on the signal. Figure 2-5 illustrates the major components of a typical transmitter, which consists of an oscillator circuit to generate the RF carrier signal and amplifier circuits to increase the power output of the oscillator to the level required for transmission by the antenna.

The voice intelligence or audio frequency signal from the microphone is added to the RF carrier by means of a modulator. The modulator uses the AF signal to vary either the amplitude or frequency of the RF carrier. If the transmitted signal is amplitude-modulated (AM), the magnitude of variation of the RF carrier signal will change with the amplitude of the audio signal

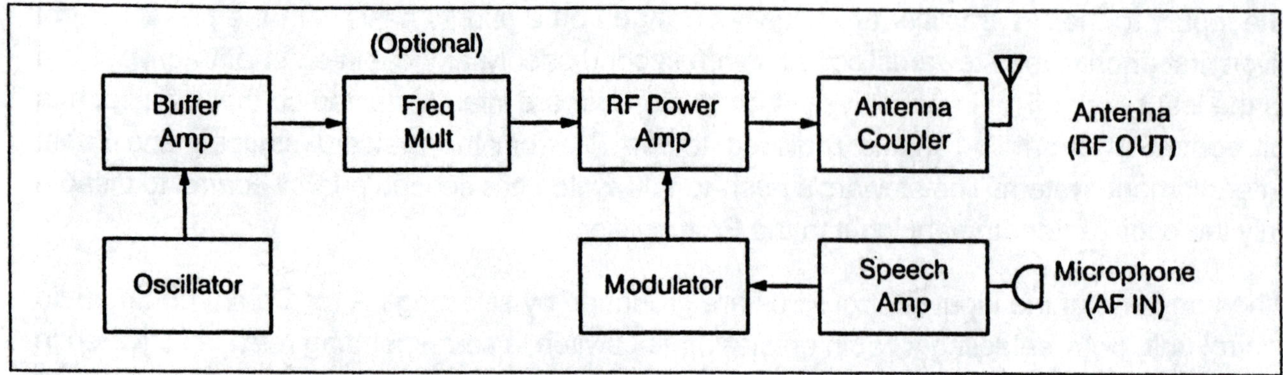

Figure 2-5. Radio Transmitter Block Diagram.

input. If the RF signal is frequency modulated (FM), the instantaneous transmitted frequency will differ from the carrier frequency by an amount proportionate to the amplitude of the applied audio signal. Figure 2-6 illustrates the waveforms for AM and FM signals.

Oscillator Circuits

The heart of every transmitter is the oscillator circuit. An oscillator may be thought of as an amplifier that employs positive feedback, which causes it to generate an AC signal. Oscillators are designed to resonate at a particular frequency determined by the charge/discharge rate of a capacitor connected across an inductor forming an LC tank circuit. The resonant frequency (Fr) of an LC tank circuit can be calculated by using the following formula:

$$Fr = 159/\sqrt{LC}$$

where:

Fr is in kilohertz, L is in microhenrys, and C is in microfarads

The AC signal generated by the tank circuit is amplified and a portion of the output is then returned to the tank circuit through a feedback loop which sustains the flywheel effect in the tank to maintain the oscillations. Various feedback coupling techniques include the use of a tickler coil in the Armstrong Oscillator, autotransformer in the Hartley oscillator, or capacitive coupling in the Colpitts Oscillator. In some cases, the oscillator may be designed to output a multiple of its resonant frequency by tuning the oscillator output coupling circuit to a harmonic of the fundamental frequency or by employing a separate amplifier to accomplish this, known as a frequency multiplier (doubler or tripler).

The oscillator circuits just described are commonly found in many typical receivers; however, transmitter oscillator circuits employ the use of a piezoelectric quartz crystal in place of the LC tank circuit in order to maintain a high order of frequency stability. When a quartz crystal is excited by the voltage applied from the feedback loop, it will vibrate at a set frequency determined by the thickness and cut of the crystal. Frequency selection of the transmitted

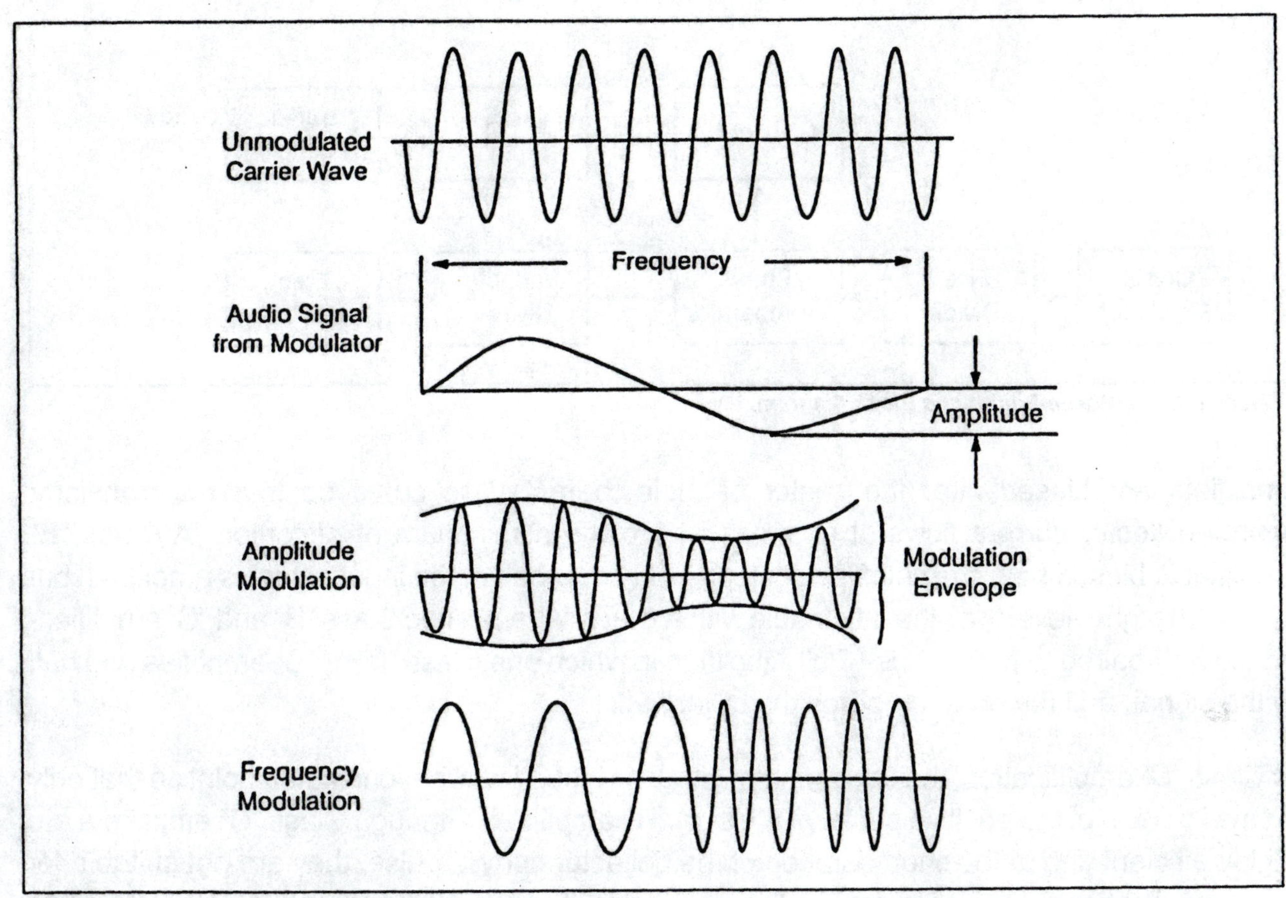

Figure 2-6. AM and FM Waveforms.

frequency can be accomplished by either switching a selected crystal from a bank of available crystals or by using a frequency synthesizer technique.

Frequency Synthesis

Frequency synthesis may be accomplished by using a mixer circuit to combine the outputs of a high-frequency and low-frequency oscillator, each with selectable crystals, to provide a wide selection of available frequencies. A more acceptable method, commonly found in avionics equipment, is the use of the phase-lock-loop circuit, shown in Figure 2-7, in which the output of a variable-frequency, voltage-controlled oscillator (VCO) is constantly compared with the frequency of a crystal-controlled reference oscillator. Any unwanted change or drift in frequency of the variable oscillator in respect to the reference oscillator is detected by the phase comparator. When a phase difference exists, the phase comparator generates a control (error) voltage which returns the VCO to the correct frequency. In this manner, only one crystal is needed to generate a large magnitude of stable frequency outputs.

Amplifier Circuits

The purpose of an amplifier is to increase the gain of the applied input signal. Amplifiers are classified according to their operating characteristics as either Class "A", "B", or "C". Class "A"

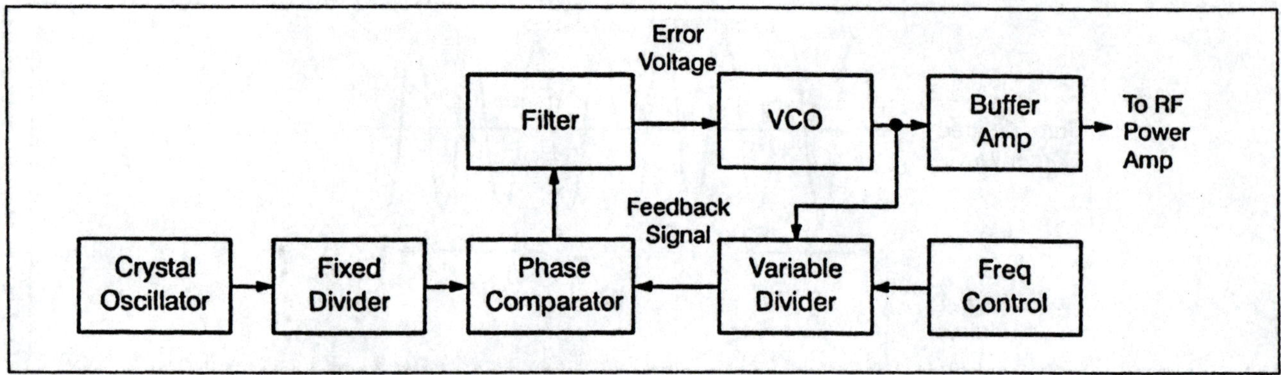

Figure 2-7. Phase-Lock-Loop Block Diagram.

amplifiers are biased near the center of their characteristic curve so that the transistor emitter-collector current flows at all times to provide a minimum of distortion. A Class "B" amplifier is biased near the cutoff point so that only one-half of the input signal is amplified, but at a much higher level than that attainable with a Class "A" amplifier. Class "B" and "C" amplifiers are usually paired to form a Push-Pull Amplifier, in which one Class "B" or "C" amplifies one half of the signal, and the other amplifies the other half.

In Class "C" amplification, the base-emitter circuit is biased well beyond cutoff point so that only a small portion of the positive peaks of the signal is amplified. Although Class "C" amplifiers are highly efficient due to the short duration of the collector current pulse, they are not suitable for audio applications since severe distortion would result from the missing parts of the signal. In RF applications, the output from a Class "C" amplifier pulses a resonant LC tank in the coupling circuit into a fly-wheel type of oscillation which reproduces the original sine-wave input signal.

A typical transmitter may use any combination of the above amplifiers depending on its application in the overall design. For example, the purpose of the buffer amplifier is to isolate and prevent loading of the oscillator in order not to affect its output frequency. Therefore, a buffer amplifier is usually operated as Class "A" so that it does not draw power from the oscillator circuit. The output from the buffer amplifier can be fed either to a frequency multiplier, if needed, or directly to the RF power amplifier.

A frequency multiplier (doubler or tripler), as previously mentioned, is basically a Class "C" amplifier whose output tank circuit is tuned to a harmonic of the oscillator frequency. The RF power amplifier, also operated Class "C", provides the needed RF power output to couple the modulated signal to the antenna. The antenna coupler usually consists of an adjustable LC circuit which matches the transmitted frequency to the resonant frequency of the antenna. The speech amplifier, which may be operated either Class "A" or "B", amplifies the audio signal input from the microphone to be applied to the modulator.

Modulator Circuits

Airborne VHF and HF transmitters amplitude-modulate the RF carrier in such a manner that the resultant RF waveform changes in accordance with the audio waveform, as previously shown in Figure 2-6. With AM, the applied audio signal must vary the amplitude of the carrier from twice the normal level during positive audio peaks to near zero level during negative audio peaks in order to fully modulate the carrier. This is known as 100% modulation. If the modulation percentage is increased above this point, overmodulation will occur causing the signal to be distorted. For this reason, an audio clipping circuit is used to prevent overmodulation.

The modulator combines the amplified audio signal from the microphone with the amplified RF carrier signal to place intelligence on the transmission. A typical amplitude modulator consists of a modulation transformer in which the primary is fed from the output of the speech amplifier. The secondary is in series with the output LC tank coupling circuit of the final RF power amplifier and the power supply. Audio signals induced across the secondary of the modulation transformer from the speech amplifier will either add to, or subtract from, the voltage applied to the RF power amplifier, thus varying the output of the RF carrier in respect to the amplitude of the applied audio signal.

Amplitude modulation results in the creation of two additional frequencies, besides the carrier and audio signals, known as sidebands. The upper sideband (USB) is higher than the carrier frequency by an amount equal to the frequency of the applied audio signal. The lower sideband (LSB) is lower than the carrier frequency by the same amount. The total bandwidth of the AM signal is the distance between the two sidebands, which is equal to twice the audio frequency, as illustrated in Figure 2-8.

Single Sideband

The control panel for a typical airborne HF communication system usually incorporates a selector switch for AM or single sideband (SSB) operation. If a SSB mode is selected (either USB or LSB), the carrier and one sideband will be suppressed resulting in the transmission of only one sideband. In SSB mode, the bandwidth of the transmitted signal is reduced to one-half of that used for standard AM operation resulting in greater range and usability under the most trying propagation conditions.

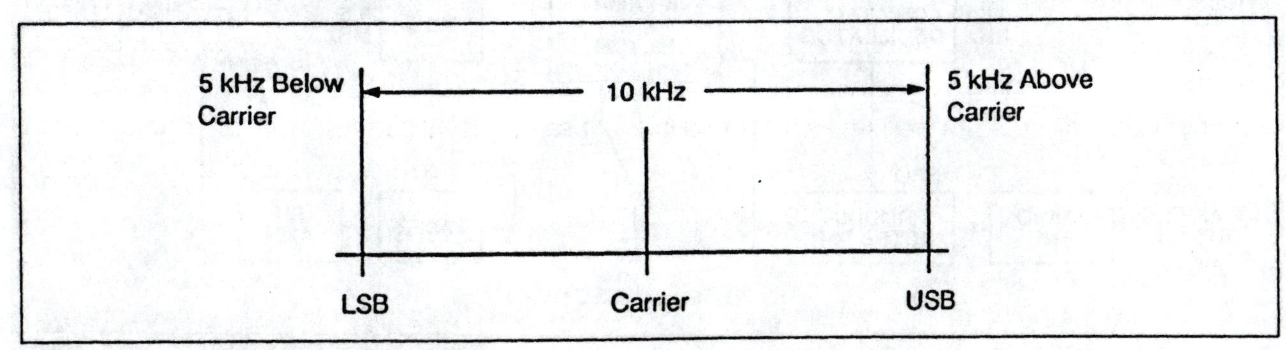

Figure 2-8. AM Sideband Generation.

A block diagram of a typical SSB transceiver is shown in Figure 2-9. The SSB transmitter uses a mixer circuit or balanced modulator, which translates the AF signal supplied by the speech amplifier into RF sidebands adjacent to the carrier frequency generated by the low-frequency crystal oscillator. In addition, the balanced modulator rejects (balances) the carrier, leaving only the two sidebands to be output to the filter which selects the desired sideband. For standard AM transmission, the filter is bypassed and the balanced modulator is unbalanced. The second mixer combines the resultant signal with the output of a high-frequency VCO. From there, the signal is amplified and sent to the antenna. The process works in reverse for SSB reception.

RADIO RECEIVER PRINCIPLES

The amount of RF energy radiated by a transmitter affects the signal strength of the electromagnetic field radiating from the antenna, and as previously mentioned, the amount of power radiated from the antenna is attenuated as the square of the distance from it. The purpose of the radio receiver is to detect and amplify the transmitted signal to a useable level. For example, if a transmitter with a one watt output sends a radio wave along a 20-mile path at an attenuation of two decibels (dB) per mile, the signal will be attenuated minus 40 dB. However, a receiver with a gain of plus 40 dB will have an output of one watt.

The primary function of the radio receiver is to select RF signals and convert the intelligence contained within these signals into audible signals for communication or visual signals for navigation. A receiver must be able to select the desired frequency from all those existing in free space and amplify the small AC signal voltage. The receiver contains a demodulator circuit to remove the intelligence from the RF signal. If the demodulator is sensitive to amplitude changes, it is called an AM detector. A demodulator circuit that is sensitive to frequency changes is known as an FM discriminator.

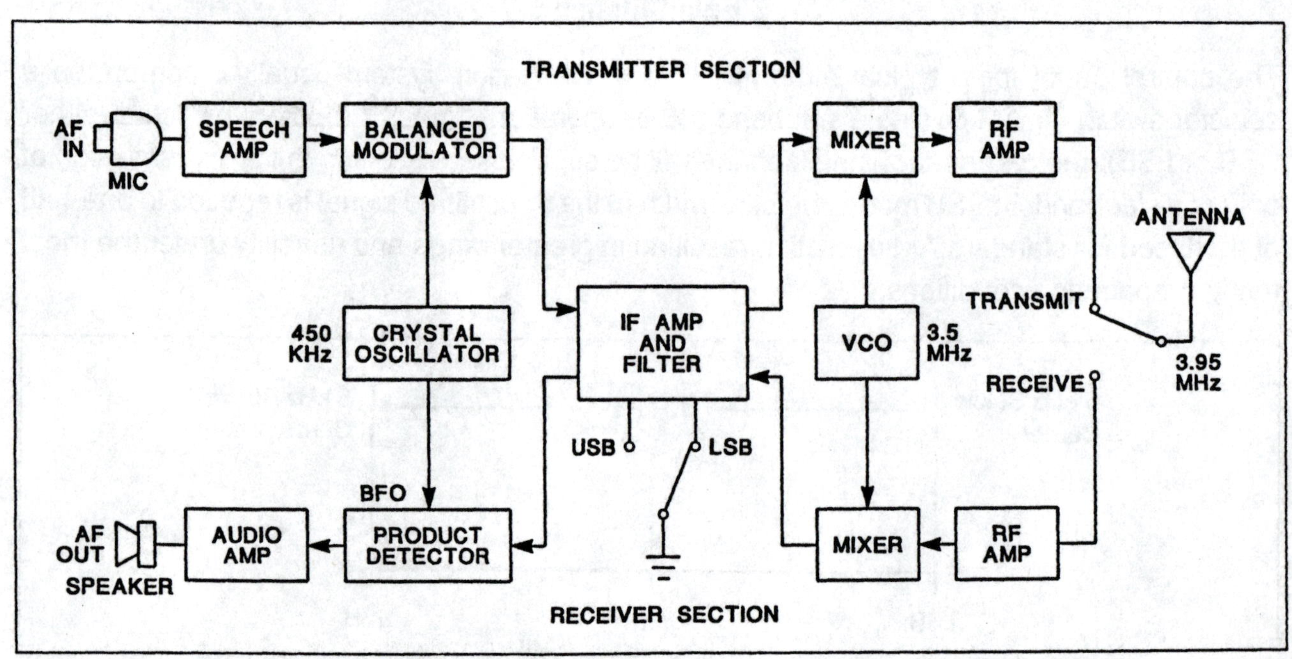

Figure 2-9. SSB Transceiver Block Diagram.

Amplifying circuits within the receiver increase the RF signal prior to demodulation to improve its sensitivity, or the ability if the receiver to detect a weak signal and distinguish it from random atmospheric noise. An audio amplifier is used after demodulation to increase the audio signal to a power level necessary to operate a headphone or a loudspeaker.

Receiver Circuit Theory

As previously mentioned, aircraft VHF and HF communication equipment transmits and receives amplitude-modulated RF signals. For purposes of illustration, a circuit diagram of the simplest type of AM receiver, known as a crystal receiver, is shown in Figure 2-10. Here, millions of radio waves of an infinite number of frequencies cut across the antenna, thereby inducing a current that is passed through the primary winding of the antenna coil (L1) to ground. This RF energy induces a voltage into the secondary winding (L2). Variable capacitor (C1) and inductor (L2) form an LC tank circuit which can be tuned to the desired resonant frequency by adjusting the value of C1. The diode detector, with the polarity as shown, only allows the positive half of the modulated RF signal to pass. Finally, the bypass capacitor (C2) filters the RF portion of the resultant signal to ground, sending the remaining audio signal to the headphones.

The Superheterodyne Receiver

Although the previous discussion on the crystal radio circuit theory is very simplistic, it is the foundation of all other types of receivers, including the superheterodyne receiver, which is used in virtually all aircraft receivers. The superheterodyne receiver operates on the principle of changing the frequency of a received signal to a lower, fixed, intermediate frequency prior to detection. This results in high selectivity, or the ability of the receiver to separate signals on closely adjacent frequencies. The process of combining two different frequencies to obtain a totally different frequency is known as heterodyning. This process, which was presented briefly in the previous section on SSB, is discussed in greater detail in the following paragraphs.

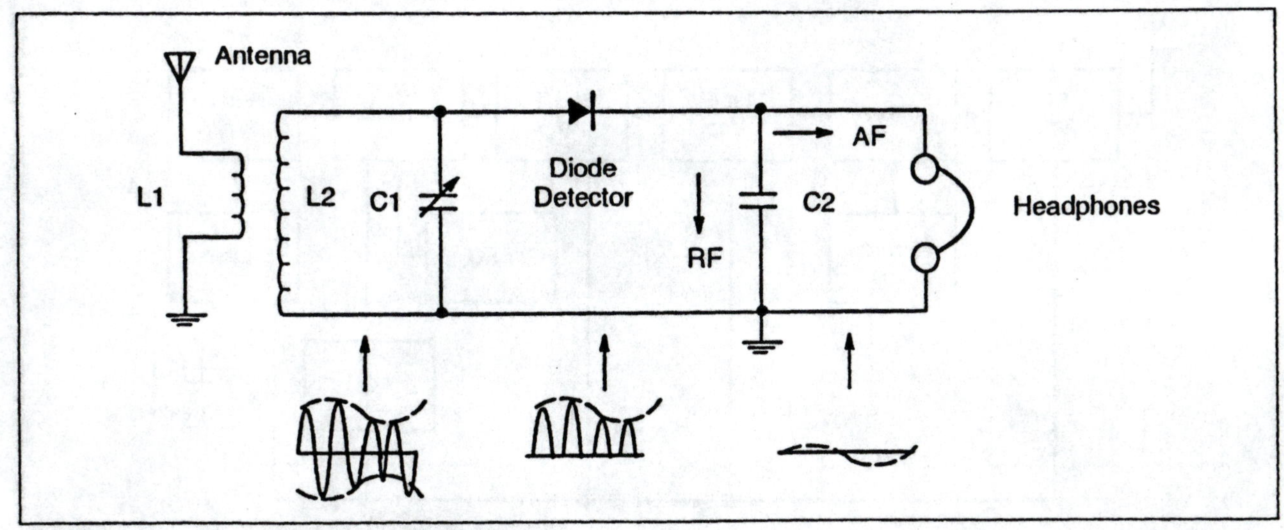

Figure 2-10. Crystal Receiver Circuit Diagram.

A block diagram of a typical superheterodyne receiver is shown in Figure 2-11. The RF signals induced in the antenna enter into the input of an RF Class "A" amplifier, which contains a resonant LC tuning circuit to select the desired frequency. The RF amplifier increases the signal amplitude of the modulated carrier and applies it to the mixer. At the same time, the local oscillator generates a frequency which is injected into the mixer. The mixer is a nonlinear Class C amplifier that combines the inputs from the local oscillator and the RF amplifier to produce two modulated intermediate frequencies, which are equal to the sum and the difference of the received carrier and the local oscillator frequencies. In addition to the two intermediate frequencies, the mixer also outputs the received carrier frequency and local oscillator frequency.

Variable capacitors in the LC tank coupling circuits, contained within the RF amplifier, local oscillator, and mixer, are mechanically ganged together so that no matter what received frequency is selected, the local oscillator frequency will change accordingly so that the lower of the two intermediate frequencies is always constant. For example, if the receiver frequency is changed from 130 MHz to 120 MHz, the local oscillator will automatically adjust its frequency output from 110 MHz to 100 MHz, so that the difference frequency of 20 MHz is always derived from the mixer regardless of the received frequency.

The carrier frequency, the local oscillator frequency, and the two intermediate frequency (IF) signals are sent from the mixer to another RF amplifier, known as an IF amplifier. The IF amplifier incorporates resonant LC tank coupling circuits which are tuned to the 20 MHz IF, thereby eliminating the other three frequency signals. The IF amplifier is designed to pass only a single frequency (20 MHz) within a specified bandwidth (typically 10 kHz) to provide a high selectivity. Usually, several IF amplifier stages are used to increase the gain of the modulated IF signal prior to detection.

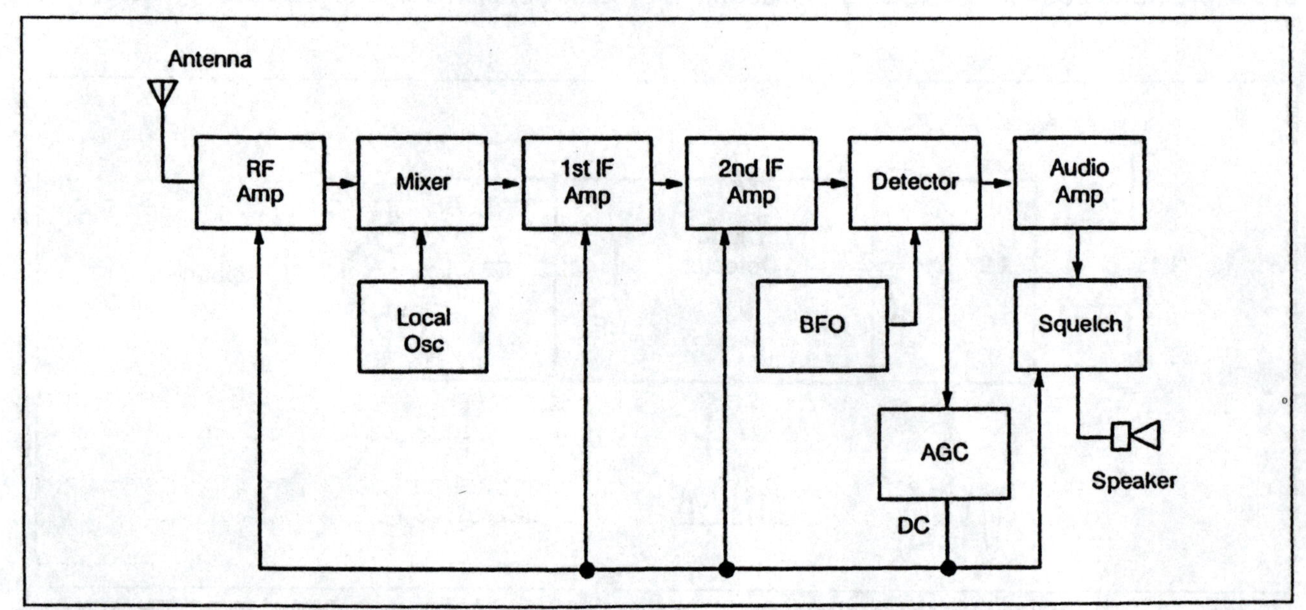

Figure 2-11. Superheterodyne Receiver Block Diagram.

Some receivers incorporate an additional oscillator and mixer between the IF stages. This process is known as double conversion. The second mixer outputs a lower IF than the first mixer to aid in the reduction of unwanted image frequencies.

The AM diode detector or SSB product detector demodulates the IF signal to extract the audio intelligence. The audio signal is then sent to both the Automatic Gain Control (AGC) detector and the audio amplifier. The audio amplifier increases the gain of the audio signal for sound reproduction by the speaker. If the receiver is designed to receive SSB transmissions, the Beat Frequency Oscillator (BFO) will inject a local carrier frequency into the detector that corresponds to the frequency of the RF carrier eliminated in the SSB transmitter to produce intelligible speech reproduction. Refer to Figure 2-9.

AGC and Squelch Circuits

An automatic gain control circuit is incorporated to automatically adjust the gain of the RF and IF amplifiers with variations in the signal strength of the received signal. This results in a near constant audio output level regardless of how close or far away (within limits) the aircraft is from the transmitting station. The AGC circuit is typically a shunt diode circuit, whereas the detector is a series diode circuit. Following the shunt diode in the AGC circuit is a long time-constant resistive-capacitive (RC) network, which filters the half-wave audio signal into a pure DC voltage. This AGC voltage is then used to control the bias of the preceding RF and IF amplifier stages to adjust the RF gain relative to the signal strength, and thereby prevent amplifier saturation which causes distortion in the output.

At times when no signal is being received, the AGC adjusts the RF gain to the maximum level, thereby resulting in the detection and amplification of atmospheric noise. Since it would be annoying for the pilot to have to listen to the background hiss emanating from the speaker when no transmissions are being received, communication receivers include an audio squelch circuit. The squelch circuit automatically disables the audio output from the receiver, except during the time that a signal is being received. The pilot should adjust the squelch control to a point just slightly above the background noise level. An incoming signal above this threshold level will open up the audio circuit so that the transmission may be heard. A typical squelch circuit uses the DC signal from the AGC applied to the gate of a Field Effect Transistor (FET), which acts as a series audio gate. The squelch threshold level is controlled by varying the signal gate voltage.

VHF COMMUNICATION SYSTEMS

Aircraft VHF communication systems consists of a VHF transceiver, control head, antenna, and an interface to the aircraft audio system for access to the microphone and cockpit speaker. In light aircraft, the transceiver is mounted in the instrument panel and contains all the necessary controls and displays. In larger aircraft, the control head, which is used for selecting the receiver and transmitter frequencies, is usually located in the center console between the pilot and copilot, and the transceiver is remotely located in the radio rack aft or below the crew station.

2 of 5 Code Table					
No.	**A**	**B**	**C**	**D**	**E**
0					
1	X	X			
2	X		X		
3		X	X		
4		X		X	
5			X	X	
6			X		X
7				X	X
8	X			X	
9	X				X
X = Ground					

Table 2-2. Two-out-of-five frequency selection.

Some control heads employ a universally accepted 2-out-of-5 frequency selection scheme. Other control heads use a digital serial data bus, such as ARINC 429, to select the desired frequency of a remotely-mounted transceiver. With 2-out-of-5 tuning, any two out of a maximum of five frequency selection inputs to the transceiver will be grounded by the selector switch in the control unit to correspond with the desired frequency selection, as shown in Table 2-2. For example, if the control head displays the frequency 121.5, the "A" and "C" 10-MHz, "A" and "B" 1-MHz, and "C" and "D" O.1-MHz frequency selection inputs to the transceiver will be grounded by the discrete signals from the control head.

The ARINC 429 serial data bus provides a balanced differential signal using nominally zero to 5-volt switching levels from the control head. ARINC 429 is also used to send digital data from the avionics equipment to the cockpit displays.

ARINC 429 messages are comprised of 32-bit data words, as shown in Figure 2-12. Each bit in the data word is set at either "0" if no voltage is present, or "1" if +5 volts DC is present. This serial data stream runs across the two-wire bus at speeds of up to 100 kilobits per second using a command-response protocol.

The message format requires that a record, consisting of up to 126 data words, begin with an initial word that notifies the receiving unit that a message is being sent, and ends with a final word that is used to test for errors in the record. Initial words and final words do not contain data in bits 11 through 29. The first 8 bits in the initial word may contain one of the following messages: Request to Send, Clear to Send, Data Follows, Data Received OK, Data Received Not OK, or Sync Lost. The first 8 bits in the final word is the file label, and bits 9 through 29 is the error control checksum, which is the addition of bits 9 through 29 in all the intermediate words in the record.

ARINC 429 does not provide for error correction, but only error detection within the serial data stream. Typically, manufacturers of avionics systems will provide not only ARINC 429, but a variation of this format, such as the Collins Commercial Standard Digital Bus (CSDB), as their own unique data bus to be used for sending and receiving data between only their brand of equipment.

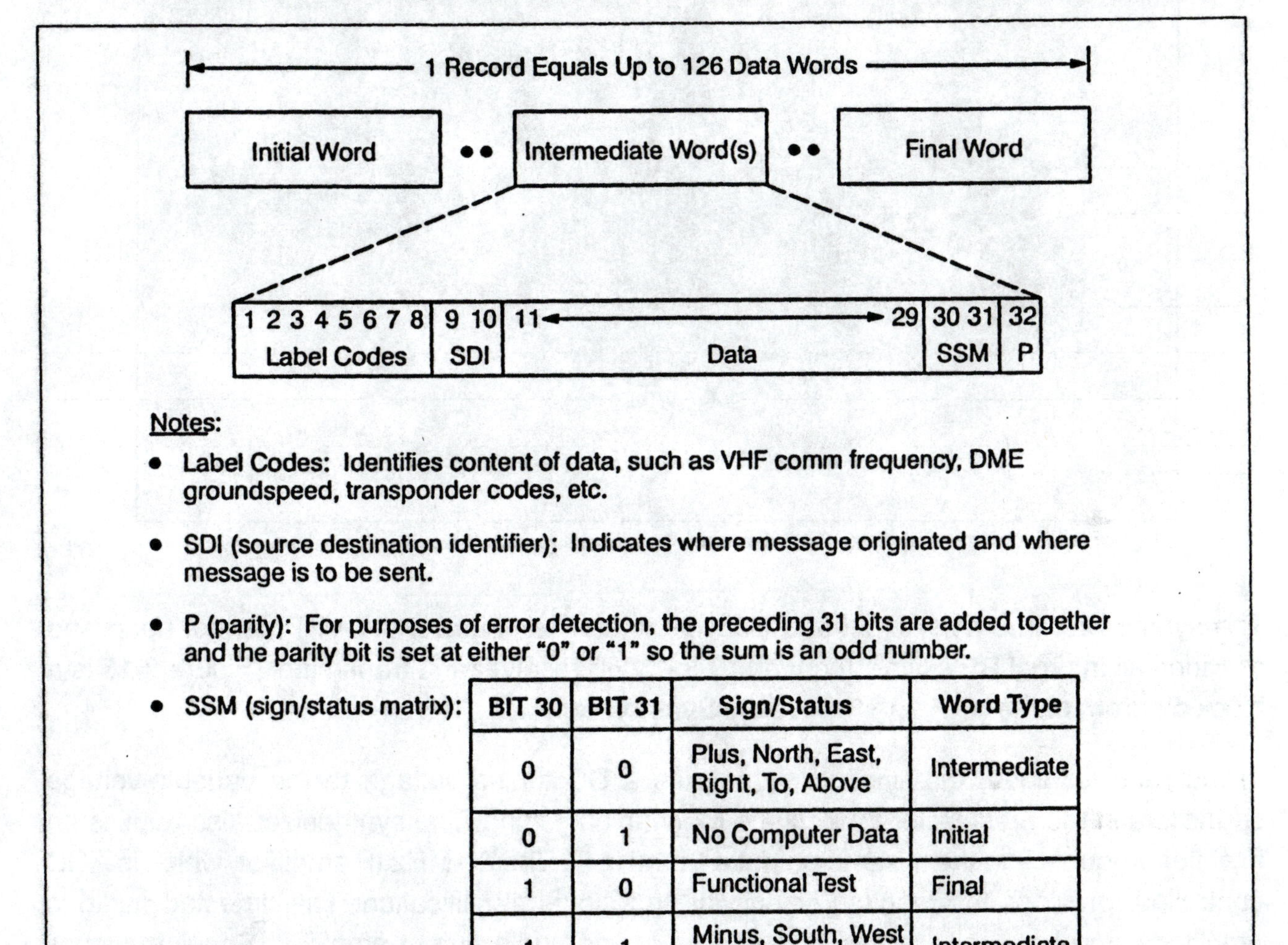

BIT 30	BIT 31	Sign/Status	Word Type
0	0	Plus, North, East, Right, To, Above	Intermediate
0	1	No Computer Data	Initial
1	0	Functional Test	Final
1	1	Minus, South, West Left, From, Below	Intermediate

Figure 2-12. ARINC 429 Data Bus Message Formats.

ROCKWELL COLLINS VHF-20A VHF COMMUNICATIONS TRANSCEIVER OPERATION

The Collins VHF-20A, a typical remotely-mounted VHF transceiver, shown in Figure 2-13, provides AM voice communication in the frequency range from 117.00 MHz through 135.975 MHz, in 25-kHz increments. The VHF-20A consists of a power supply, frequency synthesizer, receiver, modulator, and transmitter, as shown in Figure 2-14. The VHF-21/22 is an advanced microprocessor-based version of the VHF-20A that employs the ARINC 429 and Collins CSDB.

The VHF frequency synthesizer, having only one crystalcontrolled oscillator, derives accurate RF output frequencies through the use of a phase-lock-loop and solid-state switching circuits.

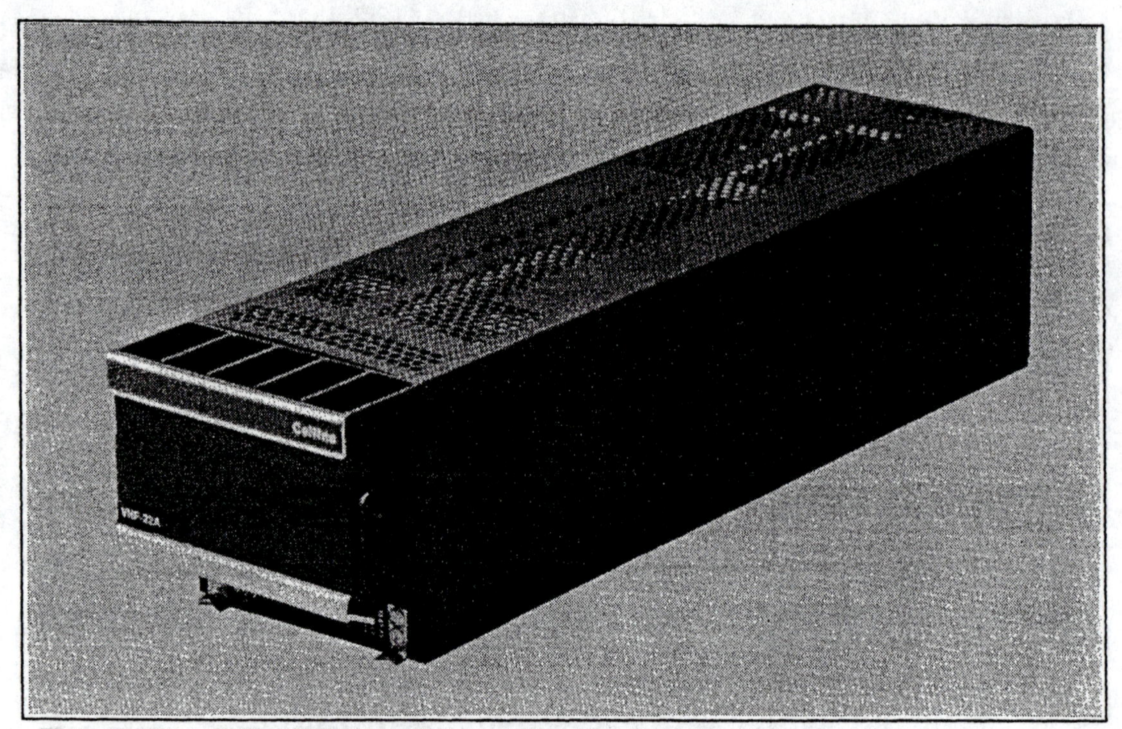

Figure 2-13. ***Collins VHF-20/21/22 Transceiver.*** *(Courtesy Rockwell Collins Avionics)*

The synthesizer interprets 2-out-of-5 frequency information from the VHF control head and provides all internal RF signals required by the VHF receiver and transmitter. Figure 2-15 is a block diagram of the VHF-20A Frequency Synthesizer.

In the receive mode, the synthesizer outputs a DC tuning voltage to the variable-voltage capacitors in the preselector to eliminate mechanical tuning. The synthesizer also applies an injection frequency to the mixer to output a 20-MHz IF. The 20-MHz IF amplifier, which is AGC controlled, provides the required selectivity and signal amplification. The detected audio is amplitude and bandpass limited and applied to the audio output amplifier. Squelch circuits disable the output amplifier if proper signal-to-noise ratio or carrier level is not present.

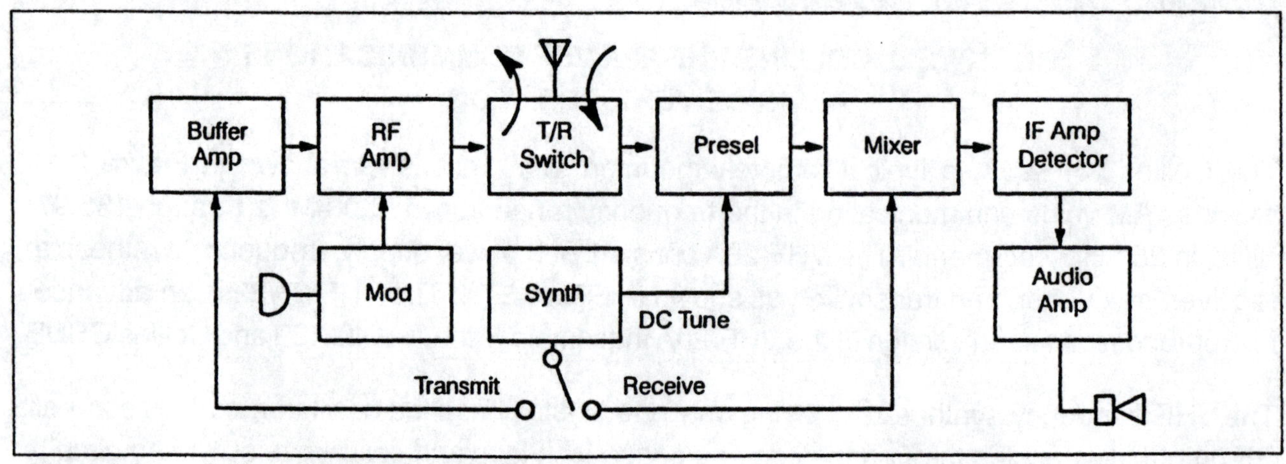

Figure 2-14. Collins VHF-20A Block Diagram.

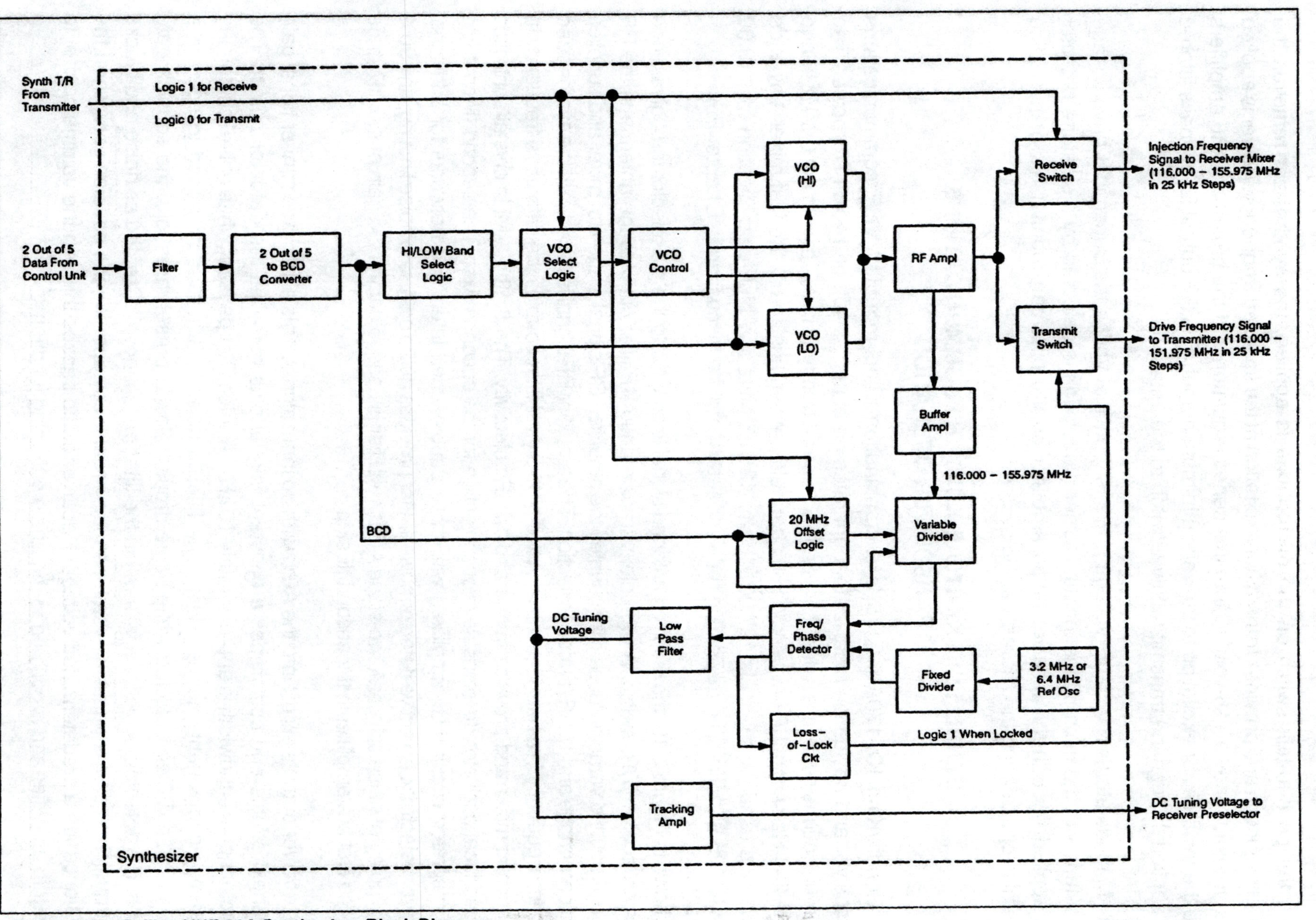

Figure 2-15. Collins VHF-20A Synthesizer Block Diagram.

When the push-to-talk switch on the microphone is applied, the synthesizer removes the receiver injection and provides transmitter excitation at the selected frequency. Power is applied to the transmitter by a +16-volt DC transmit series regulator, and the broadband RF amplifiers raise the synthesizer excitation to 20 watts minimum output. The RF output is low-pass filtered and applied through the transmit/receive switch to the antenna.

The AM modulator is a variable voltage power supply that varies the transmitter drive voltage consistent with the microphone inputs. Carrier modulation is detected by a sidetone detector and applied through the receiver audio amplifier so the pilot can monitor his or her voice transmissions through the aircraft's audio system.

BENDIX/KING KX-170A/KX-175 VHF COMMUNICATIONS TRANSCEIVER OPERATION

The Bendix/King KX-170A/KX-175 is a combination panel-mounted VHF communications transceiver and navigation receiver that operates on either 28-volt or 14-volt DC power. The communications section is a dual-conversion, superheterodyne receiver with a 9.0-MHz IF and a 861.25-kHz second IF frequency; 360 channels are synthesized at the first mixer. Low-side injection is used for channels 127.00 MHz to 135.95 MHz and high-side injection for 118.00 MHz to 126.95 MHz. The navigation receiver section will be discussed in Chapter Four.

As shown in Figure 2-16, the received antenna signal is coupled to the preselector through a diode transmit/receive (T/R) switch. A two-pole, varactor-tuned RF filter couples the antenna to the RF stage. A second varactor-tuned filter couples the amplified RF signal to the first mixer and supplies additional image and 1/2 IF spurious rejection. The amplifier RF signal is mixed with the synthesized injection frequency in a balanced mixer. A two-pole crystal filter couples the difference frequency to the second mixer and provides image and 1/2 IF selectivity. The 8.13875-MHz crystal controlled second local oscillator develops develops injection for the second mixer. The second IF contains two integrated circuit (I.C.) amplifiers with three double-tuned interstage networks for additional receiver selectivity. An active detector/noise limiter provides audio gain, rate noise limiting, and 90% AM clipping of noise spikes. A two-stage AGC amplifier is used to control the gain of the RF stage and the first I.C. amplifier in the second IF strip.

The receiver outputs six dB into the AGC with no input signal. This eliminates conventional gain threshold effects and establishes a constant "signal plus noise" at the detector output. The detector noise bandwidth is approximately 15 kHz. A noise filter passes "white noise" containing frequency components above 7 kHz. The filtered noise is amplified and used to operate a squelch gate. When detected white noise drops below a preset threshold, the squelch gate opens. If a detected audio tone falls within the filter passband, it is treated as noise and blocks the squelch. A carrier squelch overcomes this problem by opening the squelch gate when the AGC exceeds a predetermined voltage. Receiver audio passes through the volume control to an audio amplifier and is coupled to the audio summing junction.

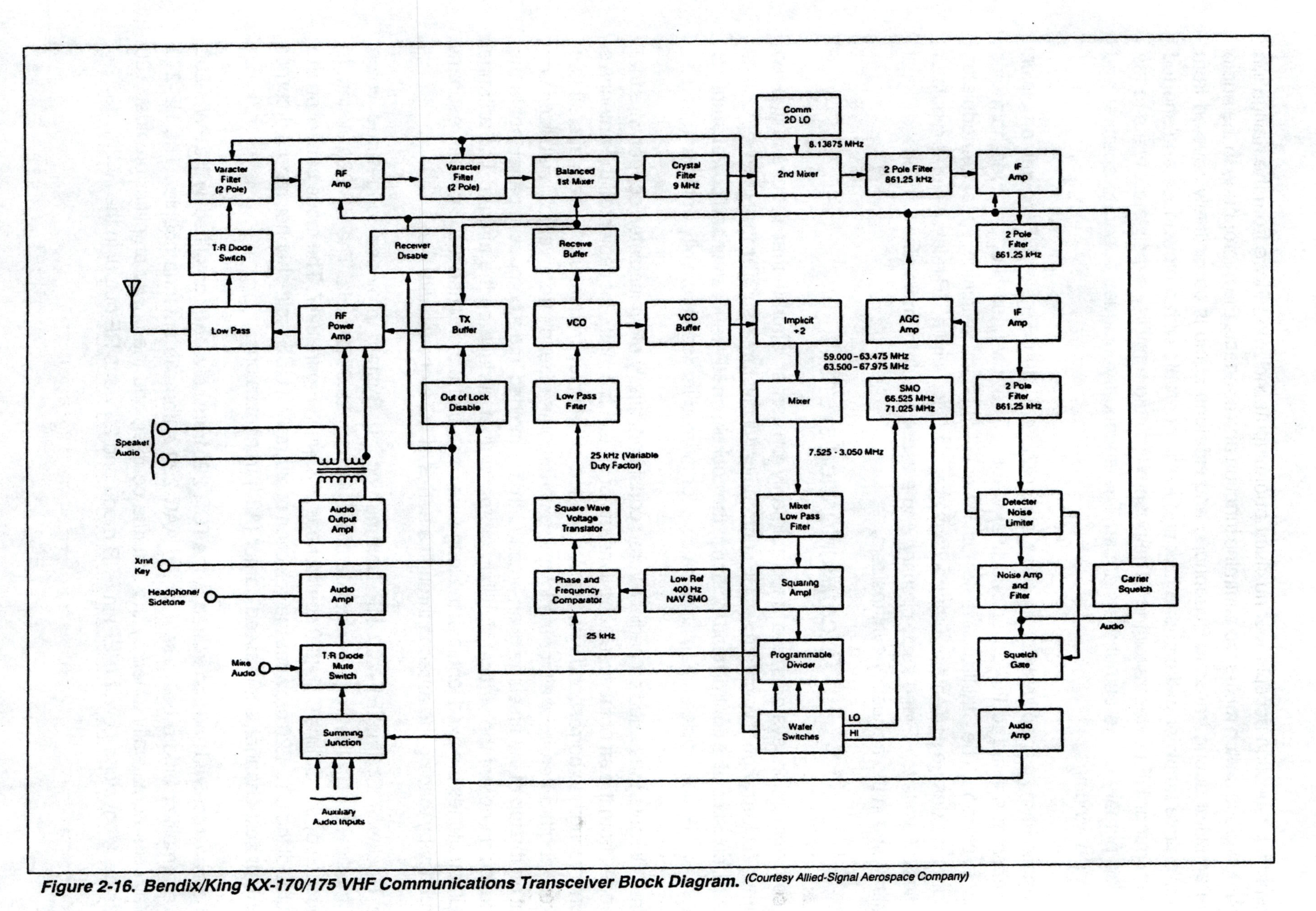

Figure 2-16. Bendix/King KX-170/175 VHF Communications Transceiver Block Diagram. *(Courtesy Allied-Signal Aerospace Company)*

The summing junction accepts four auxiliary audio inputs including those from the navigation receiver. A diode switch mutes the summing junction and connects the microphone to the audio power amplifier during transmit. Headphone and sidetone audio is capacitively coupled from the audio drive amplifier to the phone jack and is still operable with a blown fuse and shorted audio output or RF power transistors. The push-pull audio amplifier supplies six watts to the balanced primary of the output transformer. The secondary includes separate speaker and modulation windings.

The transmitter is a solid-state, four-stage, broadband, 30-dB gain, RF power amplifier. Modulation is applied to the driver and final stages. The low-pass filter provides harmonic spurious rejection. The Stabilized Master Oscillator (SMO) band switching, speaker/modulator, headphone/sidetone and antenna connections are controlled with a T/R relay. A series regulator supplies 8.5 volts to RF and audio circuitry, and a zener regulator maintains 5.0 volts to digital circuitry used in the frequency synthesizers.

HF COMMUNICATION SYSTEMS

HF communication systems are used primarily on larger aircraft that require extended communication range. Typical aircraft HF communication systems, such as the Rockwell Collins HF-220, consists of: a control head located in the cockpit, an HF transceiver and power amplifier located in the radio rack, and an antenna coupler located in close proximity of the antenna.

Aircraft HF antennas used on earlier aircraft consisted simply of an insulated copper wire that extended from the aft of the fuselage by means of an electric motor that adjusted the antenna's resonant length to correspond with the transmitted or received frequency. Modern aircraft use a fixed length antenna of either the longwire variety that extends from vertical stabilizer to a point on the top of the fuselage or an antenna that is molded into a composite panel that forms the leading edge of the vertical stabilizer or some other surface on the aircraft. The antenna coupler matches the fixed length of the long-wire or embedded antenna to the proper resonant frequency by loading the antenna through a series of LC tuned circuits.

The Rockwell Collins HF-220 HF Communication System provides long-range voice communications in the frequency range of 2.0 MHz to 29.9999 MHz. The HF-220 system is controlled by the Collins CTL-220 control head, shown in Figure 2-17. The mode control on the left side of the CTL-220 provides the selection of either AM; USB; split-channel, reduced-carrier (A3A); or split-channel, suppressed-carrier (A3J) transmissions.

Signals are received and transmitted by the TCR-220 transceiver and amplified by the PWR-200 power amplifier before being sent to the AAC-220 automatic antenna coupler. The HF-220 system is automatically tuned by the antenna coupler, and the power amplifier provides 100 watts peak envelope power (PEP) in SSB mode and 25 watts RF output in the AM mode.

Referring to the signal flow diagram of the HF-220 in Figure 2-18, the operation of the system is initiated when the on/off mode selector on the CTL-220 control is rotated clockwise out of detent. A relay in the PWR-200 is then actuated and switches +28 volts from the aircraft power bus to the PWR-200, TCR-220, and AAA-200. Each of the individual equipments have an internal power supply to provide secondary voltages. Control logic in the TCR-220 grounds the key-line whenever the microphone PTT switch is pressed. A relay in the TCR-220, actuated by the transmitter key-line, supplies power to the transmit circuits and a relay in the PWR-200, actuated by the same key-line, applies power to the power amplifier circuits.

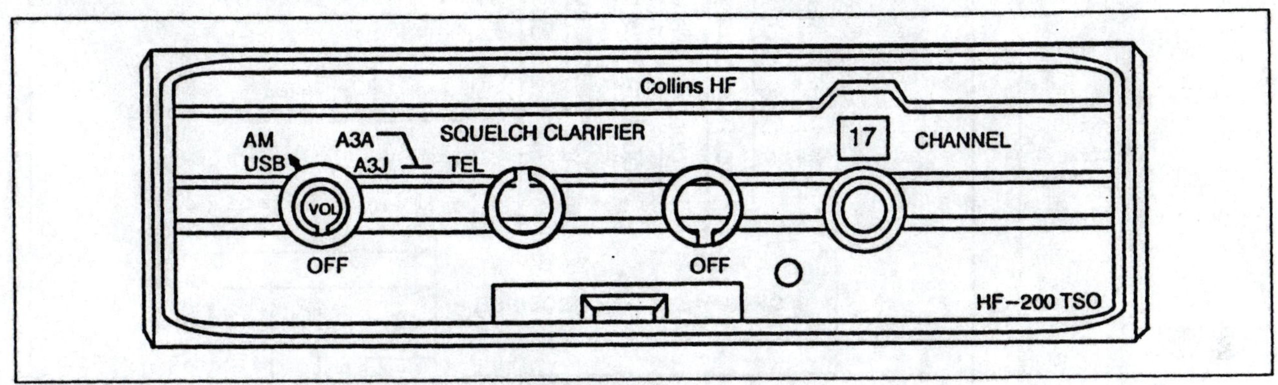

Figure 2-17. ***Rockwell Collins CTL-200 HF Control Head.***

Frequency selection involves the use of the frequency and channel controls on the CTL220, and the program card, synthesizer, and reference divider/multiplier in the TCR-220 transceiver. Direct tuning of the CTL220 control allows selection of any frequency in the 2.0-MHz to 29.9999-MHz frequency range with 100-Hz frequency spacing. Channel selection on the CTL220 allows selection of any one of 20 specified frequencies programmed on the program card. CTL220 clock and data lines send the selected frequency or channel information in serial form to the program card. The system provides either duplex operation, in which the transmitted frequency is different from the received frequency, or simplex operation, in which the transmitted and received frequencies are the same.

The HF220 system receive and transmit frequencies are established by a doubleloop circuit in the synthesizer of the TCR220. The synthesizer is frequency and phase locked to an ovenstabilized, crystalcontrolled frequency reference oscillator. The CTL220 converts the frequency or channel selected into 22-bit serial control information. The program card converts the serial control information to 22 parallel control lines that are inputs to the synthesizer. If the frequency synthesizer becomes unlocked from the crystalcontrolled frequency standard, an interrupted tone is applied to the audio output indicating that the transmitter is disabled.

In receive mode, either the USB or AM circuits in the TCR-220 are selected by the mode control switch on the CTL-220. The received signal from the antenna is applied through half-octave low-pass filters in the PWR-200 that provide protection from high-frequency, out-of-band signals. An additional bandpass filter and over-voltage diodes in the TCR-220 provide further protection for the receive

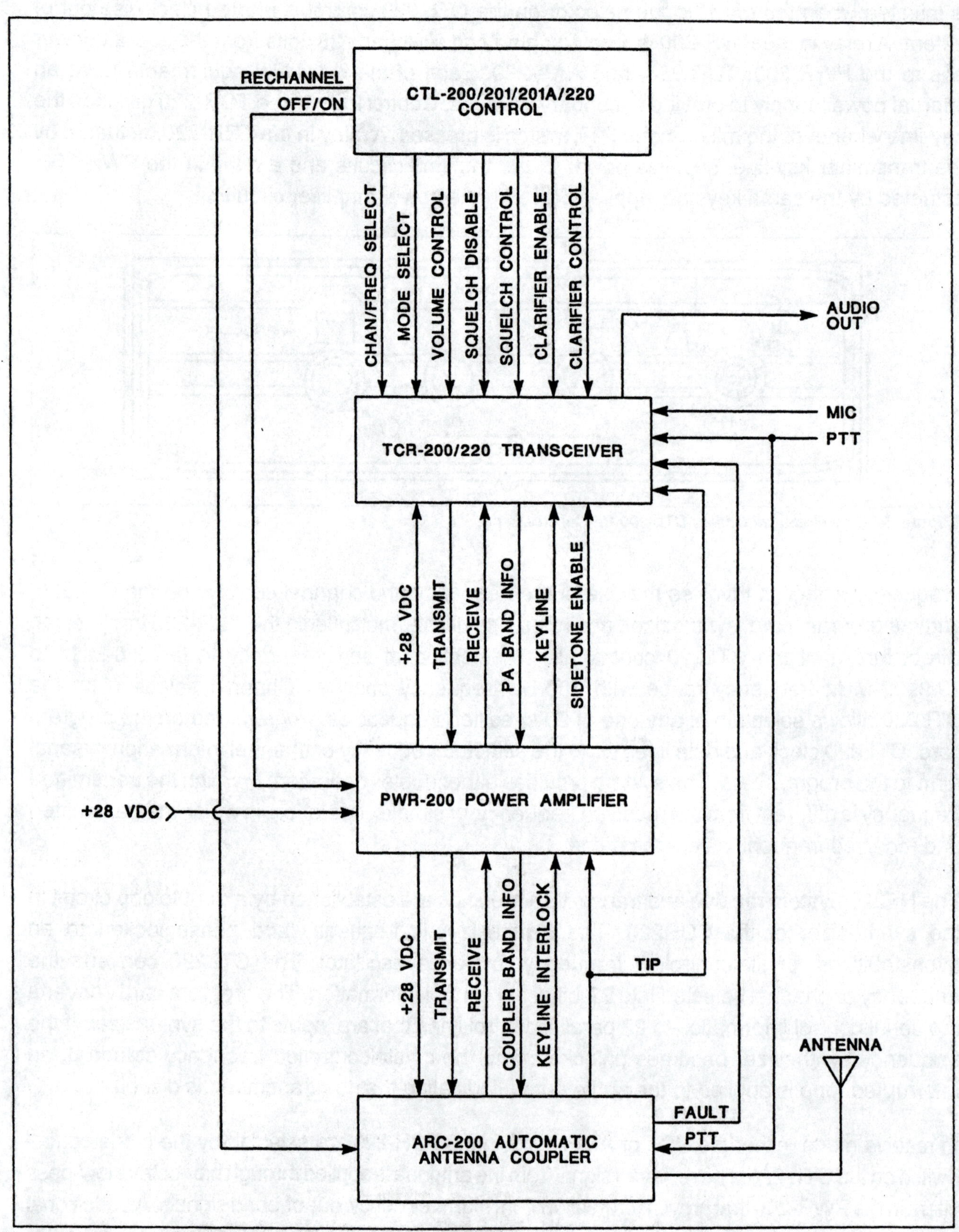

Figure 2-18. Rockwell Collins HF-220 Signal Flow Diagram.

circuits. The receive signal is then amplified, mixed to 69.8 MHz to remove spurious responses, filtered through a crystal filter, and mixed down to 500 kHz, the IF frequency. A low-noise IF preamplifier, with either the 500-kHz SSB or AM filter, is selected by control logic in the TCR-220. The signal is then amplified by the remaining IF amplifiers. Control logic in the TCR-220 actuates a product detector for SSB signals or a diode detector for AM signals. The resulting audio signal is remotely controlled by use of an electronic attenuator with the DC control located in the CTL-220. Audio is then amplified to 50 milliwatts and applied to the headphone output.

Two systems of AGC are used. The normal IF AGC operates up to an RF input level of approximately 100 microvolts. When the 100 microvolt limit is reached, the RF AGC circuits are actuated and attenuate the RF amplifier and the IF preamplifiers up to an input level of 10 millivolts. Both AGC systems continue attenuating together up to an input level of approximately 1.0-volt.

The squelch control on the CTL-220 adjusts the IF gain, resulting in smooth quieting of the receiver as the control is rotated clockwise. An audio switching circuit mutes the audio whenever the AGC drops below the threshold determined by the position of the squelch control. A receiver test function is provided by a switch on the CTL-220 which inhibits the audio muting. The CTL-220 also provides a "clarifier" control to improve the reception of off-channel SSB station. This control allows the BFO, and thus the received SSB signal, to be varied plus or minus 100 Hz from the center frequency.

The transmit circuits can be actuated in two different ways: one, by grounding the key-line (microphone PTT button), the other by grounding the tune-in progress (TIP) line. The TIP ground is automatically applied whenever the AAC200 is being tuned. The microphone PTT button is momentarily pressed, then released, to initiate the system tuning cycle.

When transmitting (PTT button pressed), a DC voltage is applied to the microphone. Microphone audio is amplified or compressed, as required, and applied to the balanced modulator. The proper sideband is selected by the mechanical filter, and further amplified by the transmit IF amplifier stage. If AM mode is selected, a 500-kHz carrier is injected into the IF amplifier stage. If the AAC-200 is tuning (TIP line grounded), the microphone audio signal is inhibited and the full carrier is injected instead. This ensures that only a phase-locked continuous-wave (CW) signal (no modulation) is used for tuning.

The same frequency translator that is used for receive is used in the opposite direction for transmit. The IF signal is mixed up to 69.8 MHz, filtered by the crystal filter, and mixed down to the desired frequency. From the translator, the signal is applied to a low-noise amplifier to keep broadband noise down. A low-pass filter for harmonics and a 33-dB, 250 milliwatt broadband RF amplifier complete the TCR-220 exciter portion of the transmitter.

The PWR-200 amplifies the RF signal from the TCR-220 to the 100-watt PEP level. This is accomplished with two stages of push-pull amplification. Following the final amplifier is a set of six one-half-octave, low-pass filters. The proper filter is selected by band decode logic in the

TCR-220 and positioned by a rotary solenoid in the PWR-200. An automatic level control (ALC) voltage is obtained in the PWR-200 from the peak detector, which limits peak envelope power to 100 watts; from the average detector, which limits average power to 25 watts; from the low-voltage detector, which limits maximum PEP during low-voltage conditions to prevent flat-topping of the signal; and from the standing-wave-ratio (SWR) detector, which reduces the RF power output if the reflected signal exceeds a 3:1 ratio (at 50 ohms) compared to the output signal. A high SWR, due to a mismatch of impedance, causes the output signal to be reflected back into the power amplifier. Therefore, it is important that the coaxial cable be terminated correctly for maximum power transfer and minimum SWR.

The RF output from the PWR-200 is applied to the AAC-200. The AAC-200 automatically matches the 50-ohm output of the PWR-200 to the 10- to 30-foot grounded wire antenna. The AAC-200 will maintain this impedance match over the system operating range of 2.0 MHz to 29.9999 MHz. Signals that provide tuning cycle control of the PWR-200 and the TCR-220 originate in the AAC-200.

HF Automatic Link Establishment (ALE)

One of the problems often encountered with the use of HF radio, is that HF depends on energy being deflected off of the ionosphere layer to achieve long-range communications. This layer changes in density and altitude depending on the time of day. As the ionosphere layer changes, the frequency at which skip waves are most effective also changes. As a result, a pilot who was experiencing good communications with a ground station may find that same frequency used an hour ago is now unusable.

The Rockwell Collins HF-9000 with Automatic Link Establishment solves this problem. The ALE processor automatically scans all the pre-programmed frequencies and selects the one that is most optimum. The HF-9000 then transmits a carrier or CW signal to a ground station equipped with ALE which locks in on the selected frequency. With 26 pre-selected frequencies stored in memory, the linking process takes approximately nine seconds before notifying the pilot, with the presence of white noise in the audio, that he or she can commence communications. ALE and SELCAL are two features that make modern HF radios very user-friendly from a pilot's standpoint.

SELCAL Decoder

The SELCAL (Selective Calling) Decoder relieves the crew from having to continuously monitor the VHF and HF receivers for incoming calls. This greatly reduces pilot workload and fatigue, especially during long-range flights.

When the ground radio operator wishes to alert a specific aircraft via the airborne decoder, he or she selects the four-tone code which has been assigned to the particular aircraft and transmits the modulated code on the VHF or HF frequency being monitored onboard the aircraft. Although the tones will be detected by all aircraft radios within range that are tuned to the frequency transmitted, only the aircraft with the SELCAL decoder set for the correct combination of tone frequencies will be alerted of the incoming call. A particular tone sequence is accomplished by setting four switches on the SELCAL decoder unit located in the radio rack.

As shown in Figure 2-19, actuating the decoder with the proper tone sequence causes a light to be illuminated in the cockpit and an alerting tone to be produced by an audible alarm. Resetting of the alarm circuit is accomplished by closing a reset switch on the SELCAL control panel or by activating the PTT switch on the microphone. The SELCAL control panel also has a selector switch to route the audio signal from the particular VHF or HF radio to be monitored to the SELCAL decoder unit.

SATELLITE COMMUNICATIONS (SATCOM)

The ability to communicate data and voice information over long distances has previously been affected by range and atmospheric conditions. To solve this problem SATCOM was recently developed. SATCOM is a system that provides reliable transoceanic communications that supports a wide range of services, such as Aircraft Communications Addressing and Reporting System (ACARS), cockpit voice, passenger voice, telex and Facsimile (FAX). SATCOM is a communication system that uses satellites as a relay station to transmit long distances, instead of HF frequencies which are susceptible to atmospheric interference. SATCOM also provides interface with ISO-8208 compatible data equipment, as well as data units such as ACARS.

Not only does SATCOM provide a more reliable medium to transmit messages, it is also capable of providing access to more data services that are present on the latest aircraft, to more users. These applications include the transmitting of data (such as ACARS for the automatic reporting of aircraft status, departure time, arrival time, or flight delay time) and voice communications (cockpit communications, as well as cabin communications such as passenger telephone). The users within this system include airlines, airports, air traffic control, and telecommunications.

System Description

The SATCOM system consists of three segments, shown in Figure 2-20: The satellite, the aircraft earth stations, and the ground earth stations. The satellite segment relays radio signals between the aircraft earth stations and the ground earth stations. The aircraft earth station (AES) is the SATCOM system carried on board the aircraft, that interfaces with various onboard communications systems and the ground earth stations (GES). The GES are fixed radio stations that interface with communication networks (through terrestrial communication links) and the AES (through the satellite segment). The system provides a variety of data and voice communication services and has the ability to provide it globally.

Satellites provide the two-way relay between the aircraft earth stations and the ground earth stations. The satellites are placed in geosynchronous orbits that provide coverage between latitudes 75° north and south. The aircraft earth stations transmit to and receive from the

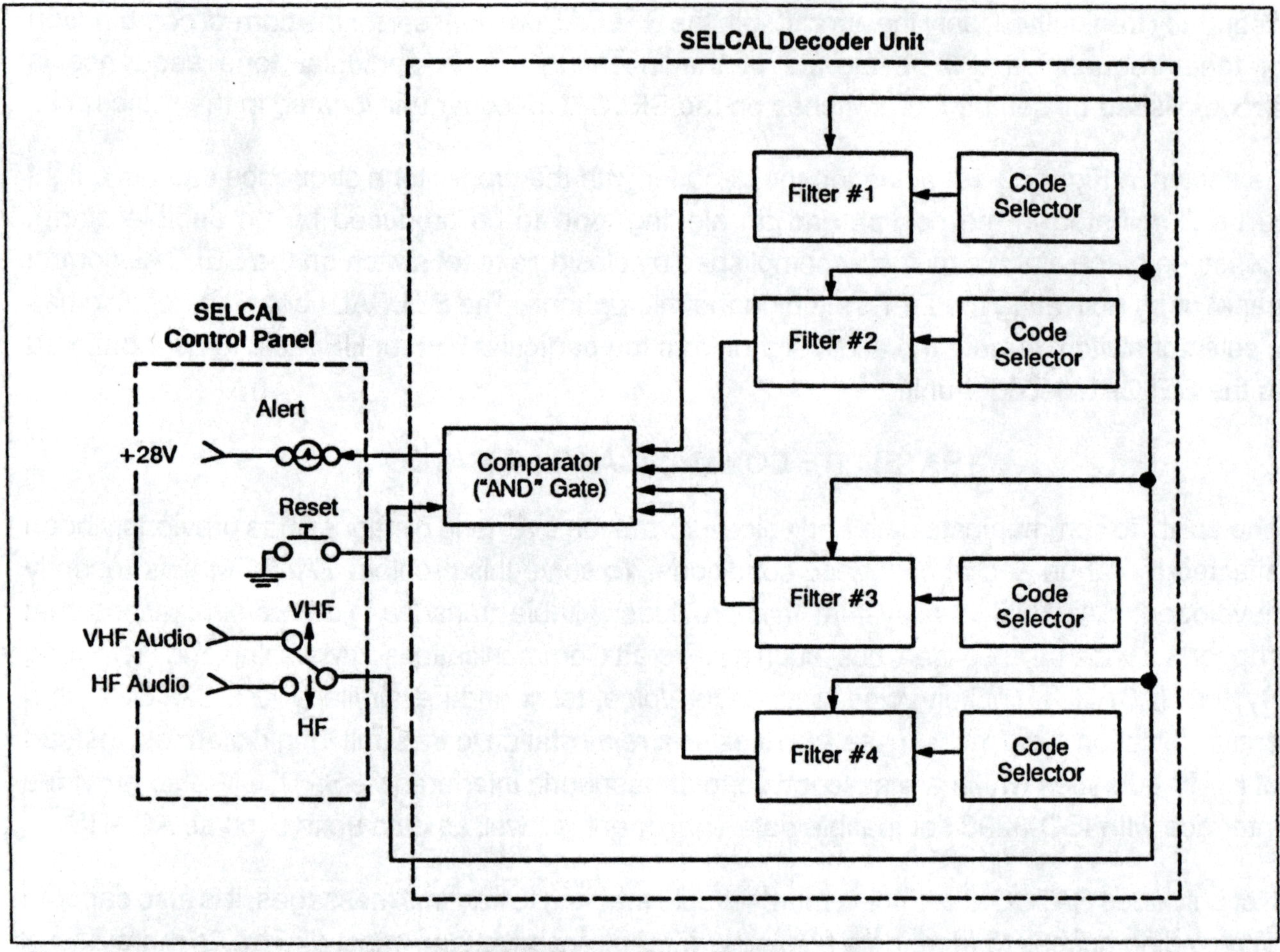

Figure 2-19. Typical Single-Channel SELCAL System Block Diagram.

satellites at L-band frequencies (1,530 MHz to 1,660.5 MHz). The ground earth stations transmit and receive from the satellites at C-band frequencies (4 GHz to 6 GHz).

The function of the aircraft earth station segment equipment is the reception and processing of signals received via a satellite operating in the L-band. Initially (Phase 1), the system will relay low speed data. When expanded (Phase 2), the system will provide voice, and high speed data allowing all aeronautical communications external to the aircraft (including passenger telephone and data services). The equipments required for this operation are described in the following paragraphs. Figure 2-21 shows a typical AES system.

The Satellite Data Unit (SDU) serves as the interface unit to all other related aircraft systems for an aviation satellite system. Within the SATCOM system, the Satellite Data Unit provides an interface to the RF Unit (RFU), the High Power Amplifier (HPA) unit, the Low-Noise Amplifier (LNA)/Diplexer(DPX), and the low gain antenna. Additional subsystem interfaces that could be added would include a full duplex ARINC 429 interface to the ACARS Management Unit (MU), an interface to an ARINC 739 Multipurpose Control Display Unit (MCDU), common message

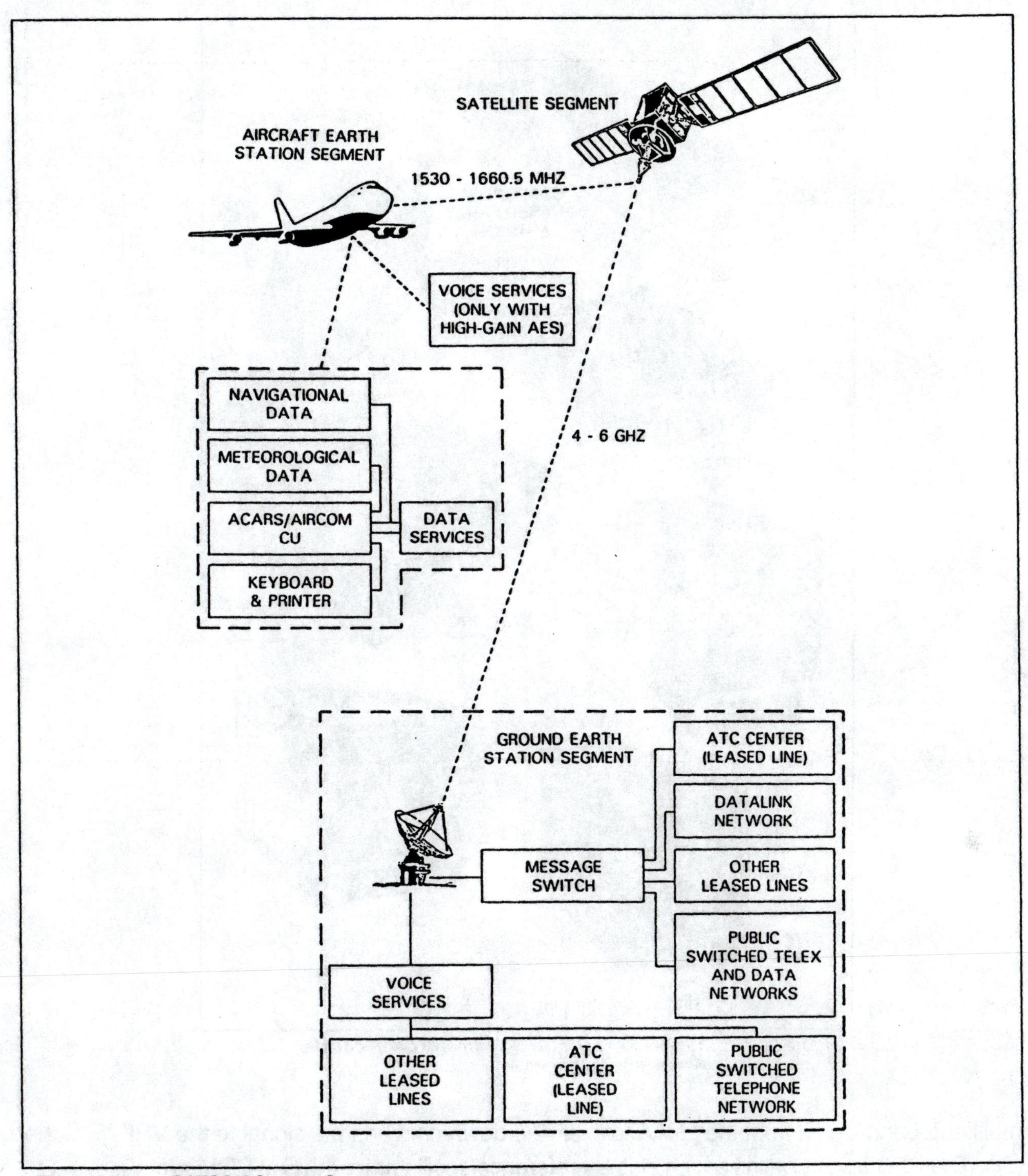

Figure 2-20. ***SATCOM System Segments.*** *(Courtesy Rockwell Collins Avionics)*

processor unit (to be shared with Mode S transponder data link), passenger telephone coder/decoder and Private Branch Exchange (PBX) units.

The HPA provides adequate RF power level, by automatic control, to the antenna. The RFU converts broadband IF from the SDU to an L-band signal applied to the HPA. The RFU also receives an

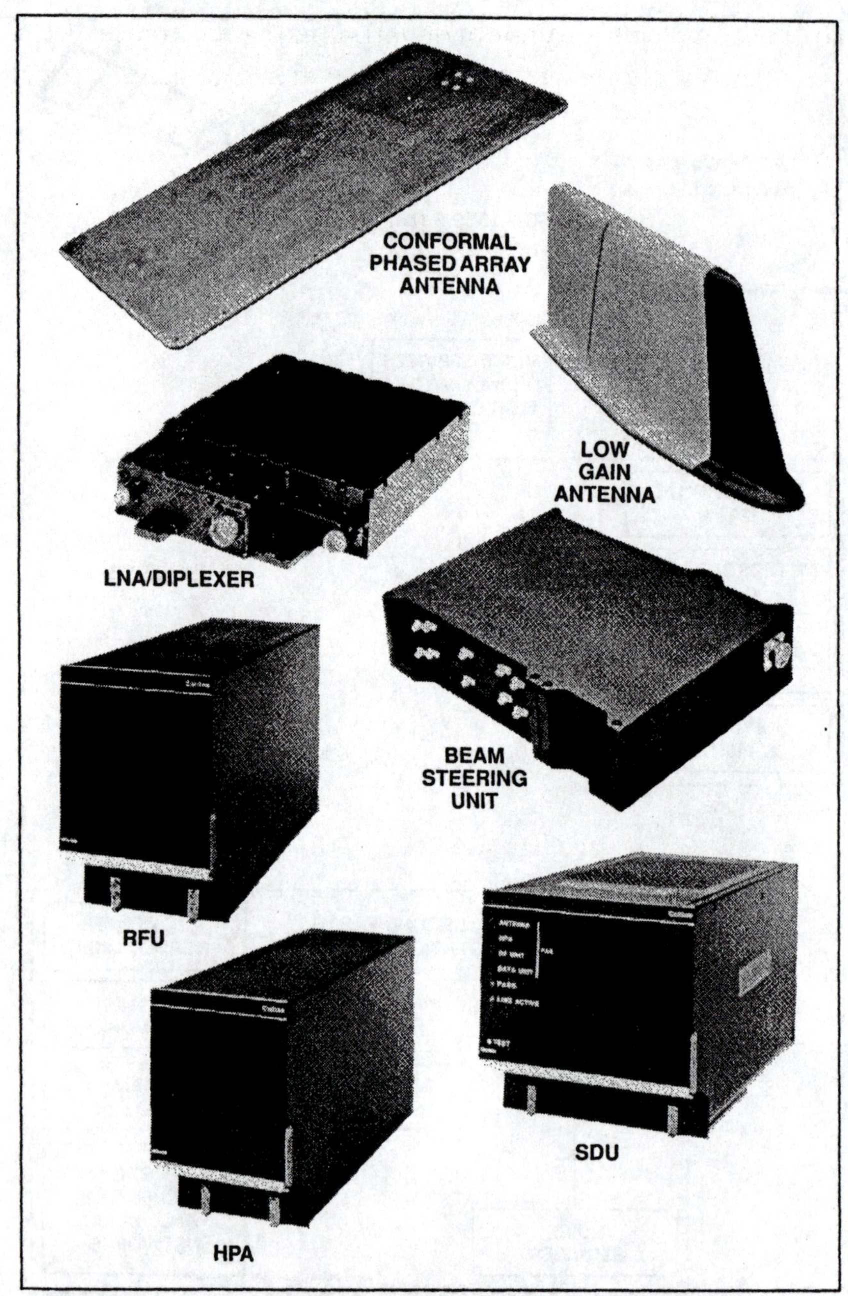

Figure 2-21. Typical SATCOM (AES Segment) Components.
(Courtesy Rockwell Collins Avionics)

amplifier L-band signal from the LNA/Diplexer and down-converts the signal to the VHF IF of the SDU. The RFU also contains a high stability reference oscillator for the SATCOM system.

An optional low-gain antenna may be used to provide low data rate communications. Figure 2-22 shows a block diagram of a low data rate system.

The diplexer and low-noise amplifier are combined into one unit for easy installation. The diplexer couples transmit and receive signals to and from the respective antenna elements to the LNA and from the HPA, while preventing transmit-frequency power from degrading the receiver system. The LNA amplifies the very low level L-band signal from its respective antenna. The LNA also compensates for transmission line losses to the RFU.

Phase two service will require a High-Gain Antenna (HGA). High-gain antennas provide at least 12 dBic of gain. Each antenna provides simultaneous transmission and reception of satellite signals for full-duplex operation.

The beam steering unit is used with electronically steered antennas to translate antenna beam position data and beam change commands received from the SDU in a standard digital format into the signals needed to select antenna elements in combinations that result in the beam pointing at the desired satellite.

The ground earth stations provide system synchronization and coordination through ground-to-aircraft transmissions. The network of ground earth stations allow the aircraft to communicate with any other user of the network. These ground earth stations are generally owned and operated by licensed organizations (nominated by the host country) to invest in and work with, INMARSAT. These organizations are responsible for the provision of communication services.

System Services

The SATCOM network provides satellite communication services for aircraft worldwide. The services required by aircraft include air-to-ground data communications, ground-to-air data communications, traffic control communications, voice communications, and passenger communications. SATCOM supplies three types of data services. These services are connectionless service, connection-oriented service, and circuit mode service.

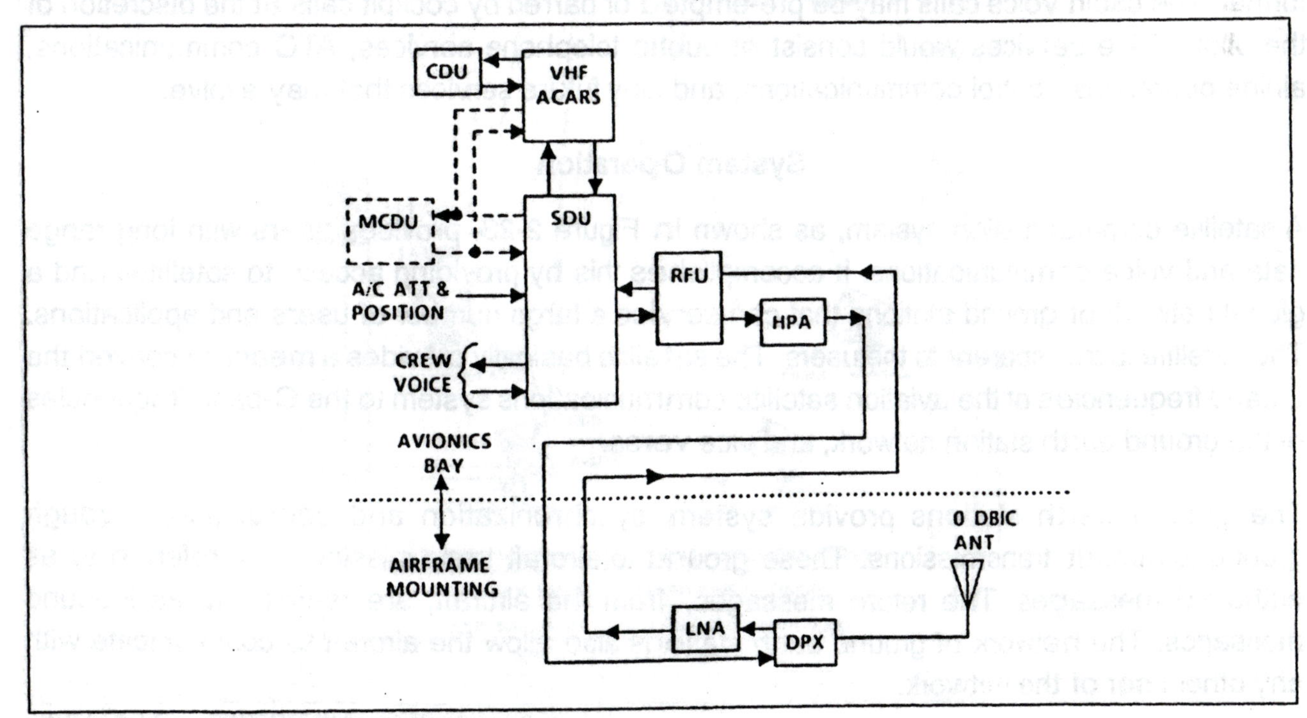

Figure 2-22. ***Low Data Rate System.*** *(Courtesy Rockwell Collins Avionics)*

data links with the operating airline operations base, an ATC control center through the log-on GES, but to operate a passenger communication through a different GES.

The AES consists of a satellite data unit, a radio frequency unit, a high power amplifier, a low-gain antenna, and/or one or two high-gain antennas. Network user information (working and emergency frequencies, satellite positions, available GESs, etc.) is stored as a system table by the AES in nonvolatile memory. This information is updated automatically by the satellite system.

The AES system, when first powered, scans a stored set of frequencies in the range of 1,530 MHz to 1,559 MHz. This is the L-band frequency range where the satellite transmission will be received. The AES locates the satellite transmission and locks on to it. The satellite transmission, as detected by the AES receiver, may be shifted in frequency due to the Doppler effect of the aircraft either approaching or flying away from the position of the satellite.

The AES then logs on to the ground earth station network. The log-on information is shared throughout the network, so that any of the ground stations can locate the AES. Once logged on to the system, communications between the AES and any user can begin. As an example, an ACARS message reporting the time of arrival is to be transmitted. The ACARS system aboard the aircraft sends the message to the satellite data unit. The satellite data unit formats the message for transmission and encodes the data to ensure message integrity. The message is transmitted at a frequency in the range of 1,626.5 MHz to 1,660.5 MHz. The AES transmitted frequency compensates for the Doppler shifting of frequencies between the aircraft and the satellite. From the perspective of the satellite, no Doppler shift occurs. This automatic frequency compensation reduces the frequency errors of the inbound RF communication signals.

The transmitted signal is received by the satellite and converted to a C-band signal and transmitted to the ground earth station network. The satellite does not perform any data manipulation or decoding. The satellite just converts the received signal frequency to the transmitted signal frequency. The ground earth station receives the C-band frequency message from the satellite and decodes the data, then reconstructs the encoded bits to produce the original message. The ACARS message is then sent to the intended user.

EMERGENCY LOCATOR TRANSMITTERS (ELT)

Although not considered an aircraft communication system, our discussion would not be complete without mentioning the Emergency Locator Transmitter. An ELT, required by law, is one device the pilot hopes he never has to use. It is usually installed in a crash survivable portion of the aircraft, such as the tailboom, and is automatically actuated when the aircraft sustains an impact resulting in 5g's or more. Upon impact, the ELT emits a 75-milliwatt tone-modulated RF signal on 121.5 MHz, the civil emergency frequency, and 243.0 MHz, the military emergency frequency. The ELT signal can be detected by search and rescue aircraft within a 100-mile radius at 10,000 feet altitude.

The ELT is completely self-contained, consisting of a transmitter, battery, and whip antenna. An AUTO/ON/OFF switch allows the ELT to be tested periodically. By momentarily (not more than three seconds according to the FARs) placing the switch in the "ON" position, an audible beeping tone can be detected on the aircraft's VHF communications receiver tuned to 121.5 MHz. During normal operation, the ELT is left in the "AUTO" position. The ELT battery will power the transmitter for up to two hours at 55° C to 48 hours at -20° C (Per FAA T50-C91) and has a shelf life of two years. Although most avionics equipment is painted black (hence the term "Black Box"), emergency avionics, such as the ELT, Flight Data Recorder, and Cockpit Voice Recorder, are painted orange to aid in recovery of this equipment for post-crash investigations.

CONCLUSION

This concludes our discussion of audio and communication systems. We began this chapter with a discussion of audio and radio frequency electromagnetic theory and then presented the principles of operation of aircraft audio systems and aircraft transceivers. Examples were given of the operation of three typical HF and VHF communication systems. We concluded this chapter with a discussion of SATCOM and the ELT.

Air-to-ground and air-to-air communications are essential in order to coordinate air traffic. However, it is equally important to be able to navigate to a desired destination. Prior to the advent of radio navigational aids, pilots would navigate their aircraft using pilotage, in which they would rely on the use of a compass and time to destination to chart their progress along a certain course. The limitation of pilotage is that it is only effective during clear daylight, since visual landmarks must be identified to validate that the aircraft is on course. Also, wind direction and speed must be taken into account during flight planning operations.

VHF TRANSCEIVER LABORATORY PROJECT

The objective of this laboratory project is to become familiar with the proper procedure for setting-up, testing, and calibrating a VHF communication transceiver prior to its installation in an aircraft. The equipment required consists of a VHF transceiver, a test panel to provide all the necessary interfaces to the transceiver, an RF signal generator, audio oscillator, RF wattmeter and load, frequency counter, modulation detector, and a Digital Voltmeter (DVM), connected as shown in Figure 2-24. The procedure to be used is outlined in the following six steps.

> ***WARNING***
>
> ***The person conducting the test must either possess a FCC Radio-Telephone license, or be under the direct supervision of someone who is licensed.***

1. Apply 27.5 volts DC (or 14 volts DC) from the power supply to the transceiver under test. Select 125.000 MHz on the control head. Connect the RF signal generator to the antenna jack on the test panel and apply a 1,000 microvolt, 125.000-MHz signal modulated 85% at 1,000 Hz. Measure the audio output voltage with the DVM and adjust the transceiver in accordance with the recommended manufacturer's setting (the VHF-20A is set at 7.75 volts RMS (root mean squared)).

2. Select 118.000 MHz on the control head. Adjust the RF signal generator for a 3-microvolt, 118.000-MHz unmodulated signal. Measure the audio output voltage to obtain a reference level.

3. Modulate the RF signal generator output to 30% with 1,000-Hz audio signal. The audio output should increase 6 dB above the reference level for proper receiver sensitivity. Repeat steps 2 and 3 with the control head and RF signal generator set at the transceiver's highest frequency. If there is less than a 6-dB receiver signal-to-noise difference at either the low- or high-end of the VHF band, refer to the applicable maintenance manual for the proper receiver alignment procedures.

4. Select 125.000 MHz on the control head. Adjust the RF signal generator for a 0-microvolt, 125.000-MHz signal modulated 90% at 8,000 Hz. Monitor the audio output while slowly increasing the RF signal generator output. If the carrier override squelch threshold is set properly, the audio output should appear when the signal generator output is between 10 to 40 microvolts.

5. Disconnect the RF signal generator from the antenna jack, and connect the frequency counter (with attenuator), modulation detector, DVM, RF wattmeter, and antenna load as

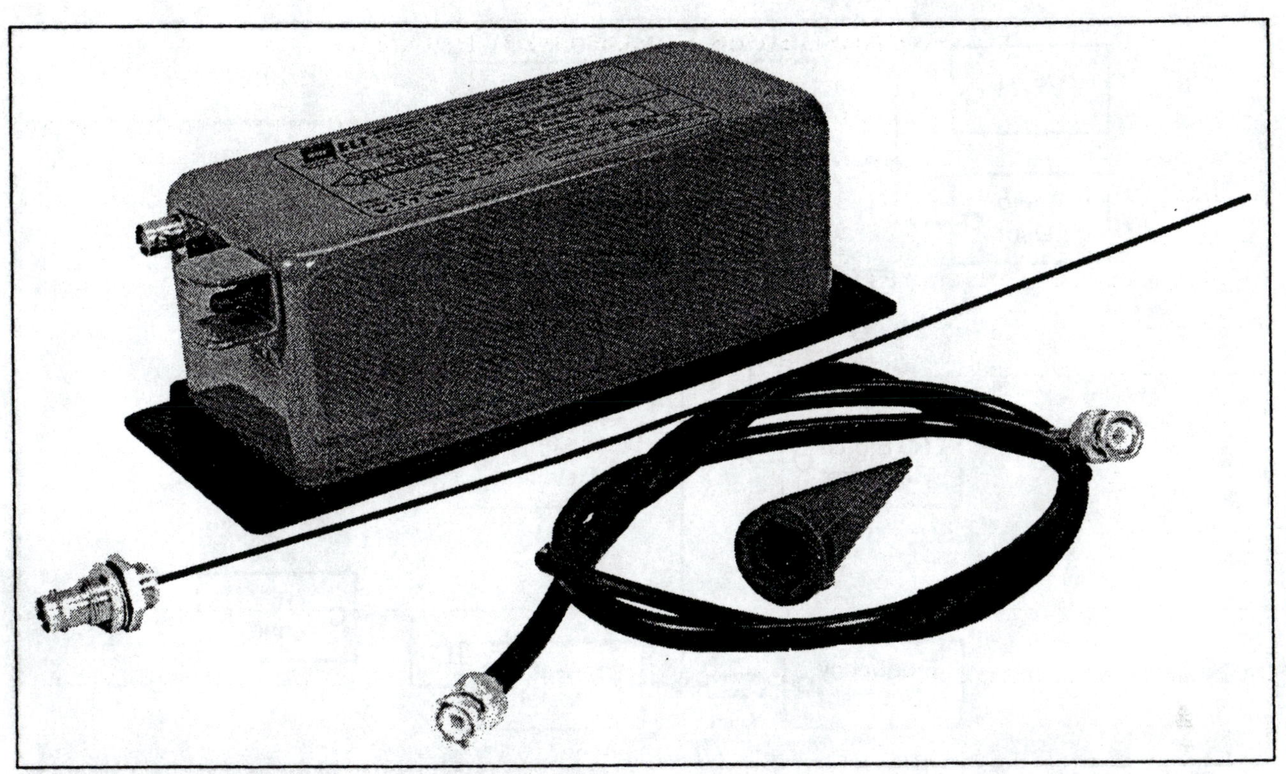

Figure 2-24. Emergency Locator Transmitter. (ELT)

shown. Close the PTT switch and measure all the available transmitted frequencies for accuracy and RF power output using the frequency counter and RF wattmeter.

> ***CAUTION***
>
> ***Do not key the transmitter for a period of more than one minute, and allow several minutes to elapse to allow the transmitter to cool before re-keying.***

Refer to the applicable maintenance manuals for the manufacturer's specifications. (The VHF20A has a frequency tolerance of plus or minus 0.0015% and an RF output between 16 to 24 watts.)

6. Apply 0.125 volts RMS at 1,000 Hz from the audio oscillator to the microphone audio input. Close the PTT switch, measure the sidetone level from the audio output voltage, and adjust if necessary (the VHF-20A is set at 3.90 volts to 7.75 volts RMS). Measure the modulation level with the DVM and adjust the transmitter modulation gain (typically 5.4 volts RMS for 100% modulation) in accordance with the procedures outlined in the applicable maintenance manuals.

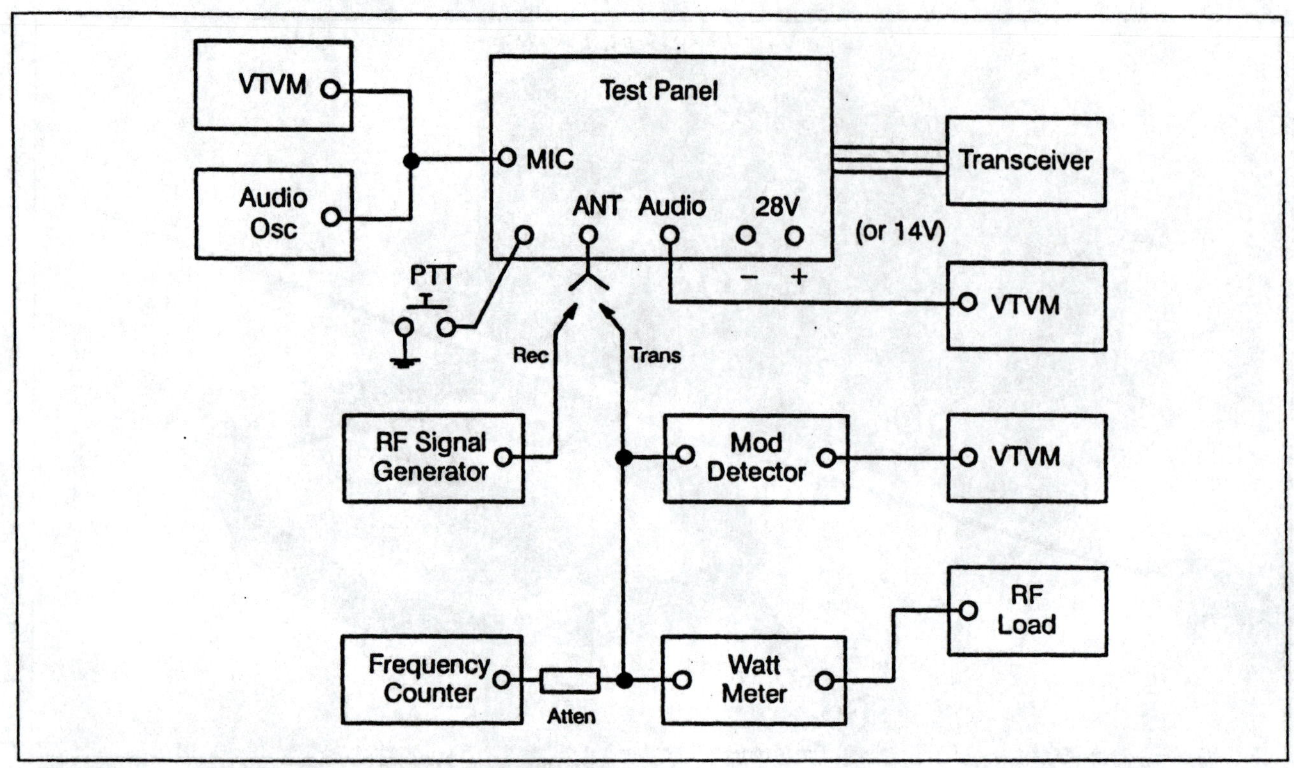

Figure 2-25. ***VHF Transceiver Bench Test Diagram.***

The preceding laboratory project describes the procedures for conducting a pre-installation test on a typical VHF transceiver. (Since these procedures may vary slightly according to the type of transceiver under test, it is recommended that the technician refer to the applicable maintenance manuals.) It is the responsibility of the avionics technician to ensure that the transceiver meets the minimum performance requirements, such as receiver sensitivity, transmitter modulation percentage, power output and frequency accuracy, and that the audio and sidetone outputs are adjusted properly prior to installation.

REVIEW QUESTIONS

1. What is the frequency range and channel spacing of a typical VHF communications transceiver?

2. What is the speed of radio wave propagation and what two components make up this wave?

3. Identify the frequency range of the different bands available in the RF portion of the electromagnetic wave spectrum.

4. Explain why HF radio waves travel at a much greater distance than VHF radio waves.

5. Calculate the frequency of an RF signal transmitted at 40 meters. What would be the resonant length in feet of a half-wave antenna designed to operate on this frequency? (Note: Length in feet = 475/frequency in MHz)

6. What are the principle components and functions of a radio communication system?

7. Identify the signal inputs and outputs controlled by the audio control unit.

8. How are the audio inputs mixed and yet isolated from one another in the audio control unit?

9. Explain the operation of the audio control emergency function.

10. Why is the cockpit speaker muted when the microphone is keyed?

11. What is the function of the sidetone output?

12. What are the different methods used for keying the transmitter?

13. What are the different interphone functions available and what is the purpose of each.

14 Explain how the priority system functions for activating the PA system.

15. What is the purpose of the PA "ground/flight" control?

16. Explain the proper procedures for shielding and grounding audio wires during an installation.

17. What are the three main functions of a transmitter and what circuits are required to provide these functions?

18. Explain the difference between AM and FM transmissions. Which type of modulation is employed in aircraft VHF and HF communication systems?

19. Calculate the resonant frequency of an LC tank circuit having a capacitance of 0.1 microfarads and an inductance of 2.2 millihenrys.

20. What is the difference between a frequency multiplier and a frequency synthesizer?

21. Explain the operation of a phase-lock-loop circuit.

22. What is the purpose of a buffer amplifier and what class is it operated?

23. Why can a Class C amplifier be used in RF applications, but not for amplifying audio signals?

24. Explain how a modulation transformer operates and what is meant by overmodulation.

25. If the USB is 4.5 kHz above the carrier frequency, what is the audio modulation frequency and the signal bandwidth?

26. What is the function of a balanced modulator for SSB operation?

27. Define receiver sensitivity and selectivity.

28. What is the difference between a detector and a discriminator? What is a product detector?

29. Describe the heterodyne process.

30. Explain the similarities and differences in the principles of operation of a simple crystal receiver to the more complex superheterodyne receiver.

31. How is the IF signal derived in a superheterodyne receiver?

32. If a singleconversion superheterodyne receiver with a 20-MHz IF is tuned to a frequency of 125.75 MHz, what is the local oscillator injection frequency to the mixer?

33. What is meant by a double-conversion superheterodyne receiver?

34. What is the purpose of the AGC, and how does it function?

35. Explain why a BFO is needed to receive SSB signals.

36. What is the purpose of the squelch control and how should it be adjusted?

37. What is the purpose of an antenna coupler?

38. What is the difference between simplex and duplex operation?

39. Why should the SWR be maintained as close as possible to a 1:1 ratio when coupling RF energy to an antenna?

40. Explain the operation of a SELCAL decoder.

41. What is SATCOM and what services does it provide?

42. Explain the operation of the three SATCOM system segments.

43. Explain how an ELT is automatically activated and what procedure is used to self-test an ELT.

44. What three avionic units are termed "Orange Boxes"?

Chapter 3
Automatic Direction Finders

INTRODUCTION

The Automatic Direction Finder (ADF) is a very important and integral part of radio navigation. The ADF provides the pilot with an indication of the direction of radio signals received from selected stations operating in the low and medium frequency range of 90 kHz to 1,800 kHz. These stations include non-directional beacons (200 kHz to 415 kHz) and standard AM broadcast stations (540 kHz to 1,600 kHz). Non-directional beacons (NDB) are identified by a CW signal modulated with a 1,020-Hz tone that transmits a three-letter identification code. Occasionally, NDBs will interrupt the CW transmission with a voice transmission to provide weather information and flight advisories. When an NDB is used in conjunction with instrument landing system markers, the beacon is referred to as a compass locator. Compass locators are identified by a continuously transmitted CW two-letter identification code. Standard AM broadcast stations are identified by voice transmission of the station call letters.

The concept of ADF navigation is based on the ability of the airborne system to measure the direction of the arrival of the received signal, and from this information, provide a relative bearing indication with respect to the centerline of the aircraft. Using the bearing information displayed on the ADF indicator, the pilot can determine the aircraft's position or can fly directly to the NDB or AM broadcast station. To determine the aircraft's position, the pilot simply plots the headings of two different stations on a navigation chart and triangulates the aircraft location at the point where the two lines intersect.

PRINCIPLES OF ADF SYSTEM OPERATION

Radio direction finders were developed in the early 1930's as the first radio navigation device to be used for airborne applications. These early devices used an indicator with a left/right needle that would center when the aircraft was pointed toward the station. The radio direction finder has since developed into an automatic system that continuously displays the direction to the station by means of a pointer on the ADF indicator.

A means is usually provided to manually or automatically rotate the compass card on the ADF indicator to the aircraft's magnetic heading so that the pointer not only indicates the direction to the station, but points to the magnetic heading the aircraft must take to fly towards the station. If the compass card is driven by a synchro, which receives heading information from the compass system, the instrument is known as a Remote Magnetic Indicator (RMI). RMIs are discussed in more detail in Chapter 5.

All ADF systems employ the directional characteristics of a loop antenna to find the direction to or from the NDB or AM broadcast station. The directional pattern of the loop antenna is such that if positioned so that the ends of the loop are in alignment with the incidence of the radio wave, the received RF signal will be maximum. However, if the loop is rotated 90° from this position, the signal will fade out. This is known as the "null" position. A non-directional sense antenna is also used in conjunction with the loop antenna to determine which of the two 180° apart null positions is the correct bearing to the station.

Early ADF systems used a rotating loop antenna and a long-wire sense antenna. Modern ADF systems use a goniometer which eliminates the requirement for the loop to rotate. ADF systems with non-rotating loops antennas are packaged in a compact module together with the sense antenna and RF amplifier to afford less drag and greater reliability.

ADF ANTENNA THEORY

As previously explained, the operation of an ADF system is based on the directional characteristics of the loop antenna to determine the direction of the incoming RF signal. The loop antenna consists of a continuously wound coil. When the magnetic lines of force from an incoming RF wave cut across the coil, a voltage is induced in the antenna. Because of the transit time of the wave, the voltage induced at the leading edge of the loop (relative to the direction of the incoming signal) will lead the voltage induced at the trailing edge. The algebraic sum of the induced voltages will result in maximum voltage when the plane of the loop is aligned to the incoming RF wave. However, as the loop is turned 90° to the direction of the RF wave, equal and opposite voltages are induced in the sides of the loop which cancel each other to result in a zero voltage output. The point of rotation where the resultant output is zero is known as the null position of the antenna. At the null position, a fairly accurate indication of the station direction can be determined.

ADF loop antennas are automatically rotated to the null position by means of a servo arrangement. The mechanical position of the shaft of the servo used to rotate the loop will reveal the bearing to the station. This shaft is mechanically coupled to a synchro, which transmits an analog three-wire signal (0 to 11.5 volts per wire) to position the shaft of another synchro which is mechanically coupled to the ADF pointer to provide bearing information.

The bidirectional fiqure-8 pattern of a loop antenna, as shown in Figure 3-1, causes it to null in two positions that are 180° apart. This condition can result in an ambiguous ADF pointer indication since the pilot would not know whether the aircraft was pointed toward the station or away from it. This problem is eliminated by the use of an omni-directional, open-wire sense antenna to provide an additional input signal which is 90° out-of-phase with the signal received from the loop antenna.

Since the phase of the loop output will always differ by 90° from that of the sense antenna, a 90° phase shift is added to the loop voltage to cause this voltage to vary with respect to the

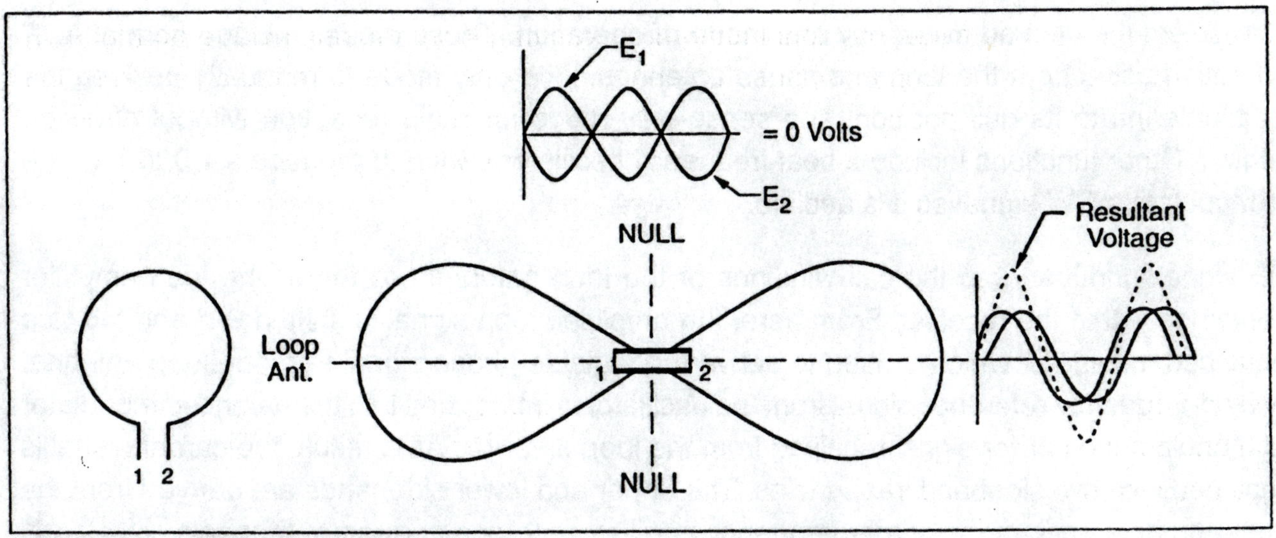

Figure 3-1. Loop Antenna Operation.

constant sense antenna voltage as the loop changes direction. By combining the loop and sense antenna voltages, a cardioid directional pattern results with only one null position.

ADF CIRCUIT THEORY

A typical ADF system, as shown in Figure 3-2, consists of a loop antenna, sense antenna, receiver, control head, and bearing indicator. The operation of the loop and sense antennas and the bearing indicator have been previously discussed. The function of the ADF control head

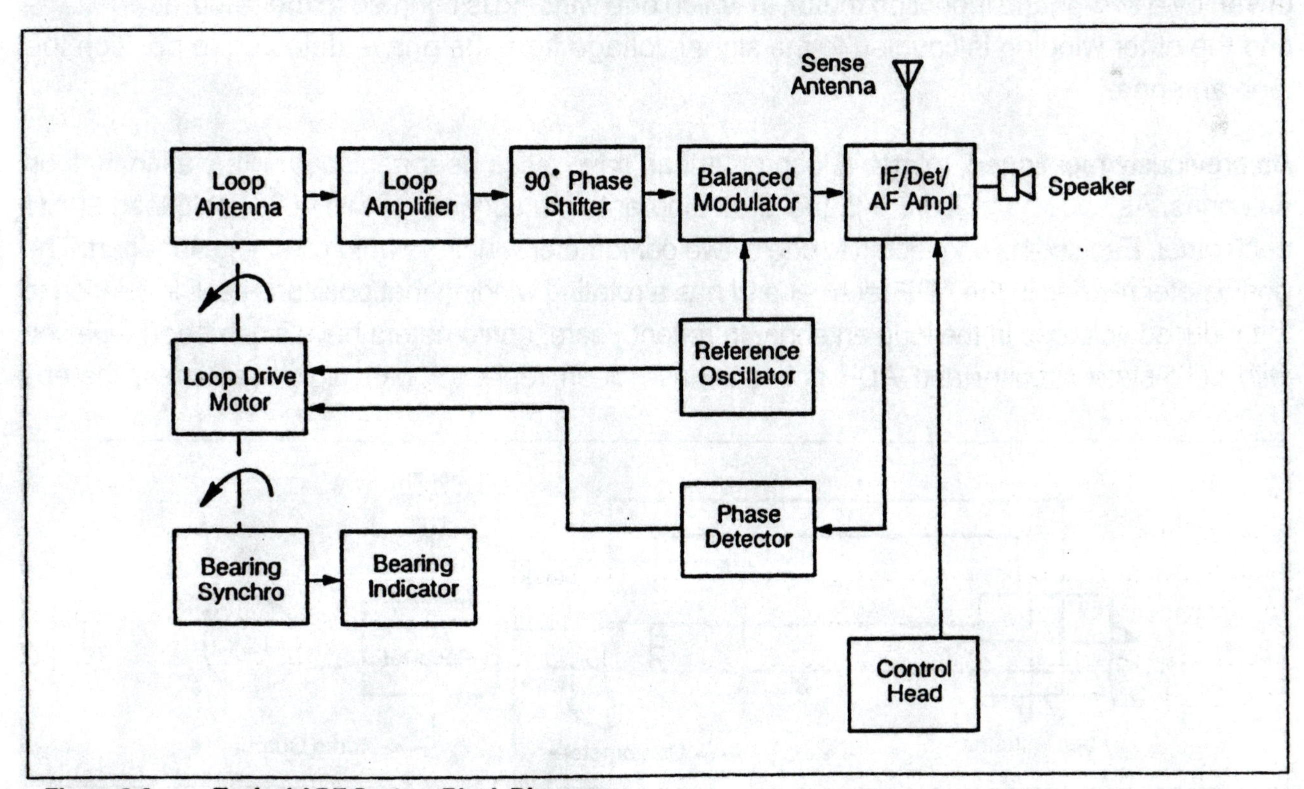

Figure 3-2. Typical ADF System Block Diagram.

is to select the desired frequency and mode of operation. These modes include normal ADF operation using both the loop and sense antennas, loop-only mode to manually position the loop antenna to its null position, and sense-only mode for radio reception without direction finding. Other functions include a beat frequency oscillator switch to produce a 1,020-Hz tone to modulate a CW signal so it is audible.

RF signals induced into the coil windings of the loop antenna are fed to the loop amplifier contained within the receiver. From here, the amplified loop signal is shifted 90° and fed to a balanced modulator which is used to derive the variable-phase signal from the loop antenna. A fixed-frequency reference signal from the oscillator is introduced into the balanced modulator to modulate the carrier signal received from the loop antenna. As a result, the carrier signal is replaced with two sideband frequencies. The upper and lower sidebands are derived from the sum and difference of the carrier frequency and the reference frequency, respectively.

These sideband products are added to the fixed-phase carrier signal received from the sense antenna. The resultant signal is detected and amplified in the superheterodyne receiver, and the modulation product from one of the sidebands is separated from the audio to be used as the loop signal. Depending on whether the station is to the right or left of the aircraft will determine if the loop signal will be in-phase or out-of-phase with the reference signal.

The loop signal is sent to the phase detector, which outputs the loop drive voltage to position the loop antenna to its null position in which the loop signal will be zero. The loop antenna is driven by a two-phase induction motor, in which one winding is coupled to the reference voltage and the other winding is coupled to the signal voltage from the phase detector, to position the loop antenna.

As previously mentioned, rotatable loop antennas have since been replaced with stationary loop antennas. As shown in Figure 3-3, the fixed loop antenna consists of two coils positioned 90° to each other. Each coil is connected to one of two goniometer windings which are also 90° apart. The goniometer resides in the ADF receiver and has a rotating winding that positions itself in relation to the induced voltages in the loop antenna. In recent years, goniometers have since been replaced with solid-state circuitry and ADF pointers have been replaced with digital readouts, thereby

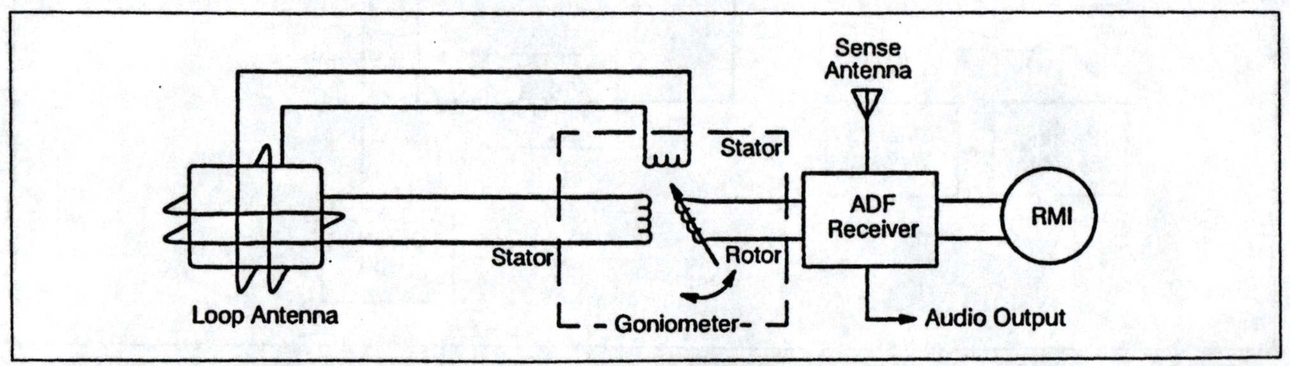

Figure 3-3. Simplified ADF Block Diagram Using a Stationary Loop Antenna.

eliminating all moving parts and increasing reliability. The following sections will discuss the principles of operation of a typical goniometer-type ADF system, such as the Bendix/King KR-86, and one that employs a solid-state antenna circuit, such as the S-TEC ADF-650.

BENDIX/KING KR-86 ADF SYSTEM OPERATION

The panel-mounted Bendix/King KR-86 ADF system contains a single-conversion superheterodyne receiver incorporating a phase-lock-loop frequency synthesizer to provide 1,551 channels. These channels are spaced at 1-kHz intervals over the frequency range of 200 kHz to 1,750 kHz. All LC tank circuits involved in tuning the receiver employ voltage-variable varistor diodes. These varistor diodes are tuned to the desired frequency by the DC output from the synthesizer. Figures 3-4 and 3-5 illustrate the operation of the KR-86.

With the KR-86 in the ADF mode, the magnetic component of the intercepted radio waves induce voltages in the loop antenna windings that lead the voltages induced in the sense antenna by 90°. By connecting the loop antenna windings to corresponding stator windings in the goniometer, the current flow resulting from the induced voltage will cause the electromagnetic field to be reconstructed in the goniometer. The voltage induced in the rotor winding of the goniometer produces a servo error signal that leads or lags the sense antenna signal by 90°.

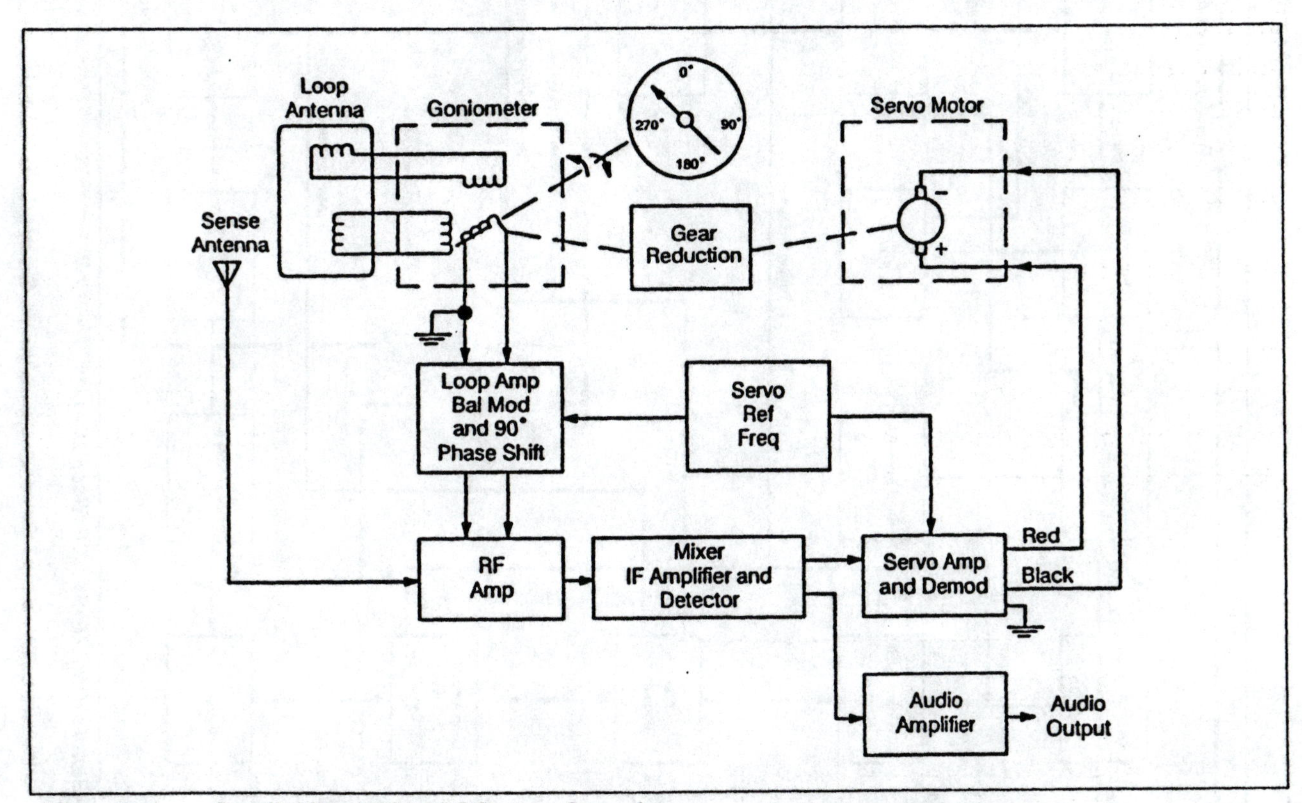

Figure 3-4. ***Bendix/King KR86 ADF System Operation.*** *(Courtesy Allied-Signal Aerospace Company)*

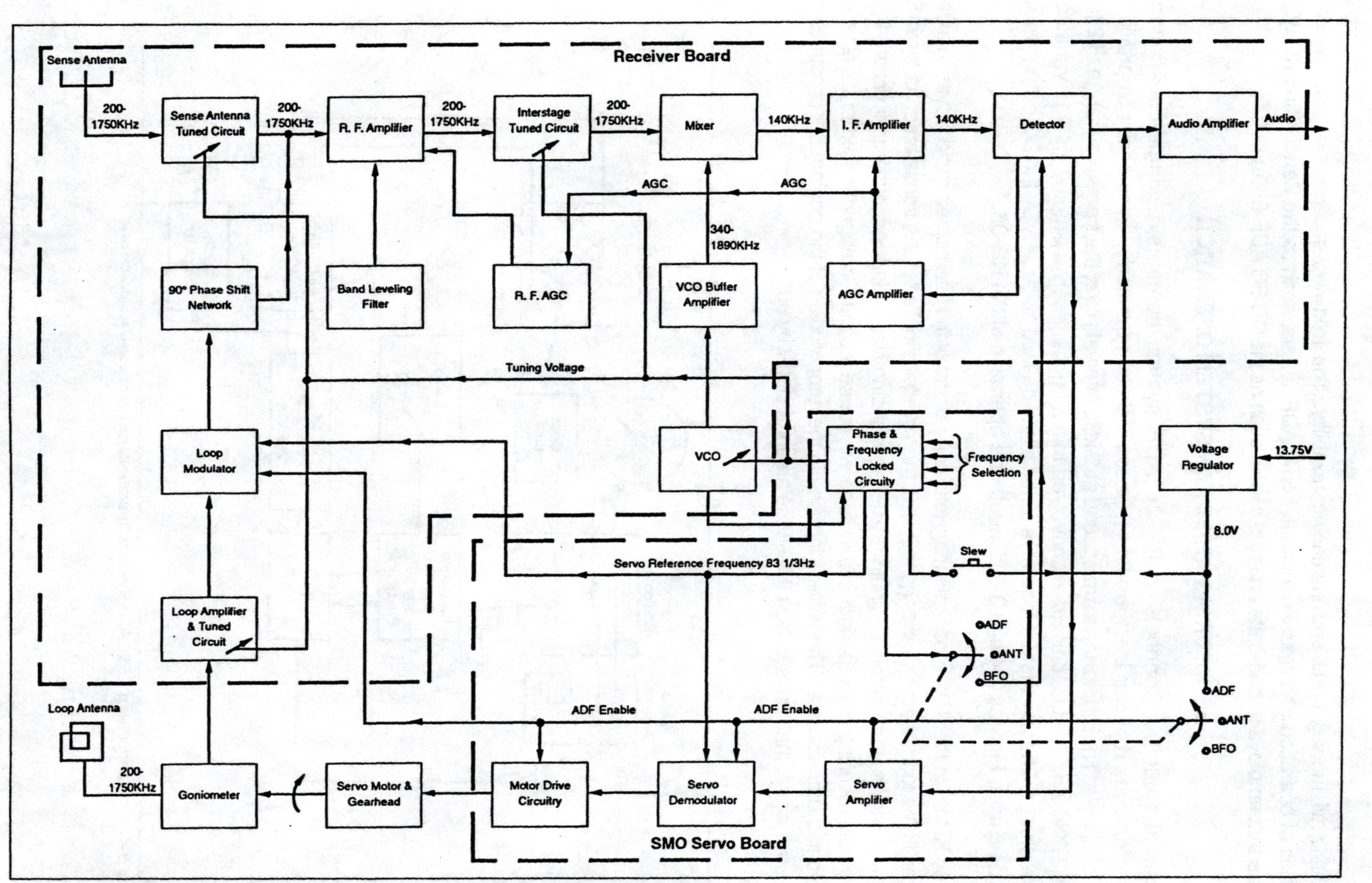

Figure 3-5. Bendix/King KR-86 ADF Block Diagram. *(Courtesy Allied-Signal Aerospace Company)*

The RF error signal developed across the rotor winding is applied to the loop amplifier input. Following amplification, the loop signal is applied to the loop modulator where the phase of the signal is alternately switched from minus 90° to plus 90° at an 83.3-Hz rate. A phase shift network retards the phase of the signal by 90°, thus putting the loop signal alternately in-phase and out-of-phase with the sense signal. This phase switched loop signal is combined with the sense signal in the RF amplifier. When the loop signal and sense signal are in-phase, they will add; and when out-of-phase, they will subtract. The net effect is to produce an 83.3-Hz AM signal. Any AM modulation of the original signal is also still present.

The modulated signal is introduced into the mixer stage along with the local oscillator signal. The resulting difference frequency of 140 kHz is passed through the IF amplifier where it is amplified prior to detection. The 83.3-Hz servo error signal and the normal audio is recovered at the detector and then filtered in the 83.3-Hz selective servo amplifier. The output of the servo amplifier is connected to the demodulator, which converts the error signal to a DC voltage to drive the servo motor. The servo motor, which is mechanically coupled to the goniometer and ADF bearing pointer, stops when the error signal is nulled. At this position, the ADF bearing indicator points to the station. When the function switch on the control panel of the KR-86 is set to ANT position, the 8-volt DC regulated supply voltage is disconnected from the servo circuit, disabling the loop modulator and servo system. Incoming signals from the sense antenna are used for audio reception only with no bearing output signal.

S-TEC ADF-650 SYSTEM OPERATION

The S-TEC ADF-650 system consists of a RCR-650 receiver, ANT-650A antenna, and IND-650 indicator. The RCR-650 is a panel-mounted ADF receiver that provides a relative bearing signal and audio output. The ANT-650A is a fuselage-mounted combined loop and sense antenna and RF amplifier. The IND-650 is a panel-mounted ADF indicator. The IND-650 indicator and RCR-650 receiver are illustrated in Figure 3-6.

The output from the receiver causes the pointer to rotate to the relative bearing between the center-line of the aircraft and the selected station. The IND-650 also provides a manual heading selector that allows present heading to be set above the lubber line. Reading the pointer against the compass card gives the pilot a direct magnetic bearing to the selected station. If a remote magnetic indicator is used instead of the IND-650, the compass card will be automatically slaved by the compass system so that the magnetic bearing of the ADF station is always displayed. Use of an RMI requires the installation of the ADA-650 three-wire synchro converter. Figure 3-7 illustrates the operation of the ADF-650.

A relative bearing indication to a selected station is accomplished by resolving voltages into two directional loop antennas and a single omni-directional sense antenna. The voltages induced into the loop antennas are amplified and applied to two balanced modulators that are driven by a low-frequency (41.6 Hz) signal. The output of the balanced modulators is summed to produce a double-sideband (DSB) signal with an audio phase relative to the 41.6-Hz clock

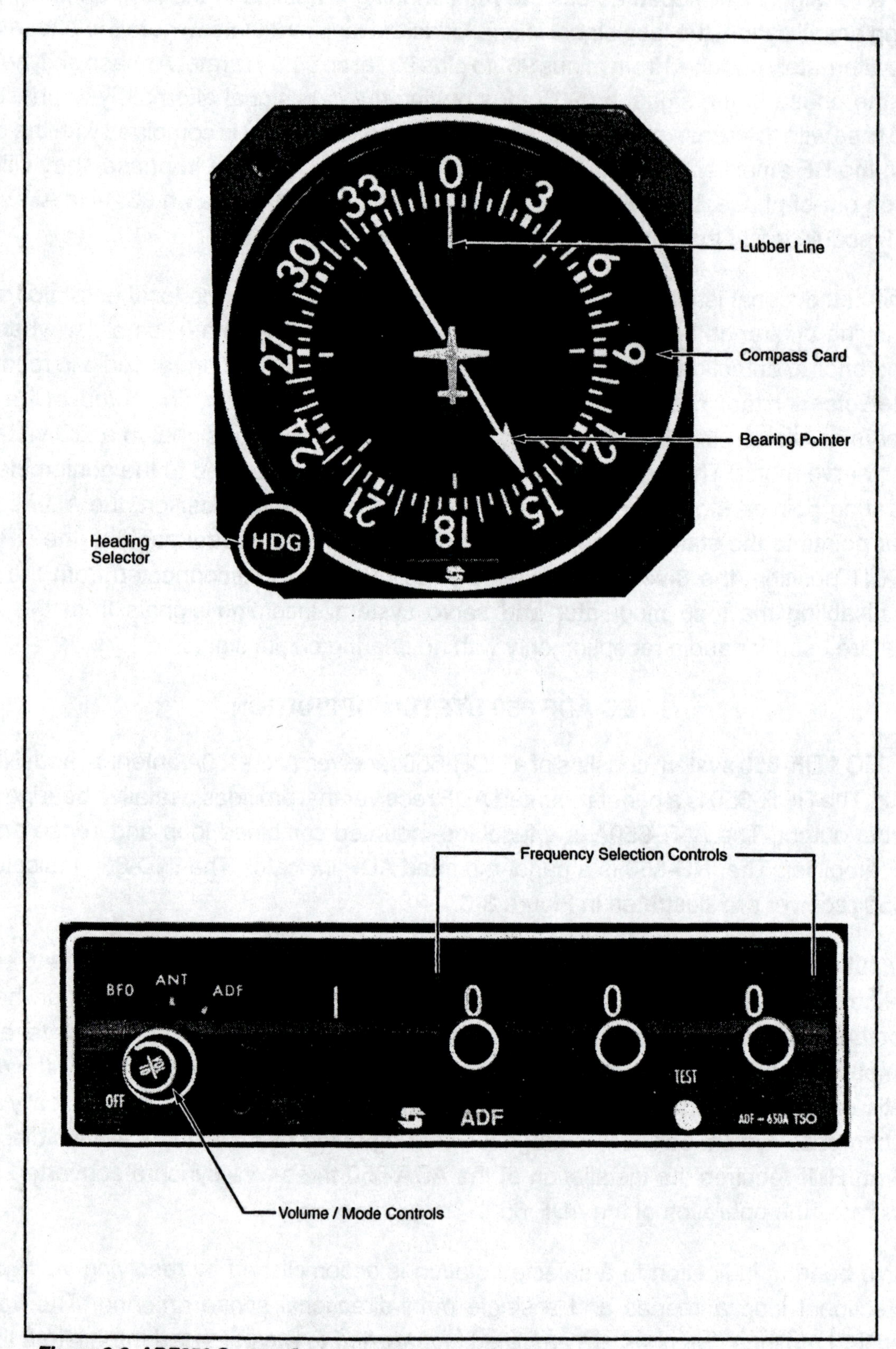

Figure 3-6. ADF650 System Components. *(Courtesy S-TEC Corporation)*

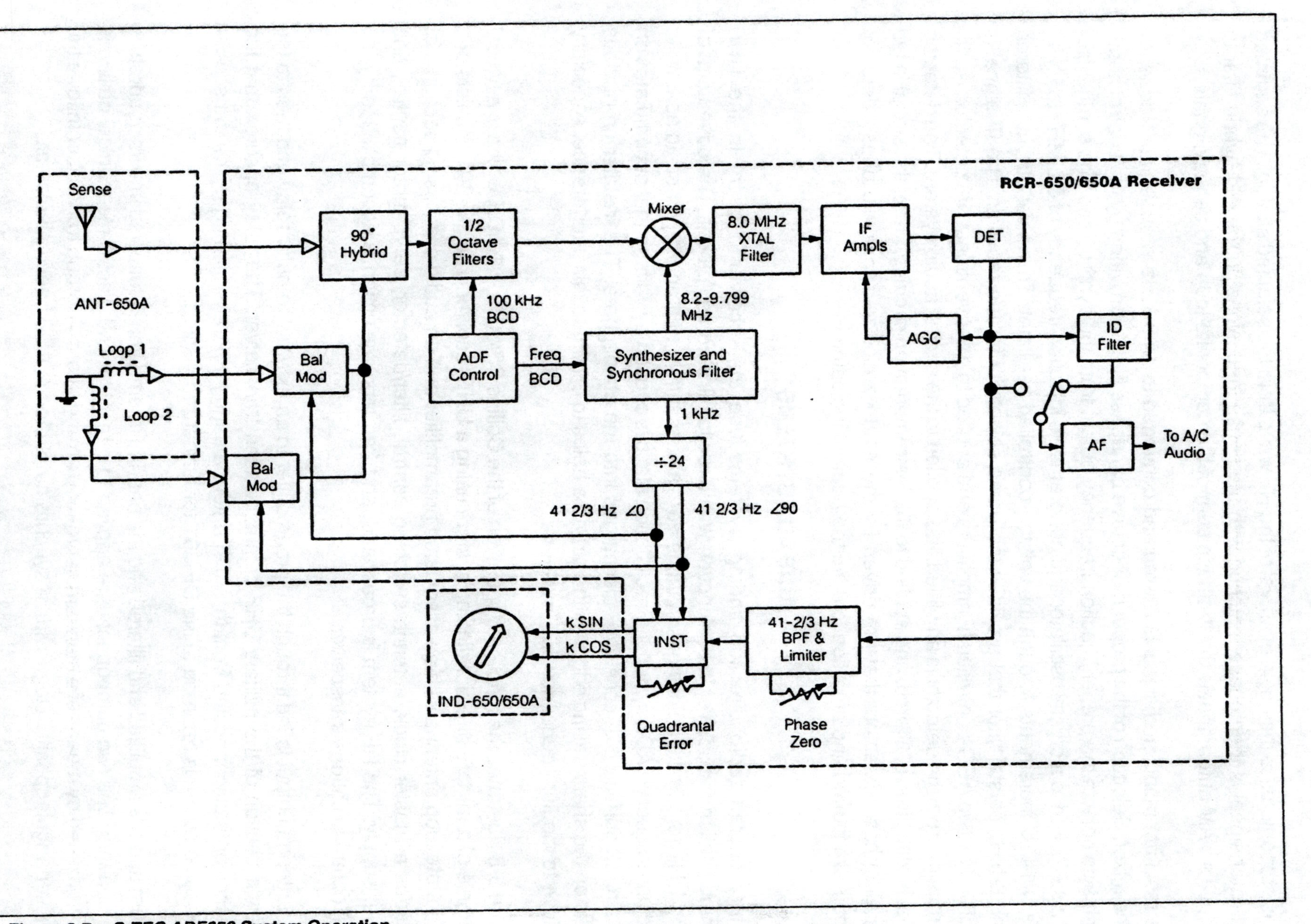

Figure 3-7. S-TEC ADF650 System Operation.

that is proportional to the angle of arrival of the RF wave from the selected station. The summed modulator signal is then phase-matched with the sense input signal in the 90° hybrid filter to produce an AM output that is the in-phase sum of the composite loop and sense signals.

This resultant modulated signal is filtered and converted to an 8-MHz intermediate frequency by the mixer. Selection of the proper half-octave bandpass filter is provided by the control head, which also selects the proper injection frequency output from the synthesizer. The synthesizer is phase-locked to a crystal oscillator and generates an injection frequency of 8.2 MHz to 9.799 MHz in 1-kHz increments through the voltage-controlled oscillator. The mixer output is filtered by an 8-MHz crystal filter that provides the prime selectivity for the receiver. The IF signal is amplified by two AGC-controlled IF amplifiers and applied to a conventional AM detector. Audio information from the detector is amplified and applied to the aircraft audio system. The detected signal is also limited, filtered, and applied to the instrumentation circuitry. Here, the signal is fed to audio phase detectors that are driven by the 41.6-Hz clocks. The audio phase detector outputs are filtered and amplified to drive the bearing indicator.

UHF ADF SYSTEMS

Military aircraft radio direction finding systems use the receiver circuitry in the UHF communication transceiver in conjunction with direction finding equipment. The advantage of using UHF for direction finding is that it is not as susceptible to precipitation static as low- to medium-frequency ADF systems. Precipitation static is caused by voltage that accumulates on the aircraft surface that constantly discharges into the atmosphere. The installation of static wicks on the sharp trailing edges of the airframe helps to eliminate this interference, especially for low- to medium-frequency ADF systems.

Figure 3-8 illustrates the functional operation of the Collins DF-301 UHF ADF, which is a typical example of a military direction-finding system using a UHF receiver. The DF-301 consists of a solid-state loop antenna with an associated preamplifier and an audio processing section that includes a master clock, antenna switching control, filter/detector, servo-motor control, and power supply. The RF system is broadband, and frequency selection is accomplished by the UHF communications transceiver.

The antenna output is fed through the preamplifier to an external transfer relay and then to the receiver section of the military UHF communication transceiver. The relay ensures that the transceiver is disconnected from the ADF antenna during the transmit mode. The transceiver ADF audio output is returned to the DF-301 for further processing.

Return audio is amplified and filtered and the bearing information is resolved into its quadrature components. The two components are applied to a resolver and electro-mechanical nulling is accomplished by means of a standard servo-loop arrangement. A torque transmitter follows the resolver angular position to provide a synchro output to the ADF bearing indicator.

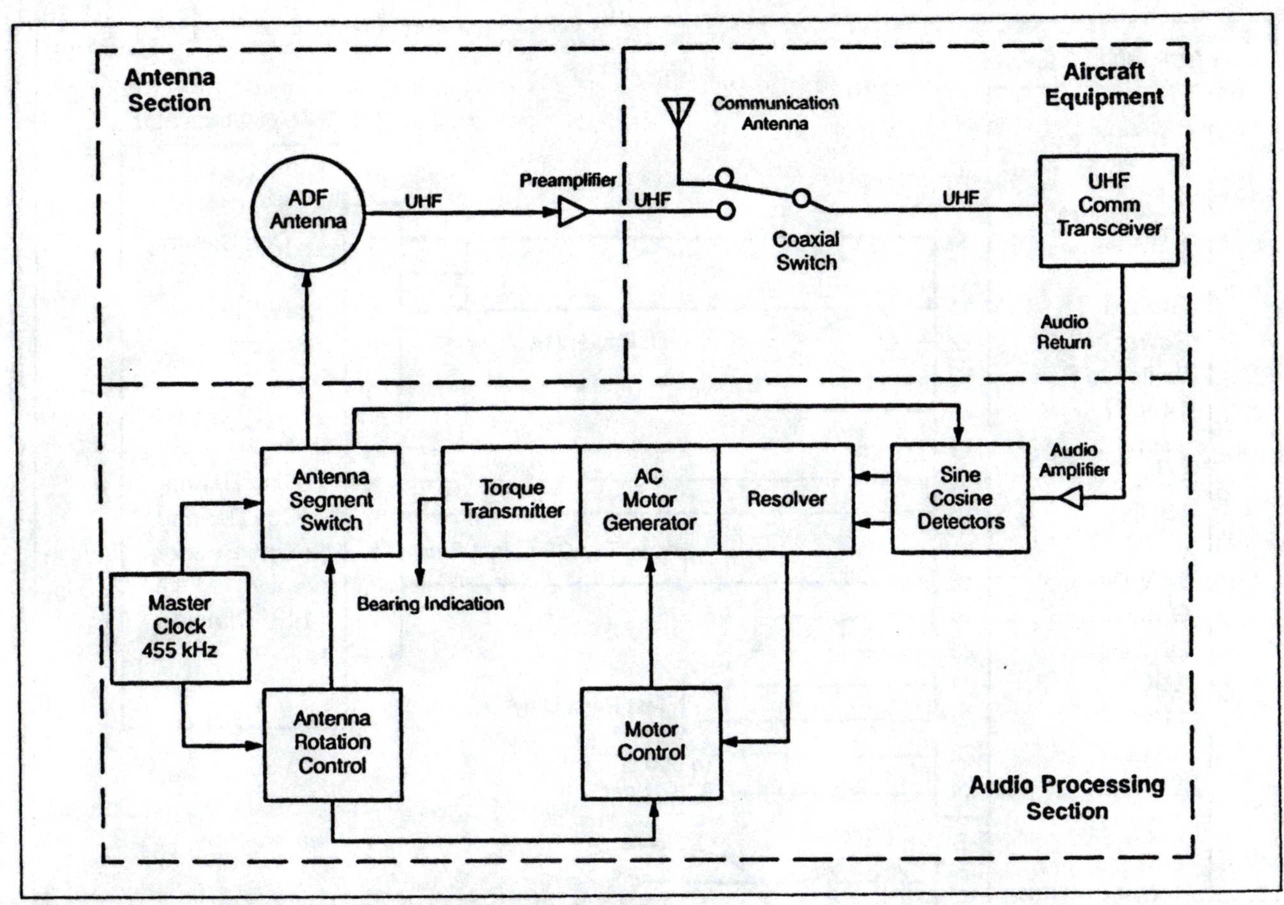

Figure 3-8. ***Rockwell Collins DF-301 System Operation.***

The DF-301 antenna amplitude-modulates the incident RF signal, and this modulation contains the bearing information. Electrical rotation of the antenna translates the spatial angle of the incoming RF signal to a phase difference between the antenna-modulated output and a reference signal synchronous with antenna rotation. From this phase difference, the bearing of the incoming signal can be recovered.

INSTALLATION TECHNIQUES

Dependable ADF system operation requires proper installation procedures. The first consideration is choosing a location for the ADF antenna that will provide the best possible signal to the antenna at all times. On low-wing aircraft, it is best to stay away from the area on the fuselage between the leading and trailing edge of the wing. This area exhibits reduced RF field strength. On high-wing aircraft, this area also exists on the top of the fuselage. In either case, optimum performance will be obtained when the antenna is located aft of the wing trailing edge. This position also minimizes quadrantal error, which is caused by the distortion of the radio wave by the aircraft structure. The quadrantal error is maximum at bearings in between the cardinal points of the nose, wingtips, and tail.

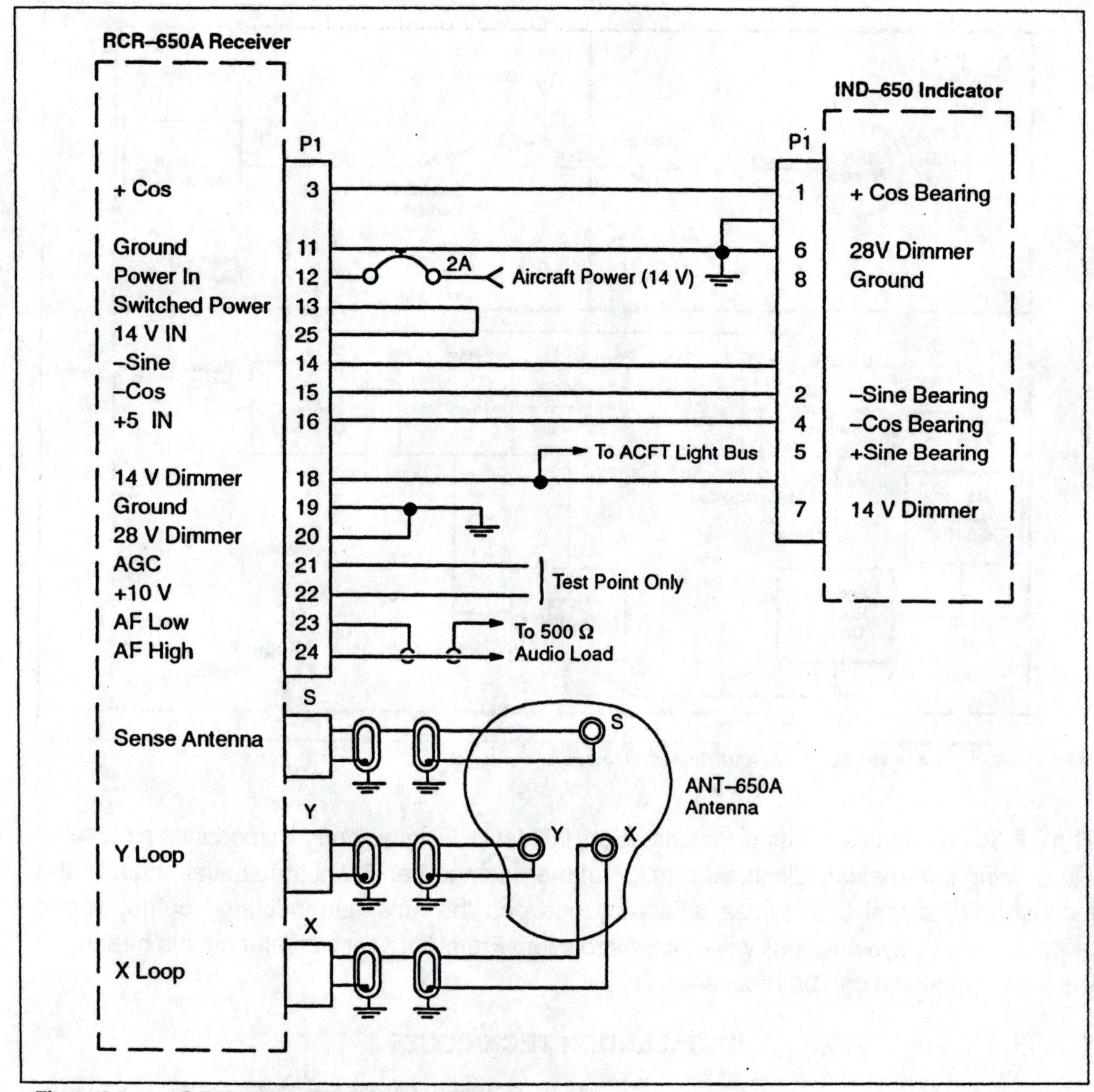

Figure 3-9. ***S-TEC ADF-650 Interconnect Wiring Diagram.***

A short metal braid or strap should be secured between the loop antenna and the fuselage to provide a good RF ground. Remove paint from the bonding surface of the loop antenna and passivate the cleaned aluminum surface with an alodine to retard corrosion. In addition, all antenna connections should be coated with silicon grease.

Noise immunity is one of the most important considerations in ADF system installation. Generators and alternators should have suitable filters installed and interference must also be suppressed from strobes, inverters, motors, and other electrical equipment. The proper number of static wicks should be installed at the trailing edges of all airframe surfaces to minimize precipitation static.

The ADF system components must be installed in accordance with the manufacturer's installation instructions and FAA Advisory Circular AC-43.13. A continuity check of the system wiring should always be performed before installing the equipment and applying power. An interconnect wiring diagram of the S-TEC ADF-650 system is shown in Figure 3-9.

PERFORMANCE VALIDATION

Once the ADF system is installed, a post-installation check must be performed to validate system performance. The aircraft compass indicator is used to determine ADF bearing accuracy in the following procedure.

Using a sectional navigational chart, select a nearby NDB or compass locator. Draw a line from the selected station to a position on the airport ramp that is at least 100 feet from surrounding buildings or other obstructions. Position the aircraft on the ramp directly toward the selected test station. Ensure that the aircraft is not grounded by tie-down chains, ground power cables, etc. Record this heading for later use.

With the engine and all other accessories turned off, apply power to the ADF receiver and position the mode switch to the ADF position. Dial in the frequency of the selected station and verify that the station is being received and audio is present at the headphones and/or loudspeaker. Adjust the heading knob on the ADF indicator until 0° on the compass card is positioned under the lubber line. The ADF indicator should show a relative bearing indication of zero plus or minus 5°. (If a slaved RMI is being used, the ADF pointer should indicate the magnetic bearing to the station plus or minus 5°.) If not, recheck the actual magnetic bearing to the test station and verify that the aircraft is properly positioned. If the inaccuracy still persists, bench test the ADF receiver to verify proper alignment.

Start the aircraft engine, turn on the generator and all radios and accessories normally used in flight and verify that the relative bearing error changes less than plus or minus 5°. If excessive bearing changes are found, isolate the cause of interference by turning off radios, accessories, generators, magnetos, etc., one at a time, until the source of interference is located. Apply the necessary suppression (ground straps, filters, etc.) to the offending interference source.

Position the aircraft at 90° increments from the original magnetic heading using the compass indicator and check to see if the relative bearing errors at 90°, 180°, and 270° are less than plus or minus 5°. If the aircraft has a slaved compass system, the proper aircraft test headings may be determined by adding increments of 90° to the original test heading of the aircraft. If the aircraft has a directional gyro, it may be set to 0° at the original test heading, and used as a reference to turn the aircraft to 90°, 180°, and 270°. Be certain to return to the original test heading to verify that the gyro has not precessed significantly during these bearing checks.

The quadrantal error adjustment does not control the error at 0°, 90°, 180°, or 270°, but does adjust the error at intermediate points, especially 45°, 135°, 225°, and 315°. This error is due to the shape of the aircraft and is slightly different for each aircraft installation. To compensate for quadrantal error, turn the aircraft 45° left from the original test heading using the slaved compass system or directional gyro for reference. Adjust the quadrantal error control so that the relative bearing indication shows 45°. Check the error at 135°, 225°, and 315° and readjust the control as necessary to split the errors at these points to get best overall tracking performance.

Verify receiver sensitivity by selecting several local and distant stations and observing the correct bearing indication. Determine that the other control head functions operate properly, including manual slewing of the ADF loop circuit (if provided). By switching from ADF mode to ANT mode, the loop circuit should be disabled causing the bearing pointer to be inoperative. Selecting the BFO mode should produce a whistle in the audio output. Pressing the test button should rotate the pointer in one direction and releasing it should cause the pointer to return to its previous position. The test function is available on some ADF receivers only when a station is being received.

The performance validation is concluded with a flight test in which the aircraft is flown directly to a fairly strong station at least 50 miles away. By selecting the ADF mode, the relative bearing should indicate 0°. The pilot then completes a 360° standard rate turn and the ADF relative bearing is compared with the compass heading indication for plus or minus 5° accuracy at 15° intervals.

OPERATING PROCEDURES

The following illustrations present typical procedures the pilot may use while operating the ADF system. The avionics technician should be aware of these procedures in order to understand the dynamic operation of the ADF while in flight. Each of the following figures illustrate the aircraft's position in relation to the received station. The compass indicator shows the aircraft's magnetic heading during the maneuver and the ADF indicator points to the relative bearing to the station.

Figure 3-10 illustrates how a pilot uses the ADF to intercept a predetermined bearing inbound from a given station. Figure 3-11 illustrates the procedure for homing to a station using the tracking method to compensate for a crosswind. Determining the aircraft's position is accomplished by alternately selecting and determining the bearing to two stations, as shown in Figure 3-12.

CONCLUSION

We learned from this chapter that the concept of ADF navigation is based on the ability of the airborne system to measure the direction of the arrival of the received signal, and from this information, provide a relative bearing indication with respect to the centerline of the aircraft. Using the bearing information displayed on the ADF indicator, the pilot can determine the aircraft's position or can fly directly to the NDB or AM broadcast station. To determine the aircraft's position, the pilot simply plots the headings of two different stations on a navigation chart and triangulates the aircraft location at the point where the two lines intersect.

This concludes our discussion of automatic direction finding systems. ADFs are still in wide use, especially for aircraft flying to and from airports that are not equipped with VHF OmniRange facilities. However, OmniRange Navigation is much more sophisticated and has definite advantages over ADF navigation, as will be discussed in the next chapter.

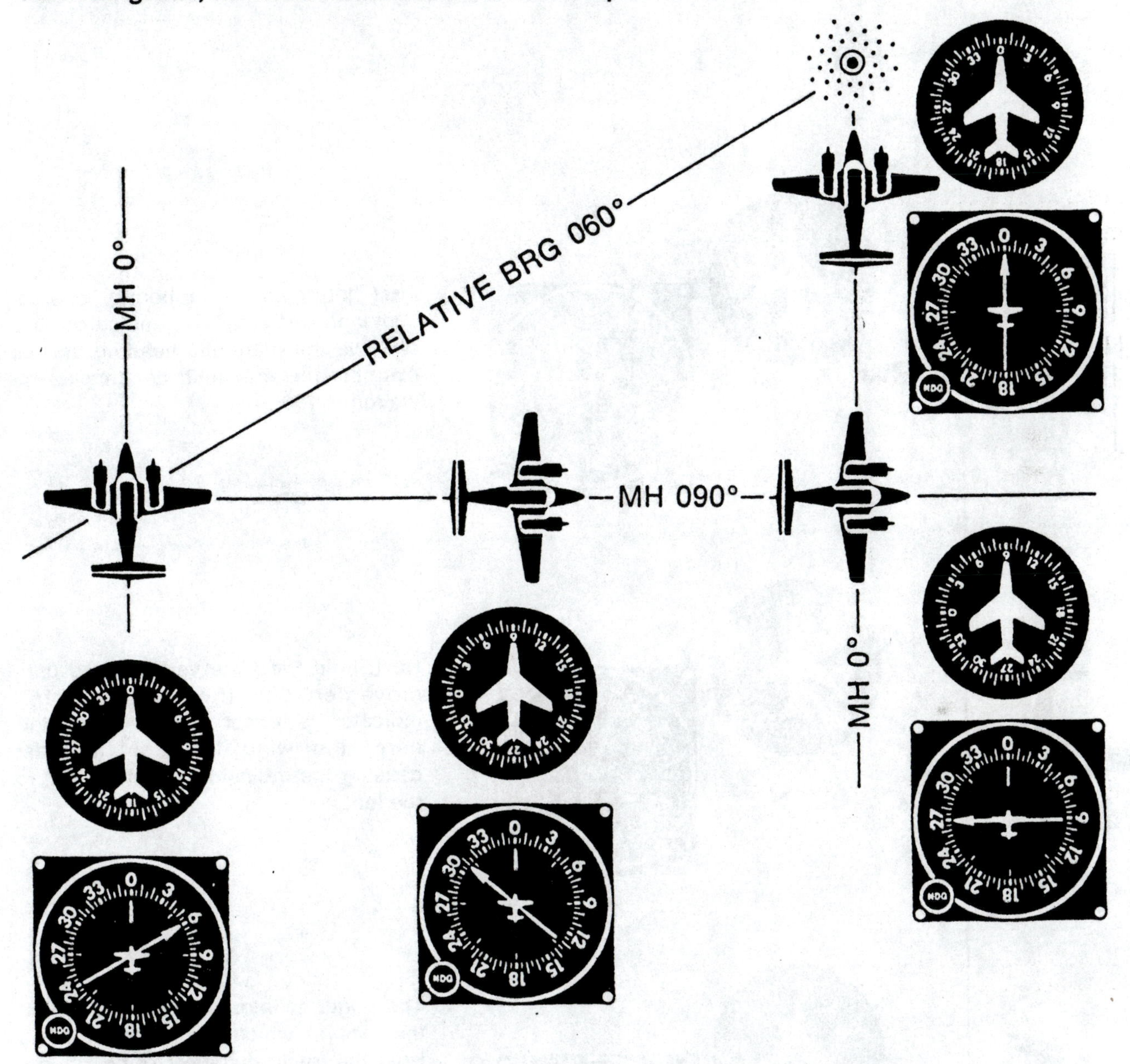

To intercept an inbound bearing, first turn the aircraft to the heading of the bearing to be intercepted. Next, observe the relative bearing to the station (in this example 60°) and double this indication to arrive at the angle of interception. Since it is not feasible to intercept a bearing at more than a 90° angle, the intercept angle should be limited to 90°. In this example the intercept angle is 90° since twice the relative bearing is 120°. Next turn the aircraft to the 90° intercept bearing and continue until the ADF bearing indicator is deflected the same number of degrees from 0 as the intercept angle. Finally, upon course interception, turn inbound and track to the station.

Figure 3-10. Interception of an Inbound Bearing.

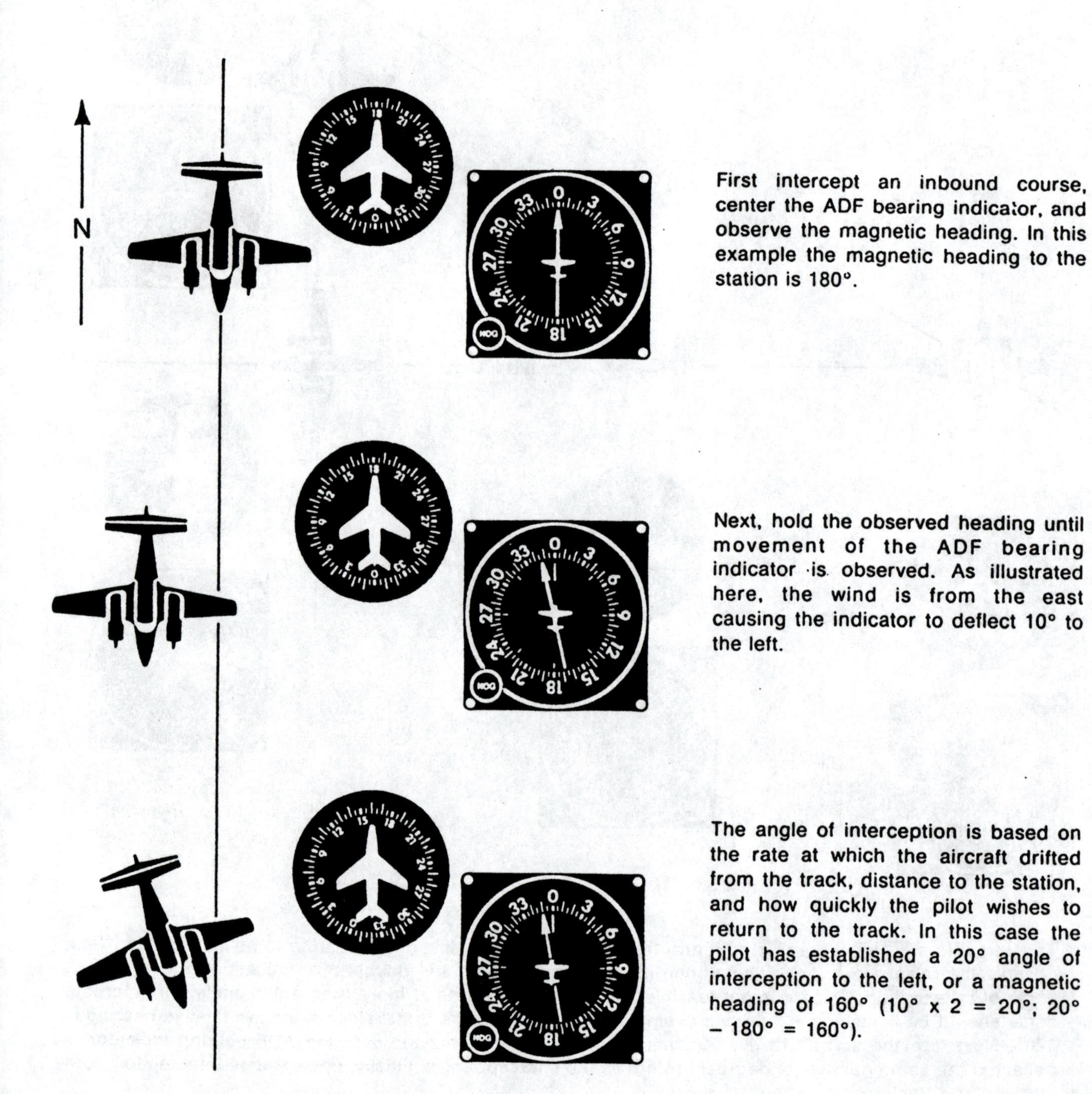

N
First intercept an inbound course, center the ADF bearing indicator, and observe the magnetic heading. In this example the magnetic heading to the station is 180°.
Next, hold the observed heading until movement of the ADF bearing indicator is observed. As illustrated here, the wind is from the east causing the indicator to deflect 10° to the left.
The angle of interception is based on the rate at which the aircraft drifted from the track, distance to the station, and how quickly the pilot wishes to return to the track. In this case the pilot has established a 20° angle of interception to the left, or a magnetic heading of 160° (10° x 2 = 20°; 20° – 180° = 160°).

Figure 3-11. ***Tracking. (Sheet 1 of 2)***

As the aircraft moves back toward the desired track, the bearing indicator will move in a clockwise direction. Rotation will continue until the pilot recognizes course interception and removes a portion of the previously established course intercept angle. Arrival on the desired track is indicated by an ADF bearing indication that equals the angle of interception. In this example a bearing indication of 20° right indicates interception of the track.

When the desired course has been intercepted, a new heading should be estimated by considering how long it took to drift off course initially. Once the course is held, the bearing indicator will be displaced the number of degrees equivalent to the crab correction angle, or in this case 10°. As long as the relative bearing remains constant, the aircraft is on the desired track. If the wind is underestimated, a decrease in relative bearing will occur. Should this happen, reintercept the track and increase the drift correction angle. If this wind is overestimated, parallel the aircraft to the desired track and drift back to the track. Upon interception turn the aircraft back into the wind at a reduced correction angle.

To continue tracking outbound from the station, maintain the correction angle established earlier by observing the tail of the bearing indicator and the DG. Variations in the wind will be observed as bearing indicator movement. Progressively make additional corrections as required until both DG and bearing indicator are stable. The main difference in tracking outbound as opposed to inbound is that the ADF bearing indicator moves further away from the 180° position as a change in heading is made toward the desired track.

Figure 3-11. Tracking. (Sheet 2 of 2)

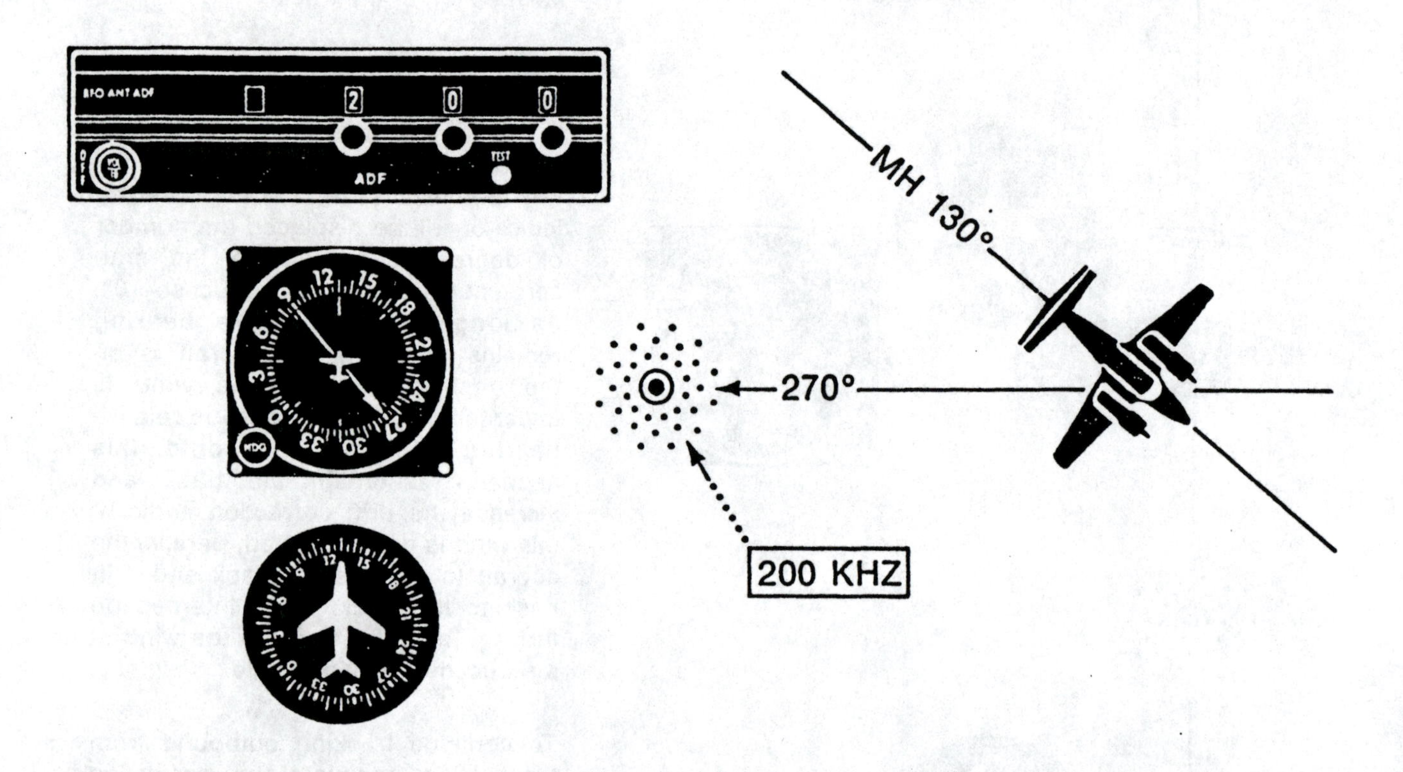

To perform an ADF fix, the pilot first selects and identifies a station in the vicinity. Once the station has been positively identified, the ADF mode is selected and the aircraft magnetic heading is observed; in this example the DG shows a magnetic heading of 130°. Using the HDG knob on the IND-650, the pilot sets the magnetic heading (130°) beneath the lubber line and observes the head of the indicator arrow. The number indicated is the magnetic bearing to the station; in this case 270°. The pilot now draws a 270° line to the station.

Figure 3-12. Determining a Fix. (Sheet 1 of 2)

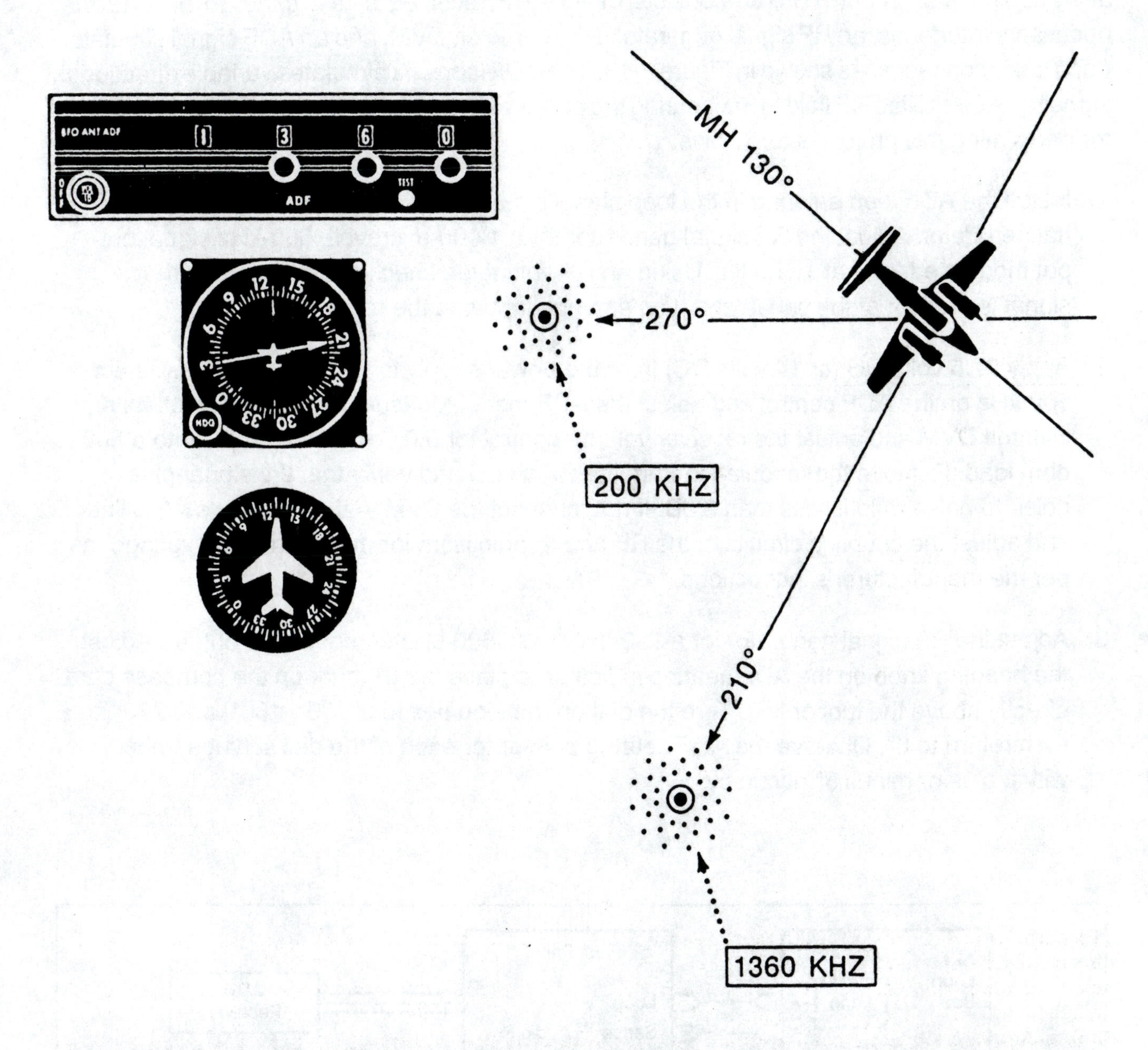

Next, the pilot selects another station in the area and takes another bearing. As shown above, the second station selected is 80° from the present heading of 130°; therefore, the magnetic bearing to the station is 210° as indicated on the IND-650 indicator. The pilot now draws a 210° line to the second station until it intersects with the 270° line drawn from the first station. The point of intersection is the approximate position of the aircraft.

Figure 3-12. Determining a Fix. (Sheet 2 of 2)

ADF RECEIVER LABORATORY PROJECT

The objective of this laboratory project is to become familiar with the proper procedure for setting-up, testing, and calibrating an ADF navigation receiver prior to its installation in an aircraft. The equipment required consists of an ADF receiver, a test panel to provide the necessary interfaces, an RF signal generator, RF voltmeter, DVM, and an ADF signal simulator (loop box), connected as shown in Figure 3-13. The ADF loop box simulates variable directional signals in a controlled RF field for measuring the performance of the ADF system. The procedure for performing this project is as follows.

1. Install the ADF loop antenna in the loop signal simulator box and connect to it an RF signal generator. Adjust the RF signal generator for a 1,000-microvolt, 500-kHz signal output modulated 30% at 1,000 Hz. Using an RF voltmeter, check to make certain that a signal is present at the sense and loop antenna inputs to the receiver.

2. Apply 27.5 volts DC (or 14 volts DC) from the power supply to the ADF receiver. Select 500 kHz on the ADF control and select the ADF mode. Measure the audio output voltage with the DVM and adjust the receiver volume control for 5.0 volts RMS output into a 500-ohm load. Remove the modulation from the RF signal and verify that the signal-plus-noise to noise ratio is less than 6 dB. If not, connect the DVM to the receiver's AGC line and adjust the coupling circuits in the RF and IF amplifiers for maximum AGC voltage, as per the manufacturer's instructions.

3. Adjust the RF signal generator for a 350-microvolt, 500-kHz unmodulated output. Adjust the heading knob on the ADF bearing indicator to place the 0° mark on the compass card directly above the lubber line. Turn the dial on the loop box to 0°, 90°, 180°, and 270°, then return to 0°. Observe the ADF bearing pointer for each of the dial settings to be within plus or minus 3° accuracy.

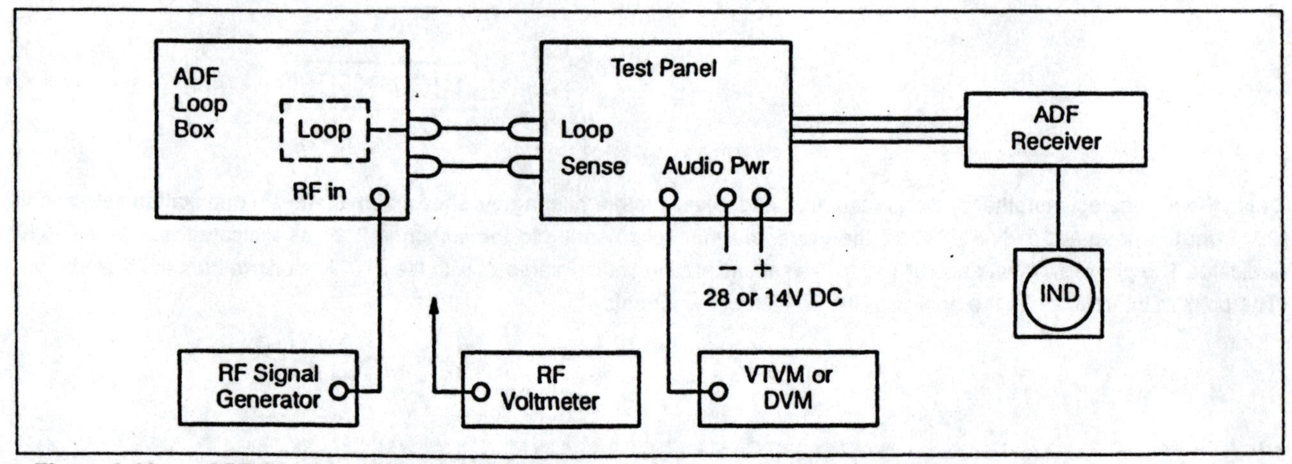

Figure 3-13. ADF Receiver Bench Test Diagram.

No response from the indicator would suggest a problem in the loop servo drive motor or possibly a missing reference signal. If the pointer continuously rotates, a defect may be present in the balanced modulator. If the pointer is unsteady, there could be a problem with the sense antenna circuit. A defect in the sense antenna circuit will also exhibit low sensitivity and a high noise level when operating in the Antenna mode.

In any case, if the ADF system does not perform satisfactorily in accordance with the required tests on the bench, refer to the applicable maintenance manual for troubleshooting procedures. If the ADF system passes the required tests on the bench, the problem can be isolated to the aircraft wiring and interconnections.

REVIEW QUESTIONS

1. What is the frequency range of the ADF receiver and how are the various stations contained within this range identified?
2. What is the null position of an ADF loop antenna in relation to the incidence of the radio wave?
3. How does the loop antenna automatically seek the proper null position?
4. Explain the three modes of operation provided for selection on most ADF control heads.
5. How is a balanced modulator similar to a standard mixer circuit?
6. What is the purpose of the phase detector in an ADF receiver?
7. Explain the operation of a goniometer.
8. Compare the major differences in the operation of the King KR86 system to the S-TEC ADF-650 system.
9. If an aircraft is flying due East and is receiving a station on its ADF receiver that is 90° to the left of it, what would be the relative bearing and the magnetic bearing to the station?
10. In the above example, where would the pilot set the compass card on the ADF indicator to provide a magnetic bearing indication to the station?
11. What is the advantage of using an RMI over a standard ADF indicator?
12. What is precipitation static and how can it be reduced?
13. How do military airborne direction finding systems differ from commercial ADFs?
14. Explain the important factors to be considered when installing an ADF system.
15. What is quadrantal error and how can it be corrected?
16. How is a directional gyro used to verify the accuracy of the ADF on the ground and in flight?
17. What is the purpose of the ADF test function?
18. How can a pilot determine the position of the aircraft by using the ADF system?
19. Why is special equipment needed to bench test an ADF system compared to a standard receiver?
20. What would happen to the operation of the ADF system if suddenly the sense antenna became disconnected?

Chapter 4

VHF OmniRange Navigation

INTRODUCTION

As mentioned in the previous chapter, ADF systems provide the pilot with an indication of the aircraft's relative bearing to the station. The only means the pilot has to determine his position using the ADF system is to plot the headings of two different stations on a navigation chart and triangulate the aircraft's location at a point where the two lines intersect. It is evident that this method, eventhough effective, can be very cumbersome. This problem was addressed in 1945 with the introduction of VHF OmniRange (VOR) navigation stations.

As shown in Figure 4-1, VOR stations appear on the U.S. Aeronautical Charts as a compass circle centered over the station and oriented toward magnetic north. Lines, known as radials, are drawn from the station out in the direction of a magnetic heading. A VOR station broadcasts an infinite number of radials. If the VOR system indicates that the aircraft is on a zero° radial, this means that it is somewhere on a line drawn from the VOR station to magnetic north. Therefore, the VOR indication is independent of aircraft heading, whereas the ADF indication changes with the aircraft heading as the ADF points to the station.

The principle of VOR operation is based on the generation of radials, or magnetic bearings, by a ground station transmitter and their reception by an airborne receiver. The receiver instrumentation unit determines which radial is passing through the aircraft position. The

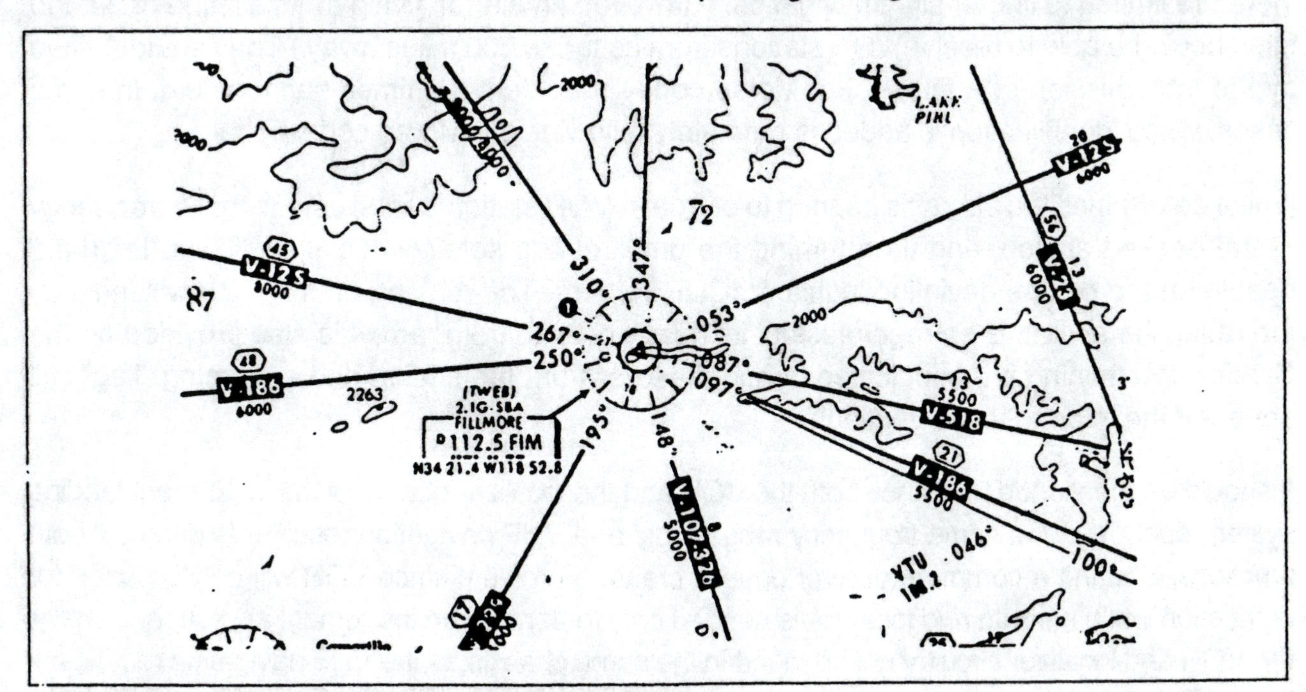

Figure 4-1. VOR Station Shown on Aeronautical Chart.

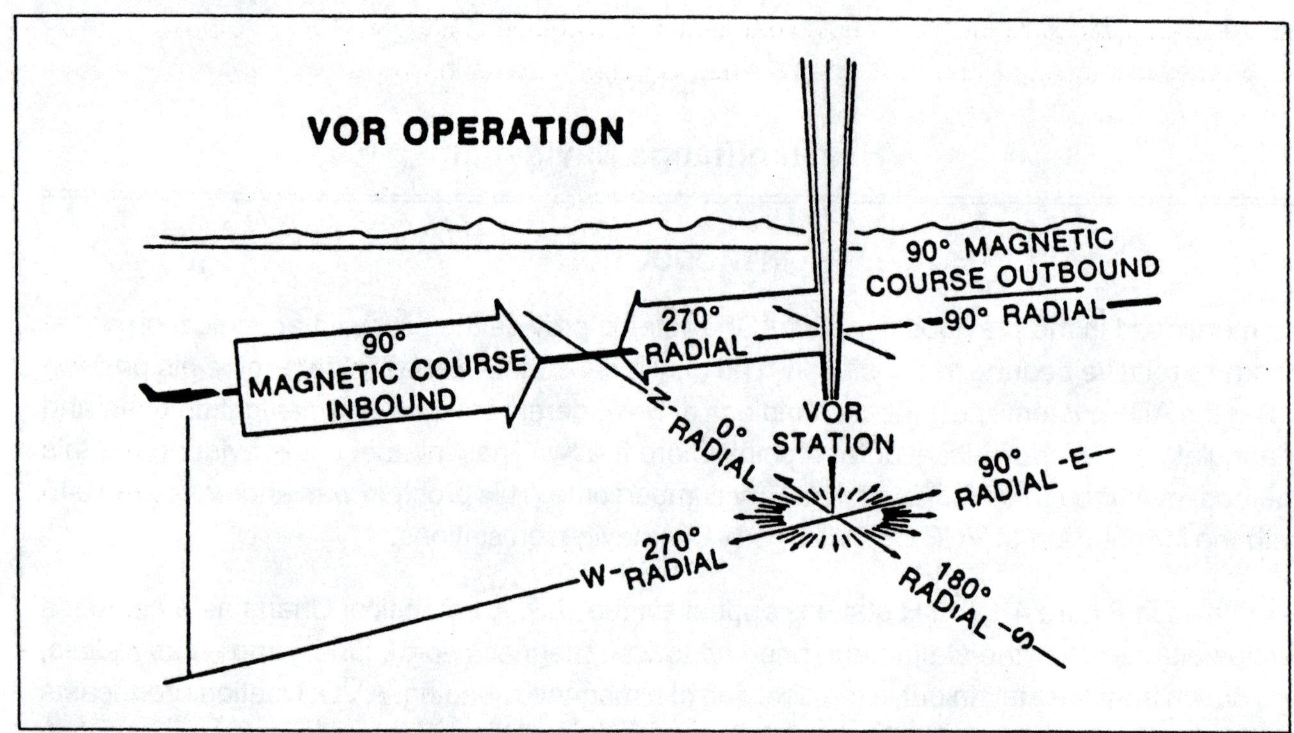

Figure 4-2. ***VOR Radials and Magnetic Courses.***

determined radial is the angle between magnetic north and the aircraft as measured from the ground station. The magnetic course from the aircraft when flying inbound to the station is the reciprocal of the radial, as shown in Figure 4-2.

VOR stations operate in the VHF frequency range from 108.00 MHz to 117.95 MHz, and are therefore limited to line-of-sight navigation. (However, an aircraft flying at an altitude of 20,000 feet should be able to receive VOR stations from as far as 200 miles away.) They are identified by the transmission of a three-letter Morse code group sent ten times each minute. In some cases, voice identification is added immediately following the Morse code signal.

A pilot determines the aircraft's bearing to or from a VOR station by first selecting the frequency of the desired station, and then turning the omnibearing selector (OBS) until the "left/right" needle on the course deviation indicator (CDI) centers. The reading on the OBS will indicate the radial the aircraft is on regardless of its heading. A "to/from" arrow is also provided on the CDI for determining if the indicated bearing is to or from the station and a warning "flag" will appear if the information is unusable.

It should be mentioned that since both the VOR and the localizer portion of the instrument landing system operate in the same frequency range, only one VHF navigation receiver is used for both functions. Sharing a common receiver doesn't create a problem since VOR will only be used for navigation while enroute and localizer is needed only to perform an instrument approach. Usually, the VOR and localizer circuitry will be found in the same chassis as the VHF navigation receiver. If the VOR and localizer circuitry is contained within the instrument, the combination is known as a converter-indicator. In light aircraft using panel-mounted equipment, the VHF navigation receiver is sometimes located in the same chassis as the VHF communications transceiver. Such is the case of the Bendix/King KX-170A/175, which will be discussed later.

VOR NAVIGATION CONCEPTS

The concept of VOR navigation is based on the ability of an airborne receiver to detect two distinct signals contained within one carrier signal transmitted from a VOR ground station and then to compare the phase difference between them to derive the bearing from the station. This concept is analogous to a tower having a sharply focused beacon light which rotates at a constant speed, and a second light which flashes in all directions (omni-directional) when the rotating beacon points to magnetic north. The bearing from the tower can thus be determined by multiplying the speed of the rotating beacon by the time required between when the fixed (reference) light flashes and the rotating (variable) light passes by the observer.

A VOR station radiates a composite electromagnetic field from two ground-based antennas on the same carrier frequency. The first antenna is omni-directional and radiates an amplitude-modulated reference signal. The modulation frequency of the reference-phase signal varies from 9,480 Hz to 10,440 Hz at a rate of 30 times per second. The resultant reference-phase signal then consists of a 9,960-Hz sub-carrier, frequency-modulated at 30 Hz, that amplitude-modulates the RF carrier.

The second antenna is a horizontal dipole which rotates at the rate of 1,800 revolutions per minute (30 revolutions per second) and produces a figure-eight electromagnetic field pattern. The RF field within one of the lobes is exactly in phase with the RF radiated from the omni-directional field. The in-phase lobe extends the omni-directional pattern on one side that results in a cardioid field pattern which rotates at the rate of 30 revolutions per second.

The airborne VOR receiver detects the RF carrier whose amplitude is varying at a rate of 30 Hz due to the rotation of the cardioid pattern. The carrier is also amplitude-modulated by the 9,960-Hz reference-phase signal, which is frequency-modulated at 30 Hz on a subcarrier to distinguish it from the 30-Hz variable-phase signal. The VOR airborne receiver detects the variable and reference-phase signals and compares their phase difference to determine the aircraft's bearing. At magnetic north, both signals appear in phase.

PRINCIPLES OF VOR OPERATION

As previously mentioned, the VOR station produces the radial pattern by transmitting 30-Hz reference signals and 30-Hz variable-phase signals for comparison by the airborne receiver. The 30-Hz variable-phase signal is an amplitude-modulated component of the VOR station RF signal. This signal is generated by rotating the transmitting antenna pattern, either mechanically or electronically, at 30 revolutions per second. The station identification code and voice transmissions are also amplitude-modulated components of the signal. The sum of all modulation components from the station results in a maximum of 90% modulation equally divided between the reference-phase component, variable-phase component, and the remaining voice or code identification component.

Magnetic north is the reference for all VOR measurements. At magnetic north, the 30-Hz variable signal is in phase with the 30-Hz reference signal. At all other radials, the 30-Hz variable signal leads or lags the 30-Hz reference signal by the number of degrees from magnetic north to the radial. For example, at 180° from magnetic north, the variable signal is 180° out-of-phase with the reference signal, as shown in Figure 4-3.

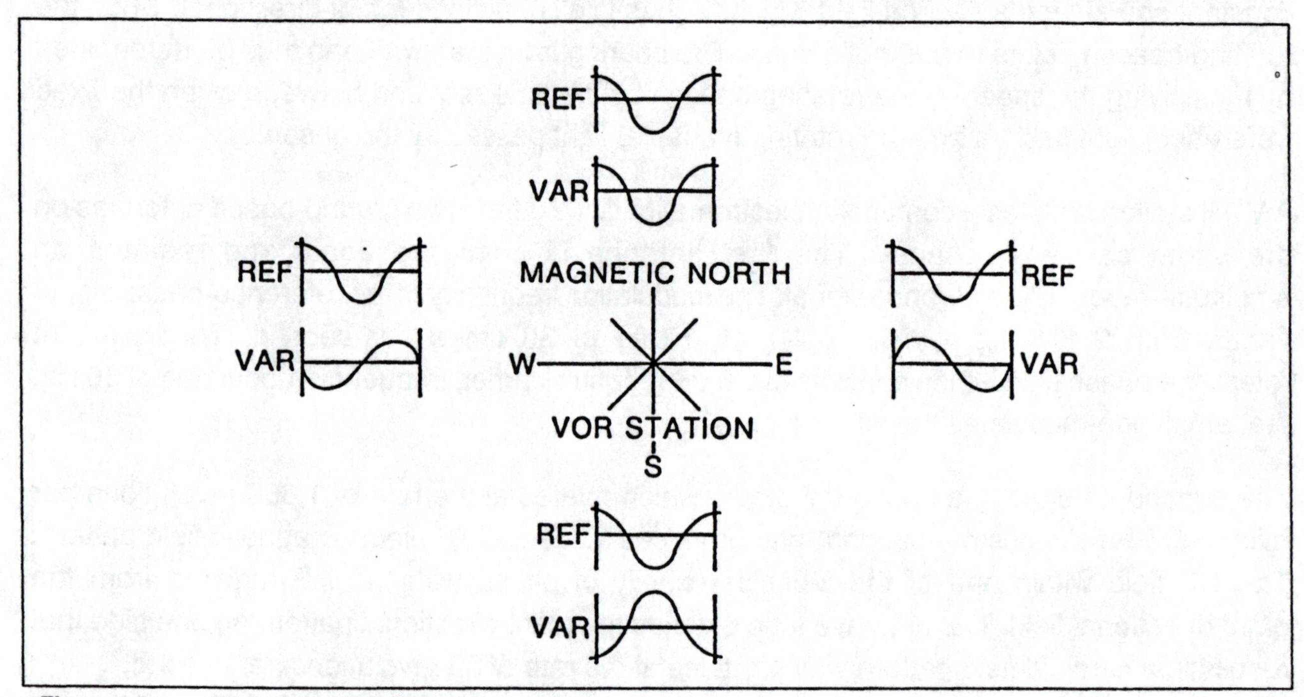

Figure 4-3. ***VOR Phase Relationships.***

Figure 4-4 illustrates the phase relationship of the 30-Hz reference and variable signals for an aircraft flying to a VOR station on the 260° radial. The 30-Hz reference component is applied to an FM detector that produces a 30-Hz reference signal that is positive when the subcarrier frequency is high (10,440 Hz) and negative when the frequency is low (9,480 Hz). The 30-Hz variable component is applied to an AM detector that produces a 30-Hz variable-phase signal that is positive when the major lobe of the radiated signal is toward the aircraft. When the 30-Hz variable and reference phases are compared, the difference is a direct measure of bearing from the VOR station to the aircraft (in this case, 260°).

The VOR circuitry derives to/from bearing information and left/right deviation from the selected course by comparing the received bearing with the setting of the omnibearing selector. As illustrated in Figure 4-5, an aircraft flying inbound to a VOR station on the 260° radial requires the pilot to slew the OBS to the 80° magnetic course setting. Since the 80° selected magnetic

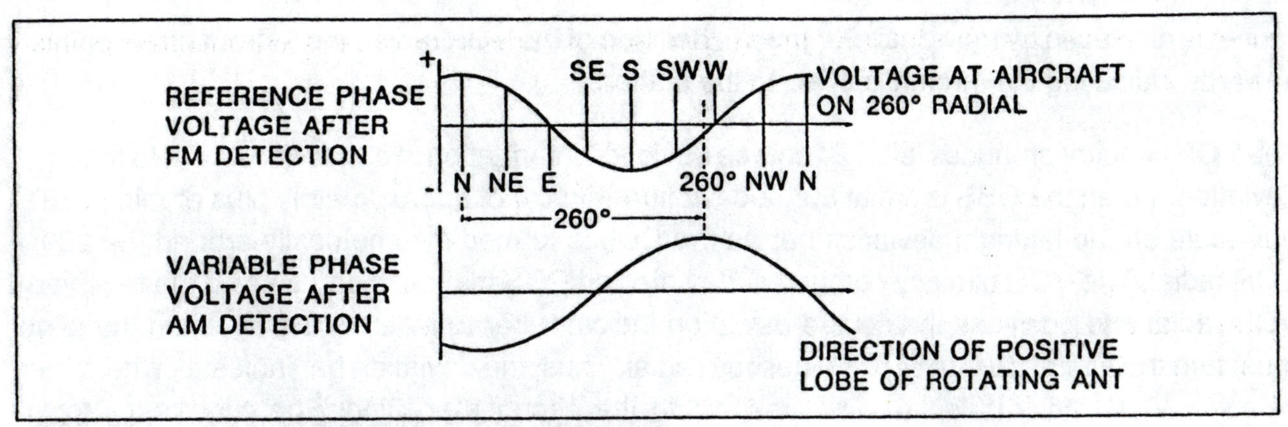

Figure 4-4. ***Phase Relationships of a 260° VOR Radial.***

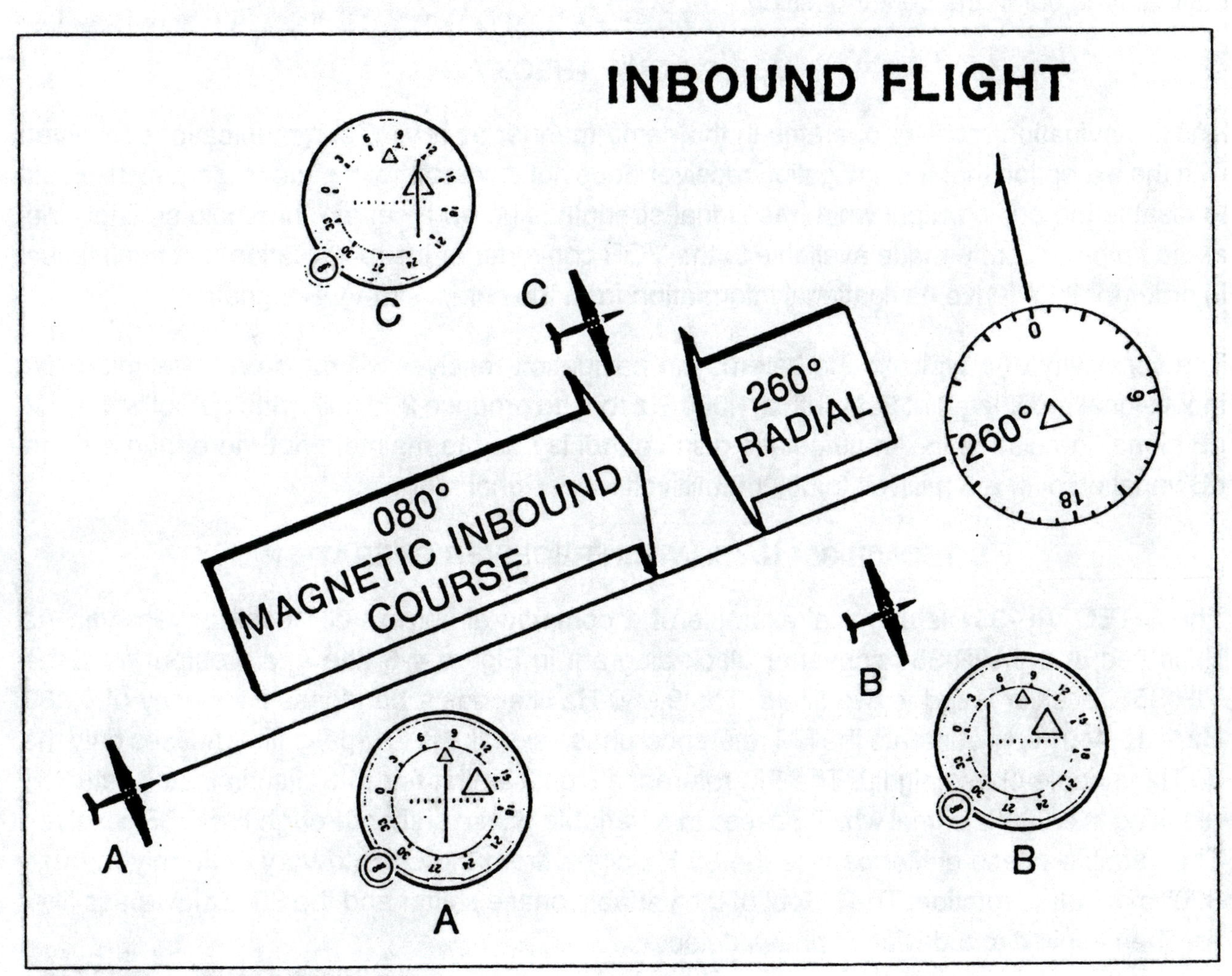

Figure 4-5. ***Flying a VOR Radial.***

course is displaced by more than 90° from either side of the VOR radial, the to/from arrow points upwards indicating the aircraft is flying to the station.

The VOR circuitry produces left/right course deviation information in a manner similar to to/from deviation. When the OBS is set at 80°, a deviation window of approximately plus or minus 10° (full scale on the left/right deviation bar on the CDI) is formed symmetrically around the 260° VOR radial. The VOR circuitry compares the selected OBS magnetic course with the received VOR radial and positions the course deviation indicator bar to indicate the direction the pilot must turn the aircraft to intercept the desired radial. Thus, the deviation bar indicates where the selected VOR radial is located with respect to the aircraft's position. The pilot keeps from deviating from the selected course by following the command of the left/right deviation bar to maintain the bar in the center position.

VOR CIRCUIT THEORY

A VHF navigation receiver operates in the same manner as a VHF communications receiver with the exception that the navigation receiver does not incorporate the use of a squelch circuit to disable the audio output when the signal strength is below a certain threshold setting. The audio output must be made available to the VOR converter or instrumentation unit at all times in order for it to derive navigational information from the composite VOR signal.

The sensitivity of a typical superheterodyne navigation receiver will cause a three microvolt input signal modulated at 30% with a 1,000-Hz tone to produce 200 milliwatts output with a six dB signal-to-noise ratio. An automatic gain control is used to maintain not more than a three dB variation over a 5 millivolt to 50,000 millivolt input signal range.

S-TEC VIR-351 NAVIGATION RECEIVER OPERATION

The S-TEC VIR-351 is a typical example of a common airborne VOR/localizer receiver. As illustrated in the VIR-351 converter block diagram in Figure 4-6, the audio output from the VIR-351 receiver is fed to two filters. The 9,960-Hz filter has a bandpass frequency of 9,480 Hz to 10,440 Hz to separate the FM reference-phase signal; the low-pass filter passes only the 30-Hz variable-phase signal. The FM reference signal is first fed to a discriminator detector, resulting in a 30-Hz signal which passes to a variable phase shifter through the OBS resolver. The variable phase shifter causes the 30-Hz signal from the OBS to vary uniformly through 360° of resolver rotation. The output of the variable phase shifter and the 30-Hz low-pass filter are then applied to a deviation phase detector.

The deviation phase detector actuates a galvanometer movement on the course deviation indicator. The polarity of the galvanometer is such that when the left/right deviation bar is centered, the setting of the resolver will indicate the bearing of the VOR station. To fly on this radial, the pilot keeps the bar centered by turning toward it. The variable-phase signal will lag the reference-phase signal when flying to the station and will lead the reference signal when

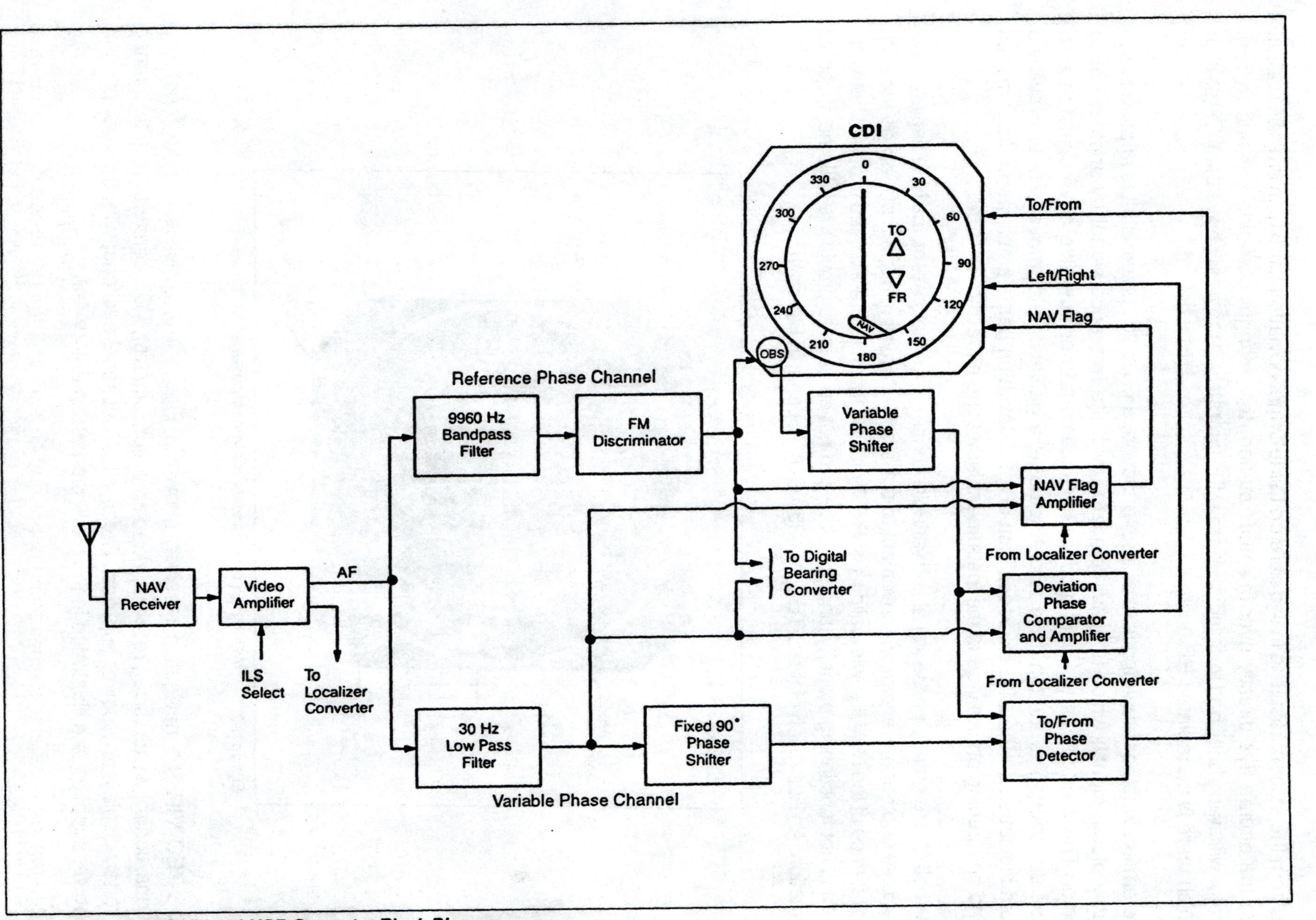

Figure 4-6. S-TEC VIR-351 VOR Converter Block Diagram.

flying from the station, resulting in two nulls on the left/right deviation bar which are 180° apart. When the aircraft flys directly over the VOR station, the left/right deviation bar and to/from indicator will swing back and forth indicating that the aircraft is in the area above the VOR station in which there is no radiated RF energy.

To determine whether a bearing is to or from a station, a fixed 90° phase shift is applied to the variable phase signal, which is again compared with the reference signal by another phase detector. The fixed 90° phase shift produces a maximum to or from flag drive signal at zero course deviation. The to/from drive signal actuates a galvanometer movement to provide a to/from indication on the CDI. However, instead of a deviation bar, a small metal "flag" moves behind an opening on the face of the instrument indicating either "to" or "from".

When both the reference-phase signal and variable-phase signal are present, the NAV flag amplifier will actuate another galvanometer movement to pull the NAV warning flag out of view on the CDI. The presence of the NAV flag with valid VOR RF signal reception indicates to the pilot that the VOR system is not functioning properly, and therefore can not be used to provide reliable navigational information. An illustration of the S-TEC IND-350 CDI, appears in Figure 4-7.

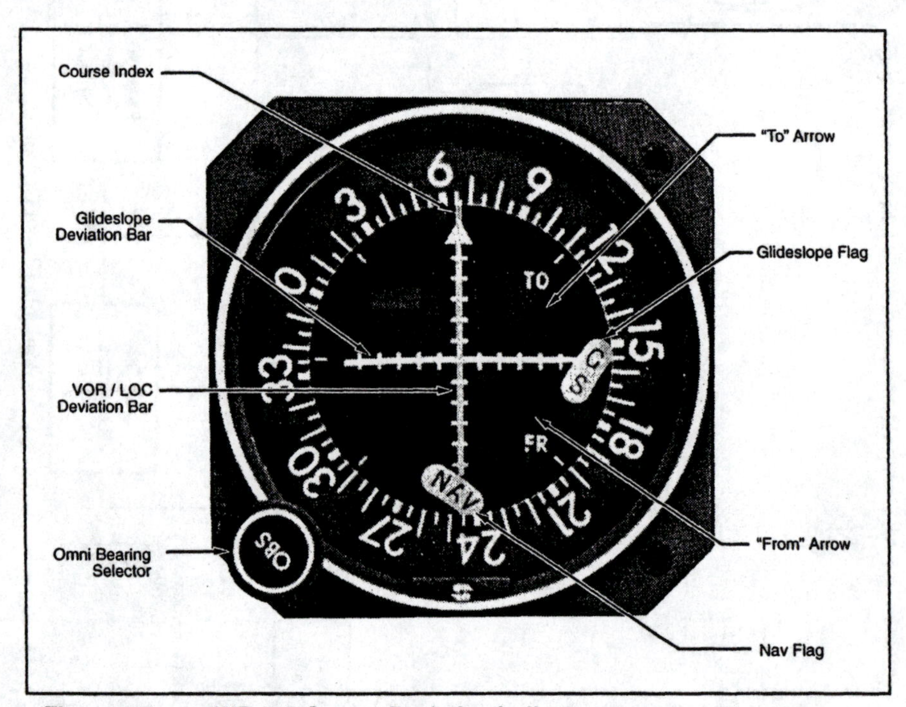

Figure 4-7. ***IND-350 Course Deviation Indicator.*** *(Courtesy S-TEC Corporation)*

The S-TEC VIR-351 navigation receiver, shown in Figure 4-8, consists of nine functional sections: control, synthesizer, receiver, video amplifier, localizer, VOR converter, digital bearing converter, display, and power supplies. These nine functional sections, with the exception of the power supplies, are illustrated in the block diagram in Figure 4-9.

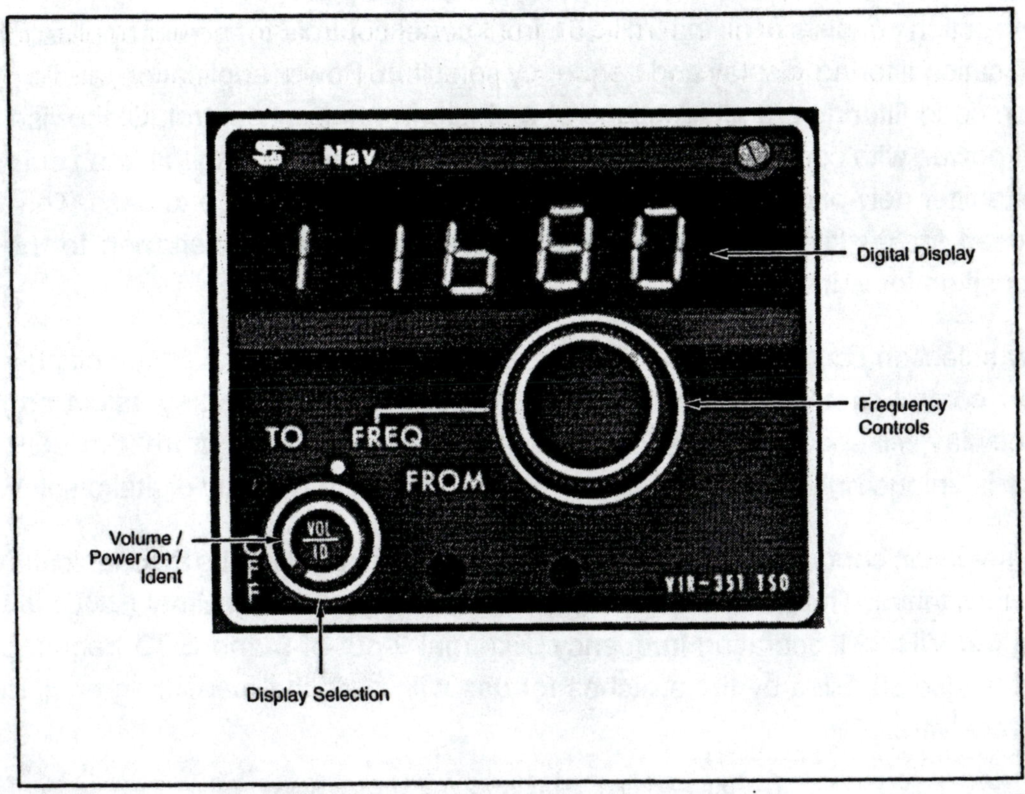

Figure 4-8. ***VIR-351 Navigation Receiver.*** *(Courtesy S-TEC Corporation)*

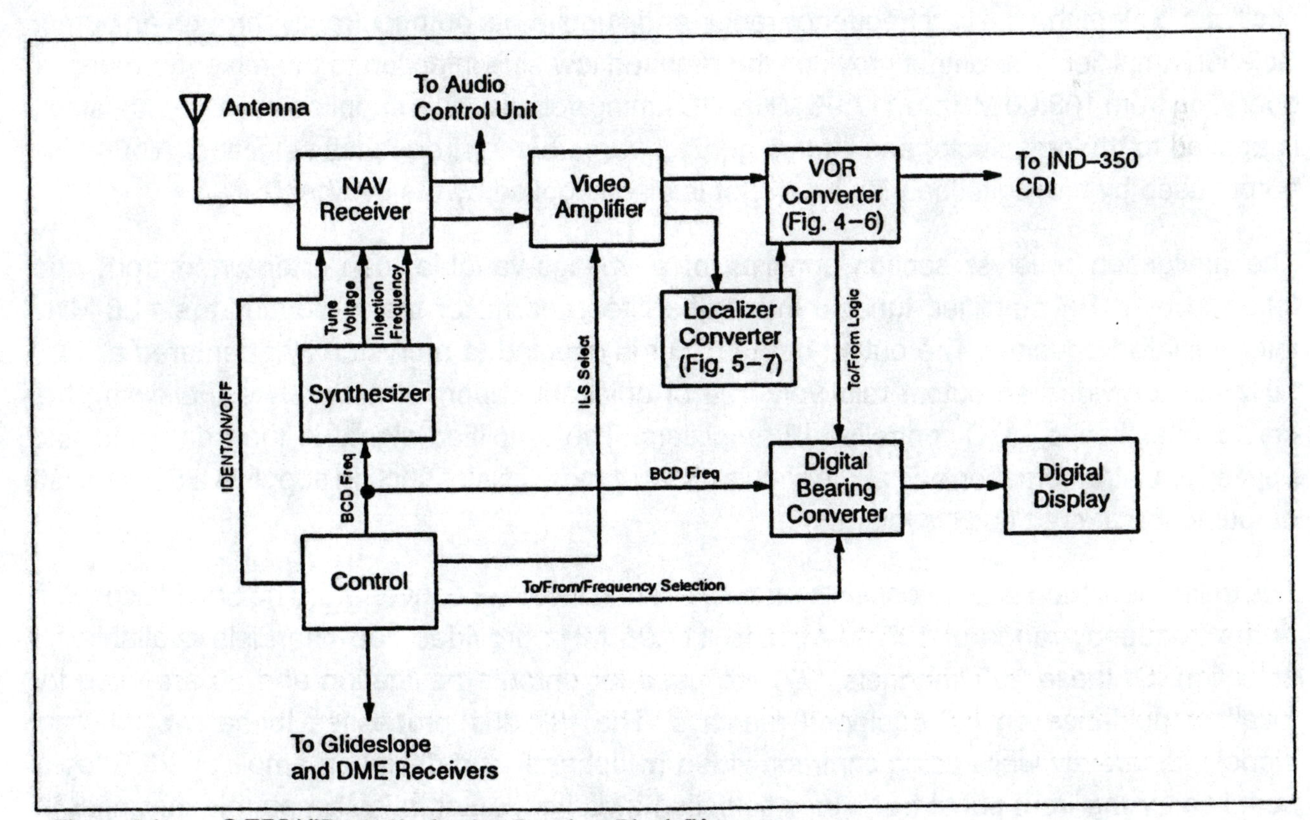

Figure 4-9. S-TEC VIR-351 Navigation Receiver Block Diagram.

The control section consists of all the VIR-351 front panel controls for power application, audio gain, identification filtering, display and frequency selection. Power application, audio gain, and identification code filtering are all functions of a single front panel control. Clockwise rotation applies unit power, with continued rotation increasing audio gain. Pulling the knob out removes a portion of a filter network, thereby allowing the CW identification code to pass. This function is initially used for positive station identification, then the switch is returned to the normal operating position for voice reception.

The display selection control consists of a three-position switch that supplies display logic to the localizer control circuits. Depending upon the position of the display select control, the electronic display will show either frequency, bearing "to", or bearing "from" the VOR station. This feature is unique only in VOR receivers that incorporate electronic digital displays.

Frequency selection controls consist of a pair of concentric control knobs mechanically coupled to five wafer switches. The wafer switches produce a binary-coded decimal (BCD) format that determines the VIR-351 operating frequency. External 2-out-of-5 and BCD frequency-select information is also supplied by the switches for use with distance measuring equipment and glideslope receivers.

Frequency control of the VIR-351 is accomplished by a digital frequency synthesizer that is phase-locked and frequency-locked to one stable crystal oscillator that generates a receiver injection frequency of 97.0 MHz to 106.95 MHz in 50-kHz increments. A single voltage controlled oscillator covers the desired frequency range and supplies its output directly through an output isolation amplifier. The output provides the required low-side injection to the mixer for receiver operation from 108.00 MHz to 117.95 MHz. DC tuning voltage, also supplied by the synthesizer, is applied to the preselector and interstage filter varactors for frequency selection. A 100-kHz output used by the digital bearing converter is also supplied by the synthesizer.

The navigation receiver section consists of a voltage-variable, capacitor-tuned front end followed by a RF amplifier, tunable interstage filter, and mixer that produces the 11.0-MHz intermediate frequency. The output of the mixer is coupled to a crystal filter centered at 11.0 MHz that provides an output relatively free of adjacent channel interference. Following the crystal filter is two AGC-controlled IF amplifiers. The amplified signal is then detected and applied to instrumentation and audio circuits. The audio amplifier section supplies a 50-milliwatt output to the aircraft audio system.

The instrumentation section consists of the VOR and localizer converters. The 50-kHz spacing in the frequency range of 108.00 MHz to 117.95 MHz provides 200 channels available for selection. Of these 200 channels, 160 are used for enroute navigation and 40 are used for localizer guidance on ILS-equipped runways. The VIR-351 processes these two different signals separately while using common video (wideband) and deviation amplifiers. ILS logic, supplied by the front panel frequency selection switches, automatically controls the receiver mode of operation (VOR or localizer). Selection of an ILS frequency automatically disables operation in the VOR mode and sends the detected signal to the localizer converter. The localizer portion of the VIR-351 receiver is discussed in the next chapter.

The detected VOR signal, supplied by the receiver section, is amplified by a video amplifier and applied to reference and variable-phase channels for processing. The composite VOR signal is applied to a 30-Hz bandpass filter, where the variable signal is extracted, and to a 9,960-Hz bandpass filter and FM discriminator, where the 30-Hz reference signal is derived. The 30-Hz reference signal excites the OBS resolver rotor. The output of the OBS resolver is applied to a variable phase shifter network, making the OBS phase shift of 30 Hz vary uniformly through 360° of resolver rotation. The outputs of the variable phase shifter and the 30-Hz variable-phase bandpass filter are applied to a deviation phase comparator that produces a chopped 30-Hz output signal used to drive the deviation amplifier. The output of the deviation amplifier is a filtered DC voltage used to drive the left-right needle of the course deviation indicator.

The to/from output is produced by applying the output of the variable phase shifter and the 30-Hz variable signal to the to/from phase comparator. The 30-Hz variable signal is first phase-shifted 90° to produce a maximum flag drive signal at zero deviation. The comparator then drives the to/from flag above, below, or at the five volt reference level as a function of the compared inputs.

Both VOR and localizer signal validity are monitored by the flag circuit. The presence of a flag may be due to one or more conditions. Typically, variables such as reception of an unreliable signal, attempting to tune an out-of-range station, ground station failure, or a malfunction in the airborne VOR equipment will cause the flag to come into view. In the VOR mode, the 30-Hz reference and 30-Hz variable signals are applied to the flag circuit. The absence of one or both of the 30-Hz signals will result in the appearance of the NAV flag on the CDI.

The digital bearing converter measures the phase difference between the 30-Hz reference and 30-Hz variable signals and provides a digital output that is proportional to the phase displacement between the two signals. Depending upon the position of the front panel selector, the VIR-35l will display the digital bearing either to or from the selected VOR station. When the from position is selected, the digital bearing will be followed by the letter "F" to distinguish between bearing indications.

Control logic to the electronic display is supplied by the digital bearing converter circuits. The input logic is decoded by four BCD-to-seven-segment converter/drivers and applied to incandescent filament-type displays. Automatic dimming of the displays is accomplished by an internal photocell dimming circuit that senses cockpit ambient light. Manual adjustment is provided through a front panel access hole to match the display intensity with the other electronic displays on the instrument panel.

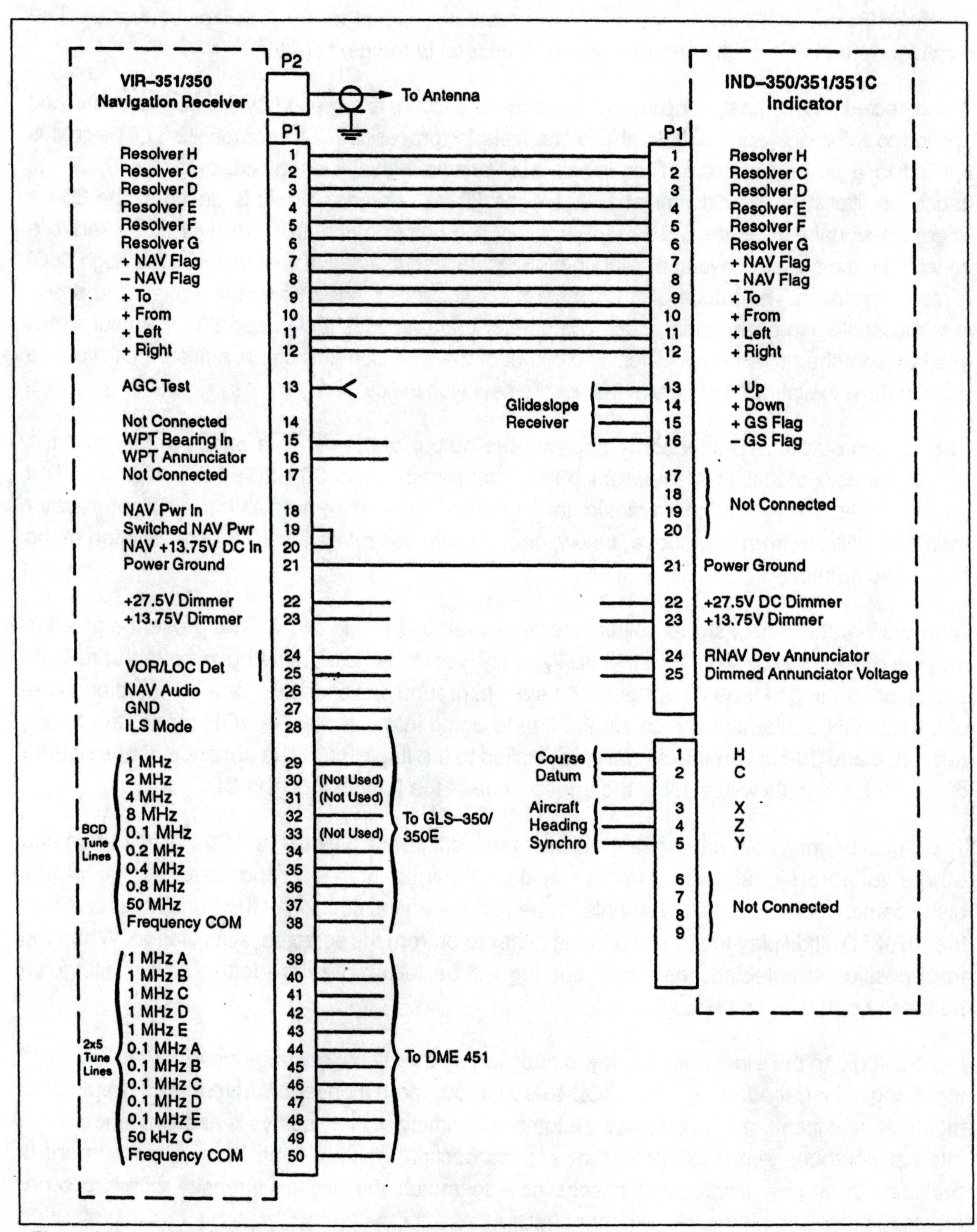

Figure 4-10. ***VIR-351 Interconnect Wiring Diagram.*** *(Courtesy S-TEC Corporation)*

There are two distinct power supplies contained in the VIR-351, both of which are supplied the primary 14-volt DC input power. A +10-volt DC regulator is included along with a -12-volt and a +5-volt supply. The later supply consists of a switching regulator that produces +5-volt DC for use throughout the radio. The -12-volt DC output is produced by a step-up transformer with a half-wave rectified output. The electronic display driver circuit also operates off the +5-volt DC output. Display drive is controlled by the front panel mounted photocell. The +10-volt DC regulator consists of a transistor with an active error feedback circuit.

Figure 4-10 is an interconnect wiring diagram of the S-TEC VIR-351 navigation receiver and S-TEC IND-350 course deviation indicator.

BENDIX/KING KX-170A/175 NAVIGATION RECEIVER OPERATION

The KX-170A/175 is a combination panel-mounted VHF navigation receiver and communications transceiver. The communications transceiver portion was discussed in Chapter Two. The navigation receiver section is a dual-conversion, superheterodyne receiver with a 15.1875-MHz first IF and a 1.1857-MHz second IF frequency. The 200 channels are synthesized at the first mixer and low-side injection is used for all channels.

As shown in Figure 4-11, a two-pole, varactor-tuned, RF filter couples the antenna to the RF stage. A single pole, varactor-tuned filter, couples the amplified RF signal to the first mixer and supplies additional image and 1/2 IF spurious rejection. The amplified RF signal is mixed with the synthesized injection frequency in a balanced mixer.

A crystal filter couples the difference frequency to the second mixer and provides image and 1/2 IF selectivity. The 14.0018-MHz, crystal-controlled, second local oscillator develops injection for the second mixer. The second IF contains two IC amplifiers with three double-tuned interstage networks for receiver selectivity. An active detector provides audio gain. An emitter follower couples the detector signal to the noise limiter and the AGC amplifier. The noise limiter clips noise peaks corresponding to greater than 90% AM modulation and provides rate limiting. A two-stage, AGC amplifier is used to control the gain of the RF stage and the second IF strip.

The receiver outputs 6 dB into the AGC with no input signal. This eliminates conventional receiver gain threshold effects. The audio signal from the rate noise limiter passes through a panel switch controlled 1,020-Hz ident filter, the volume control and then to a audio amplifier providing a 50-milliwatt, 600-ohm output.

PERFORMANCE VALIDATION

Once the VOR navigation system wiring is installed in the aircraft and a continuity check of the system wiring has been accomplished, the equipment may be installed and power may be applied. The following paragraphs describe the procedures for performing a ground or airborne post-installation check to validate VOR system performance.

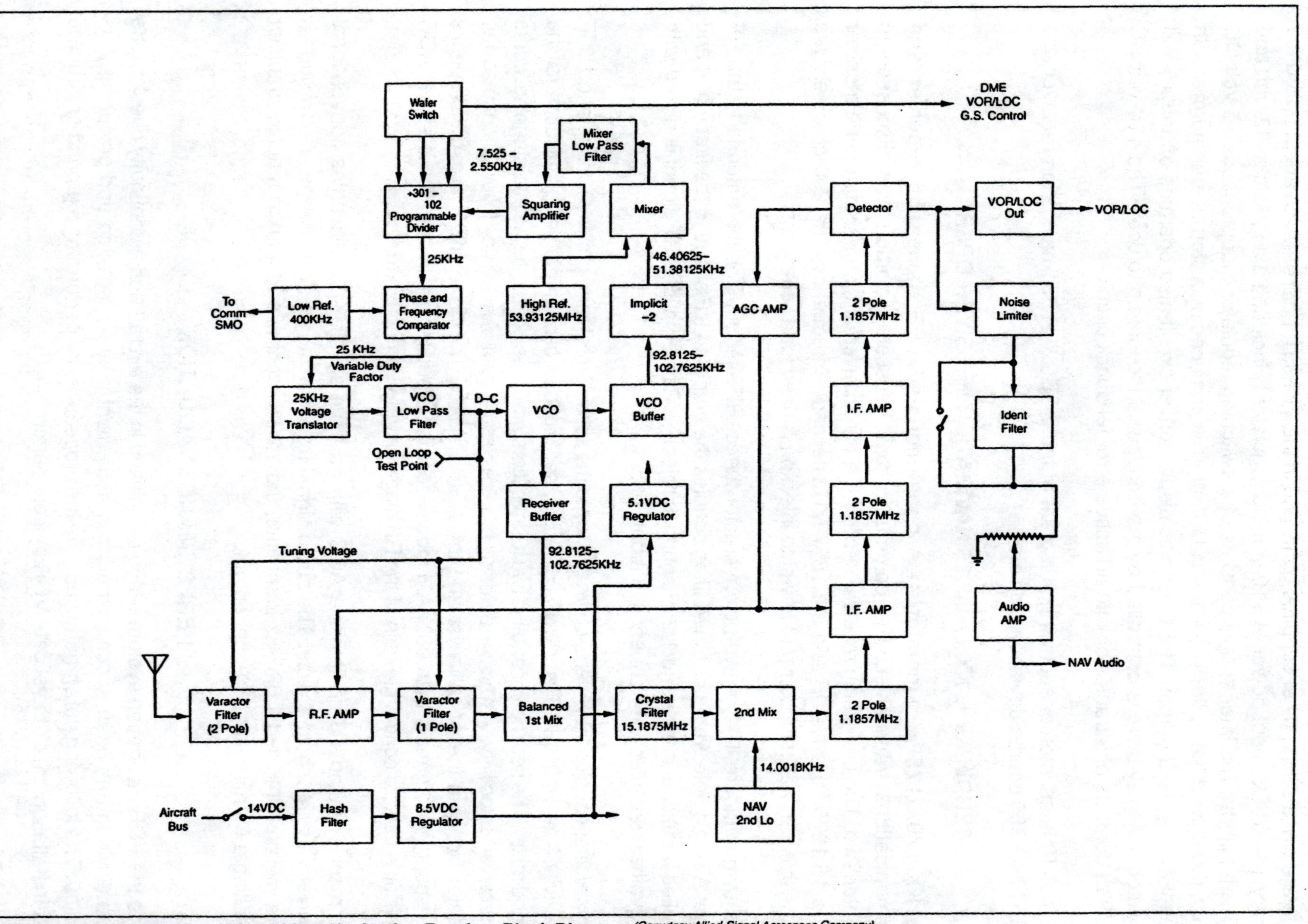

Figure 4-11. Bendix/King KX-170A/175 Navigation Receiver Block Diagram. *(Cozurtesy Allied-Signal Aerospace Company)*

The first step to validate performance is to apply power to the VOR system and channel the receiver to a VOR or VOT (VOR Test) station in the immediate area. Verify that the signal is reliable by listening to the identification code and ensuring that the NAV flag is out-of-view on the CDI. Adjust the OBS knob until the left/right deviation bar is exactly centered and a "from" indication is obtained. The bearing indication above the course index or lubber line should indicate the bearing indicated on the aeronautical chart (0° for a VOT) within plus or minus 5°. Adjust the OBS knob until the left/right deviation bar is exactly centered and a "to" indication is obtained. Observe that the bearing indication above the course index reads 180° plus or minus 5° difference from the published bearing.

To test the deflection sensitivity, record the indicator bearing when the left/right deviation bar is centered. Slowly rotate the OBS knob until the bar is positioned over the fifth dot on the right side of the scale. Note the indicator bearing at this point and compare it to the previously recorded bearing when the bar was centered. The new bearing indication should differ by 10° plus or minus 3°. Rotate the OBS knob counterclockwise until the left/right deviation bar is positioned over the fifth dot on the left side of the scale. The resultant bearing indication at this point should differ again by 10° plus or minus 3° from the bearing when the bar was centered.

If the VOR navigation system fails either one of the above tests, remove the receiver and indicator and bench test the system for proper operation. Additional tests, using a ramp signal generator, are provided in the next chapter. Calibration procedures may be obtained from the applicable manufacturer's maintenance manuals.

OPERATING PROCEDURES

The following discussion describes the three most common procedures for operating a VOR navigation system while airborne: VOR orientation, interception and tracking, and VOR crosscheck.

VOR orientation consists of determining aircraft position in relation to a station radial and/or determining the inbound course that must be flown to guide the aircraft from its present position to the VOR station. To determine the aircraft's present position on a radial, the pilot first selects and positively identifies the desired station by listening to the station identification code or voice transmission. The pilot then notes the to/from indication displayed on the course deviation indicator.

If a "from" indication is displayed, the OBS knob is rotated in the opposite direction of the left/right deviation bar deflection until the bar is centered. When the deviation bar is centered, the aircraft position will be on the radial displayed above the course index. If the CDI shows a "to" indication, the OBS knob is rotated in the direction of deflection until a "from" indication is obtained and the deviation bar is centered. The radial again will be displayed above the course index on the CDI.

To determine the bearing to a selected station, the pilot turns the OBS knob until the to/from arrow indicates "to" and the left/right deviation bar is centered. The pilot then reads the "to" bearing indicated above the course index and maneuvers the aircraft to fly the magnetic course

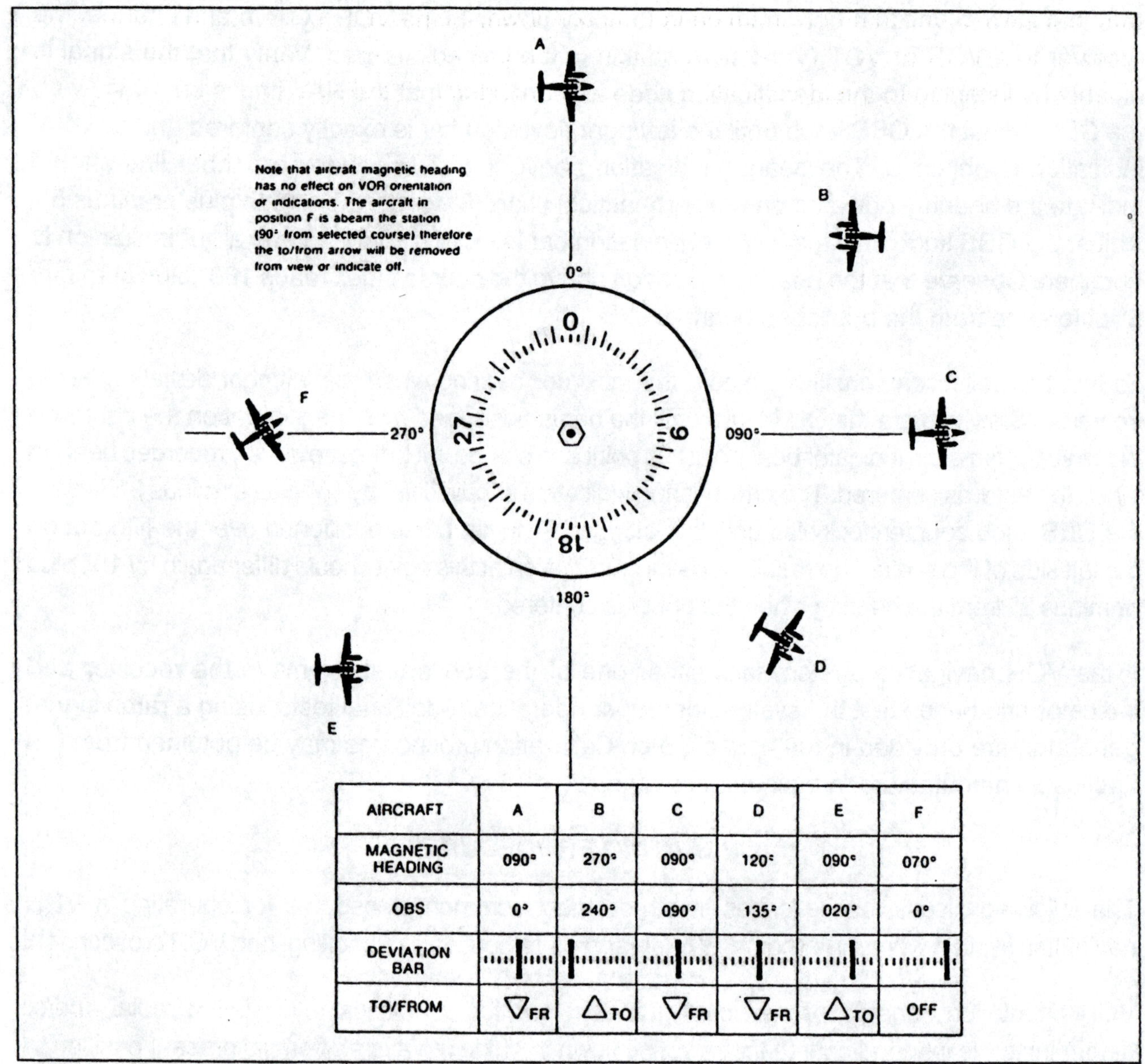

AIRCRAFT	A	B	C	D	E	F
MAGNETIC HEADING	090°	270°	090°	120°	090°	070°
OBS	0°	240°	090°	135°	020°	0°
DEVIATION BAR						
TO/FROM	▽FR	△TO	▽FR	▽FR	△TO	OFF

Figure 4-12. ***VOR Orientation.***

to the station. An example of how the VOR is used in this mode of operation is illustrated in Figure 4-12.

To intercept and fly along a selected course to the VOR station, the pilot uses the OBS knob to set the desired course above the course index and maneuvers the aircraft to establish the correct intercept angle. When an on-course condition has been established, the left/right deviation bar will center and the to/from indicator will show to. Once the deviation bar is centered, the pilot must maintain the heading corresponding to the selected course.

As shown in Figure 4-13, crosswind components will be observed by noting deviation bar drift. The direction of bar drift shows the direction of the crosswind. If the bar drifts to the left, the wind is from the left. To maintain an on-course condition, the pilot must apply correction and fly in the direction of the deviation bar deflection. With the proper drift-correction angle established, the deviation bar will remain centered until the aircraft is close to the station.

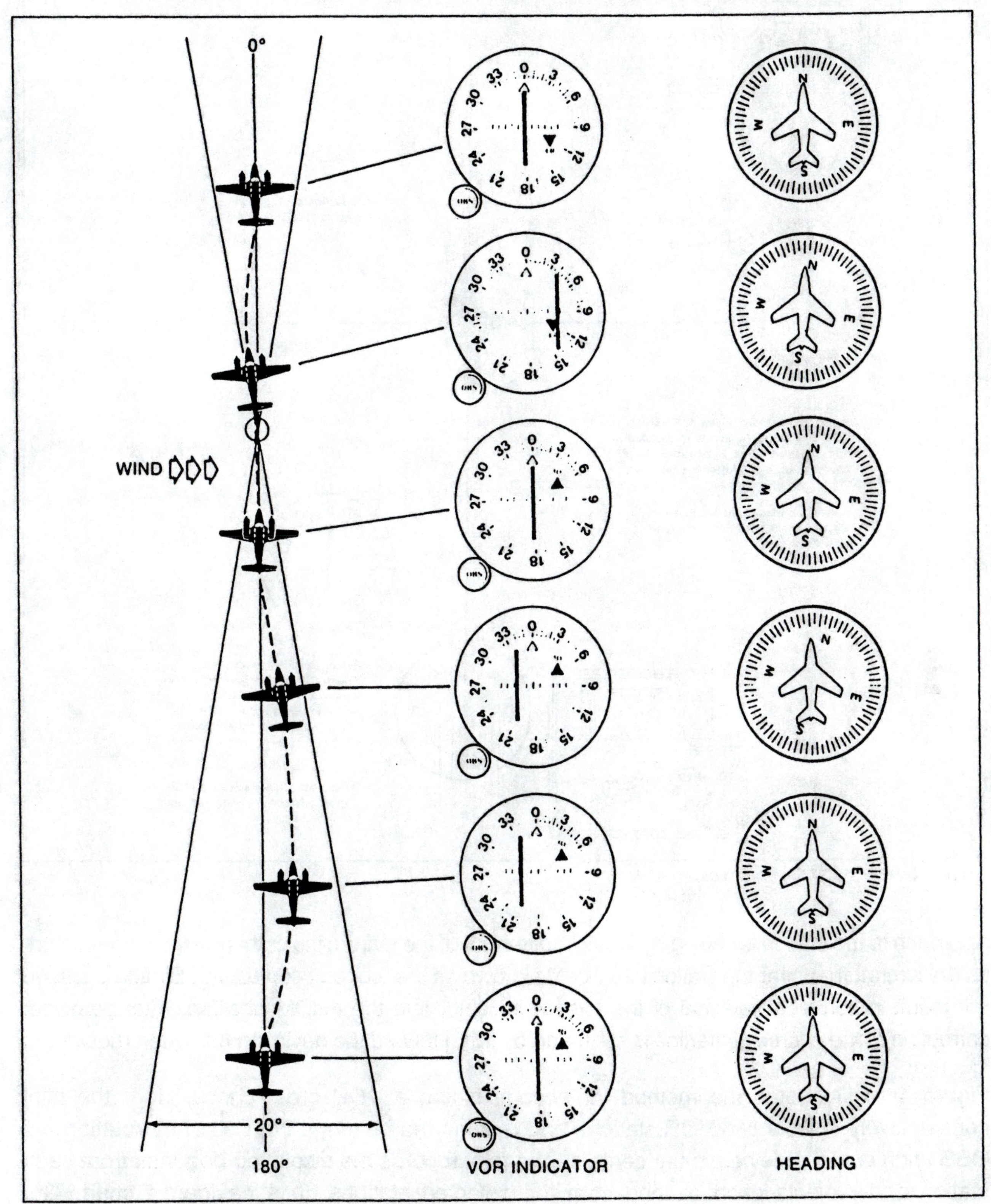

Figure 4-13. *VOR Interception and Tracking.*

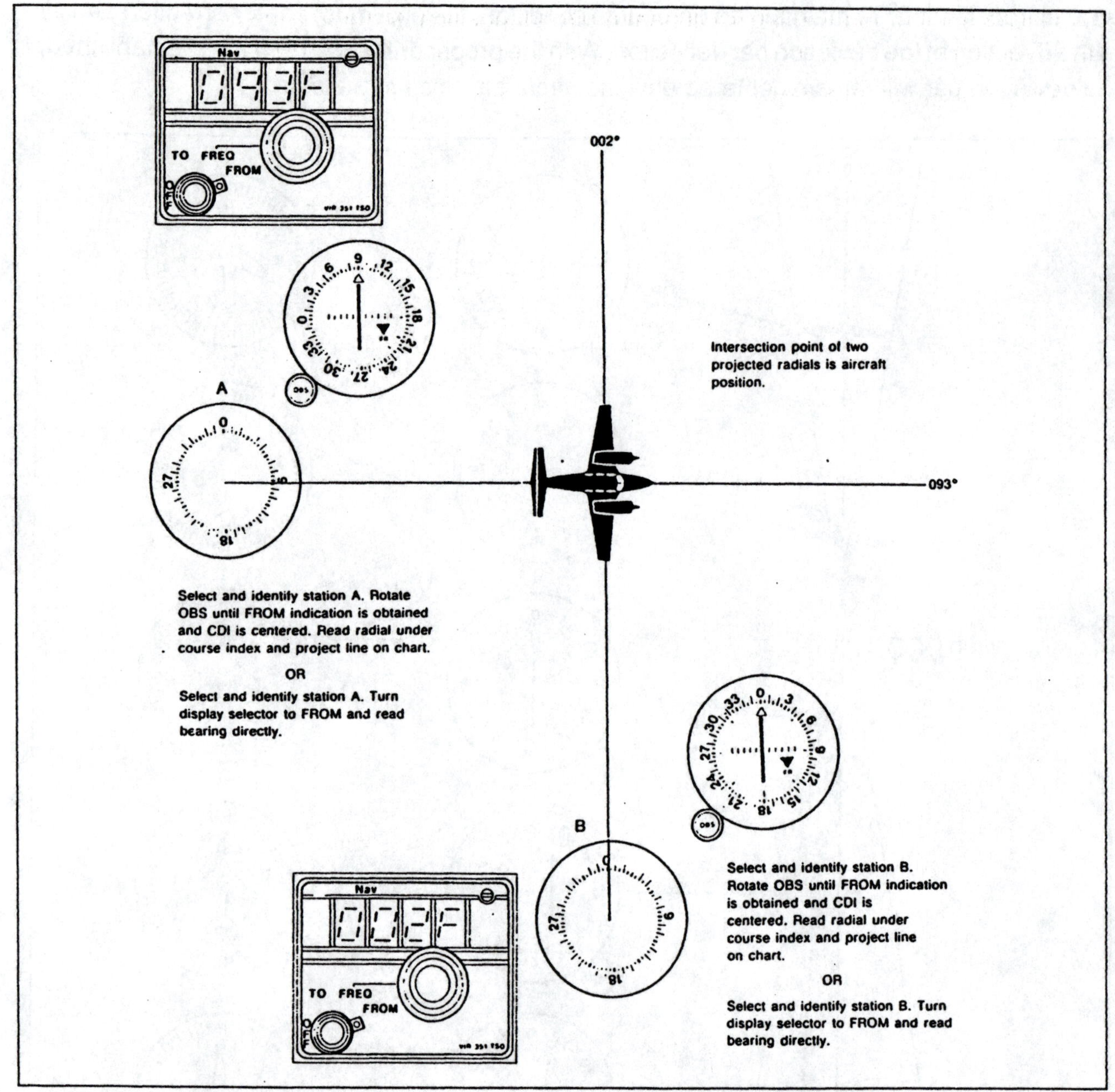

Figure 4-14. VOR Cross-Check.

Approach to the station is shown by erratic movement of the to/from indicator and the deviation bar as the aircraft flies over the station into what is known as the "cone of confusion". Station passage will result in complete reversal of the to/from indicator into the FROM position. After passage, correction to the course centerline is still made by flying toward the deviation bar when required.

Figure 4-14 illustrates the method for accomplishing a VOR cross-check. Here the pilot consecutively selects two VOR stations and obtains the bearings from each by rotating the OBS knob until the deviation bar centers. The pilot records the displayed bearings from each station and projects each bearing from the selected stations on a navigation chart. The intersecting point of the two lines will be the aircraft position. If the VOR system is equipped with a digital bearing display, rotation of the OBS will not be required since the bearing can be read directly from the display.

CONCLUSION

As can be seen from the previous discussion, VOR navigation provides a means for determining the aircraft's position while enroute to a given destination, even during times of zero visibility. However, a VOR system is of no use in directing a pilot to touch down on the runway once he or she reaches the final destination. Therefore, instrument landing systems were developed in 1942 to provide the pilot with an indication of his deviation left or right of the runway along a predetermined glide path upon final approach. The next chapter discusses the operation and maintenance of airborne ILS systems.

REVIEW QUESTIONS

1. What is the principle advantage of using VOR navigation over ADF navigation?
2. Define the term "radial". How many radials does a VOR station broadcast?
3. What is the frequency range and spacing of VOR navigation channels?
4. Describe the three components of the composite RF signal radiated from a ground-based VOR station in terms of modulation type, frequency, and percentage.
5. How is the rotating cardioid field pattern generated?
6. What is the purpose of the 9,960-Hz subcarrier?
7. If an aircraft is flying due north towards a VOR station, what radial is it on before and after it passes the station?
8. What is the phase relationship between the variable and reference signals when an aircraft is flying inbound to a VOR station on a 90° magnetic course?
9. When are the variable and reference components of the detected VOR signal both positive?
10. What information is obtained by comparing the received bearing with the OBS setting on the CDI?
11. If the course deviation bar is displaced full scale to the left when the OBS is set to 260°, approximately what radial is the aircraft on and what must the pilot do to keep from deviating from the selected course?
12. Draw a simplified block diagram of a typical VOR converter.
13. What is the function of the OBS resolver?
14. What circuit element is used to demodulate the 30-Hz reference signal from the receiver output?
15. How is "to/from" information derived and displayed?
16. Explain the procedure for testing "left/right" deflection sensitivity.
17. What is the procedure for determining the inbound course that must be flown to guide the aircraft from its present position to a VOR station?
18. How is crosswind correction applied to maintain the aircraft on the selected VOR radial?
19. What happens when the aircraft passes over the "cone of confusion"?
20. How does a pilot use the VOR system to determine the aircraft's present position?

Chapter 5
Instrument Landing Systems

INTRODUCTION

In the preceding chapter, it was mentioned that the majority of airborne navigation receivers incorporate instrumentation circuitry for both VOR enroute navigation and for Instrument Landing System (ILS) approaches while sharing a common indicator for both functions. This chapter will deal specifically with Instrument Landing Systems, which electronically guide the aircraft down the approach path to intersect the center of the threshold of the runway prior to landing by means of visual cues provided to the pilot on the Course Deviation Indicator (CDI).

Instrument landing systems were developed in 1942 in response to the need for pilots to locate and fly to an airport during times of poor visibility. Thus, with the use of the ILS, the pilot is able to fly the aircraft, without outside visual reference, along the approach path to a pre-determined decision height (usually a minimum of 200 feet altitude) before establishing visual contact with the runway approach lights. If visual contact is not established at this point, the pilot must either select an alternative airport in which to land the aircraft, or circle the airport until such time that visibility improves.

The ILS transmitters, located at or near the airport facility, radiate directional signals which are received by the airborne receiver. These signals are then processed to provide the pilot with a visual horizontal directional reference, vertical reference, and distance reference along the approach path. The localizer transmitter provides the horizontal reference signal, which is displayed by the "left/right" needle on the course deviation indicator. The vertical reference signal, obtained from the glideslope transmitter, results in an "up/down" needle indication on the CDI. Marker beacon transmitters, located at pre-determined distances from the end of the runway, provide an audio (tone) and visual (light) indication along the approach path. The pilot flys an ILS approach by centering the left/right and up/down needles on his CDI while monitoring the marker cues to determine progress along his or her descent.

PRINCIPLES OF LOCALIZER (LOC) OPERATION

A typical VHF localizer station is located 1,000 feet from the end of the runway and is offset 300 feet from the runway centerline. It has a horizontally-polarized antenna array that radiates approximately 100 watts of RF power on one of 40 available channels spaced alternately at 50 and 150 kHz apart in the VHF frequency range of 108.10 MHz to 111.95 MHz. The localizer station is identified by the transmission of a four-letter identification code modulated at 1,020 Hz and also by voice identification.

The localizer station radiates a beam of information indicating the horizontal centerline of the runway. This beam is produced by two transmitters operating on the same channel frequency but amplitude-modulated with different audio signals, one being 90 MHz and the other 150 Hz. The transmitting antennas are aligned to radiate two lobes that intersect the runway centerline. Viewed from the approach end of the runway, the 15-Hz modulated lobe is on the right and the 90-Hz lobe is on the left.

When the aircraft approaches the runway, the navigation receiver demodulates the received signal and the localizer instrumentation unit compares the amplitude of the 90-Hz and 150-Hz signal levels to provide a deviation signal output to the CDI which is proportional to the balance or unbalance of the 90-Hz and 150-Hz signals. The aircraft is centered with the runway when the 90-Hz and 150-Hz localizer signals are equal in amplitude as indicated by zero deflection of the left/right needle on the CDI. This concept is shown in Figure 5-1.

The course width, which is the amount of off-course distance for a given needle indication, is set at 2.5° off-center for full-scale needle deflection. For example, if the aircraft is five miles from the runway, a full left or right needle deflection on the CDI would represent a distance of approximately 1,500 feet from the left or right of the runway centerline. Naturally, as the aircraft flys closer to the runway, the sensitivity of the left/right deflection increases.

LOCALIZER CIRCUIT THEORY

Typical airborne localizer instrumentation circuits utilize the same VHF receiver that is used for VOR navigation. In some cases, a common VHF receiver operating in the frequency range of 108.000 to 135.975 MHz is used for VHF communication (118.000 to 135.975 MHz), VOR navigation (108.000 to 117.950 MHz) and LOC operation (108.100 to 111.950 MHz). The 40 localizer channels are located on odd-tenths frequencies to distinguish them from the even-tenths VOR frequencies. The navigation control head, which the pilot uses to select the desired VHF frequency, outputs a control signal which switches the instrumentation unit from the VOR mode to the LOC mode when an odd-tenth frequency below 112 MHz is selected. The LOC signal is horizontally-polarized, allowing the VOR antenna to also be used for localizer reception.

As illustrated in Figure 5-2, the audio output from the VHF navigation receiver is fed to two filters. One filter passes the 90-Hz modulation component, and the other separates the 150-Hz component from the detected localizer signal. The filtered signals are then rectified to produce a DC signal, which energizes a galvanometer movement to form a left/right needle indication on the CDI relative to the localizer course. LOC steering information is obtained by comparing the difference in output levels from the two bandpass filters.

In ILS mode, the output of these filters are summed together to provide the voltage necessary to pull the "NAV" warning flag on the CDI out-of-view. The NAV flag appears from beneath a cutout on the face of the indicator when the output from either one or both of the filters falls below a useable level. Since the to/from indication on the CDI is intended only for VOR

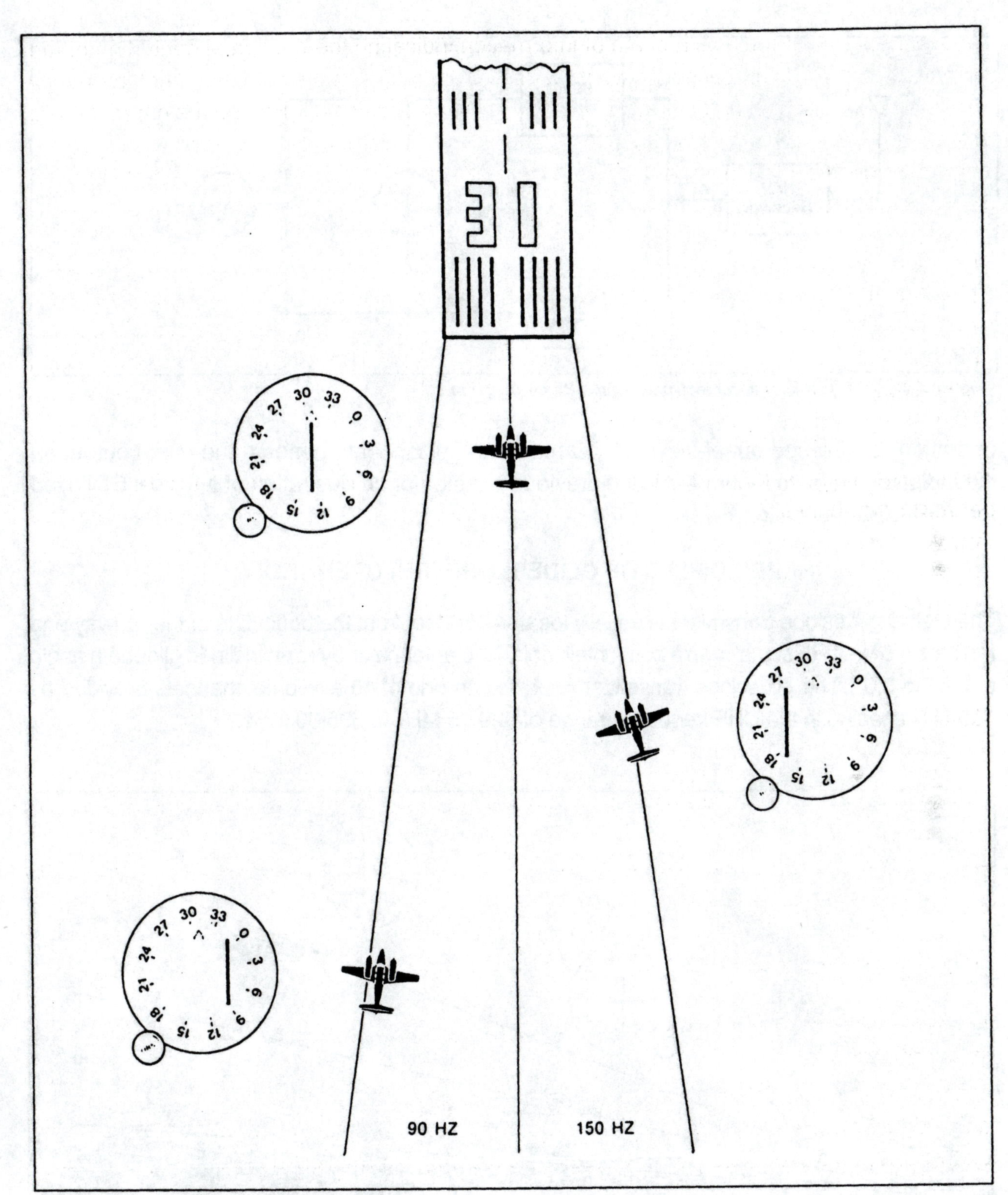

Figure 5-1. Flying a Localizer Course.

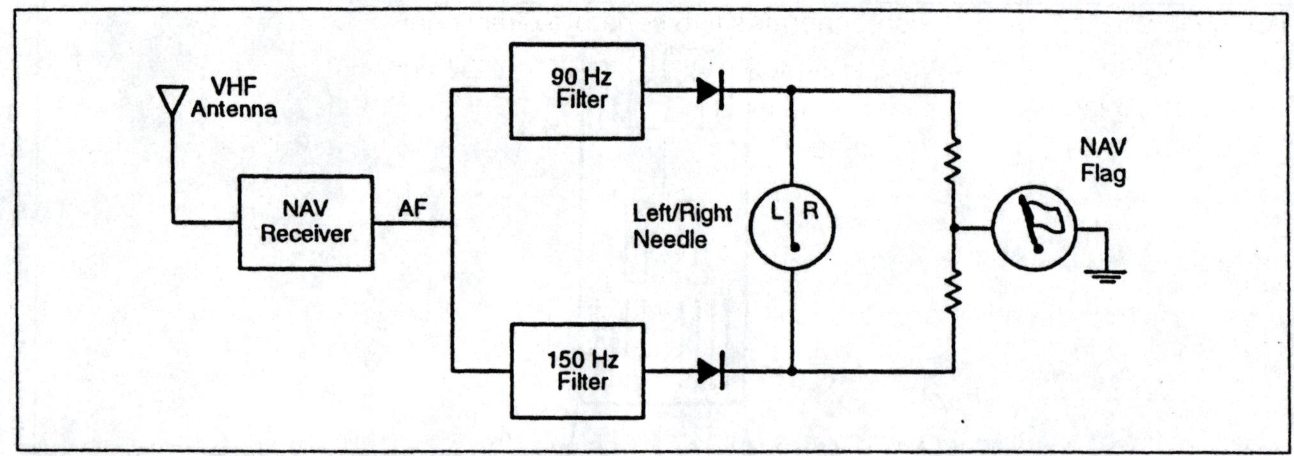

Figure 5-2. ***Typical Localizer Instrumentation Block Diagram.***

operation, it is biased out-of-view in LOC mode and disappears beneath the metal cutout on the indicator. Refer to Figure 4-7 in the previous chapter for an illustration of a typical CDI used for VOR/LOC navigation.

PRINCIPLES OF GLIDESLOPE (GS) OPERATION

The typical glideslope transmitter is usually located 750 feet from the beginning of the runway and radiates a 5-watt RF signal from a horizontallypolarized antenna array at an inclined glidepath angle of 2.5° to 3.0°. The glideslope transmitter operates on one of 40 available channels provided by 150-kHz spacing in the UHF frequency range of 329.15 MHz to 335.00 MHz.

Figure 5-3. ***Flying the Glidepath.***

Like the localizer station, it also generates two separate directional lobes on the same carrier frequency. The lobe radiated above the glidepath angle is amplitude-modulated at 90 Hz, while the lobe below is 150 Hz amplitude-modulated. The glideslope receiver compares the amplitude of the 90-Hz and 150-Hz modulation components to determine the aircraft's position in the vertical plane. When these two signals are equal, the aircraft is centered on the glidepath, which is indicated by zero deflection of the up/down needle on the CDI, as shown in Figure 5-3.

The glideslope needle is much more sensitive than the localizer needle in that its course width is set at 0.7° off-center for full-scale deflection. Thus, at five miles from the end of the runway, a full-scale deflection of the up/down needle would correspond to a distance of roughly 250 feet above or below the centerline of the approach path.

An illustration of a typical localizer course and glidepath used for an ILS approach is shown in Figure 5-4. Aircraft "A" is centered on the ILS approach path where the localizer and glideslope 90-Hz and 150-Hz signals are equal in amplitude causing the localizer left/right needle and glideslope up/down needle to be centered on the course deviation indicator. Aircraft "B" is flying above the glidepath and to the right of the localizer course. In this case, the 90-Hz glideslope signal dominates, causing a "down" needle deflection on the CDI; while the "left" needle deflection is the result of the 150-Hz localizer signal. These steering needles command the pilot to fly the aircraft down and to the left to intercept the center of the ILS approach path to the runway.

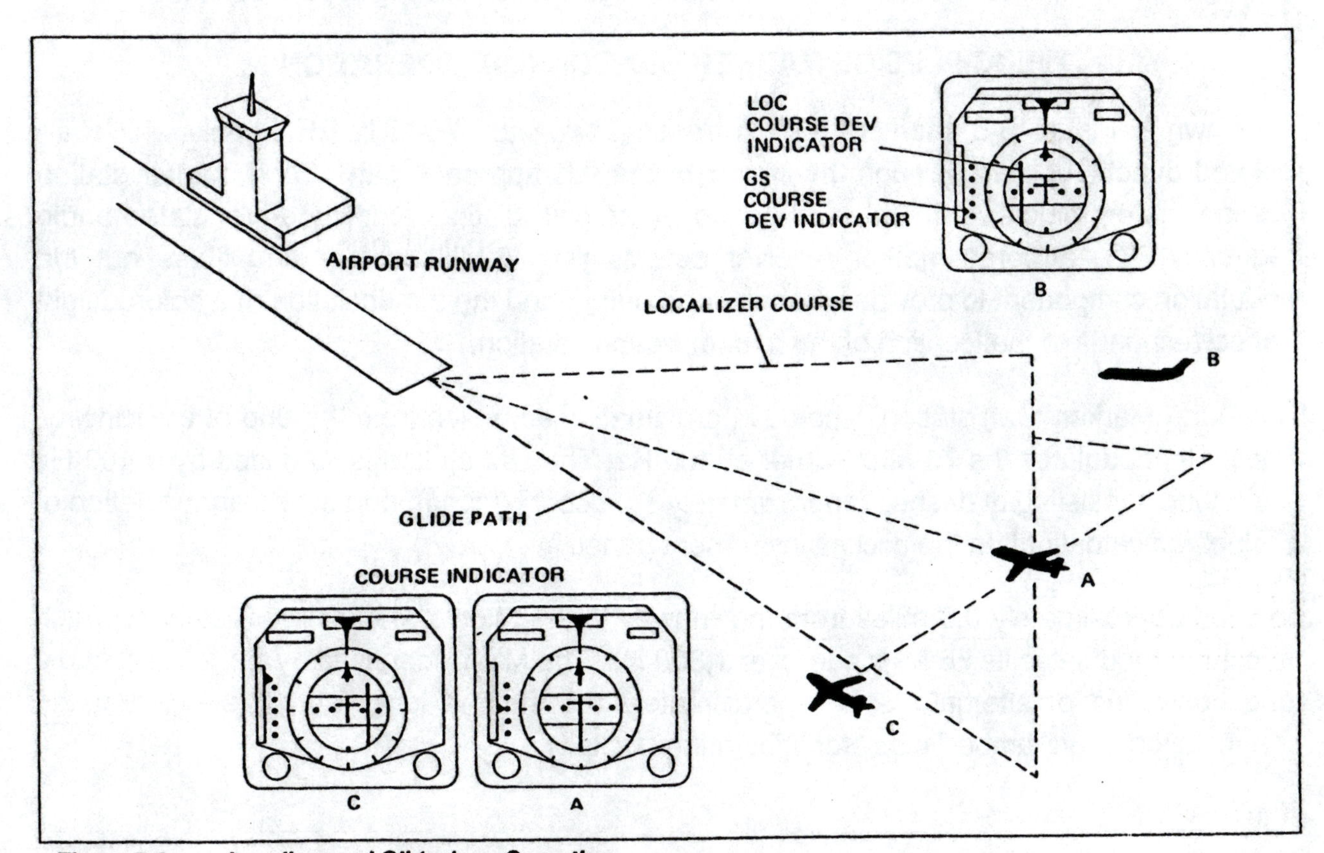

Figure 5-4. Localizer and Glideslope Operation.

As illustrated in Figure 5-4, the localizer course and glidepath signals intersect at a common point along the center of the ILS approach to provide an "X" and "Y" axis, respectively, along the path of descent. Marker Beacon transmitters are located at predetermined locations beneath this path to provide the pilot with an indication of the aircraft's distance from the end of the runway.

GLIDESLOPE CIRCUIT THEORY

As previously mentioned, glideslope receivers operate in the UHF portion of the RF frequency band between 329.5 MHz and 335.0 MHz. Since the glideslope is designed to work in conjunction with the localizer to provide the pilot with both a vertical and horizontal reference along the ILS approach, it also has 40 channels which are paired to the 40 localizer channels. To select the appropriate glideslope channel, the pilot merely tunes in the desired localizer frequency on the navigation control head and the glideslope frequency is automatically channeled to the corresponding localizer frequency. Table 5-1 shows the paired glideslope and localizer frequencies.

Superheterodyne receivers are used to receive the UHF glideslope signals and the instrumentation circuitry for glideslope operation is identical to that used for the localizer. The detector supplies the 90-Hz and 150-Hz modulation components to two bandpass filters, which separates the signals to provide an output to drive the up/down needle on the CDI. A "GS" flag is also provided to indicate an unusable glideslope signal when the flag appears on the indicator.

PRINCIPLES OF MARKER BEACON (MB) OPERATION

As shown in Figure 5-5, marker facilities transmit two-watt, 75-MHz RF signals which are radiated directly upward through the center of the ILS approach path. Each marker station location (outer, middle, or inner) is identified by its own distinct amplitude-modulated audio frequency. The airborne marker receiver detects the 75-MHz carrier and filters out the modulation component to provide an audio tone output and the annunciation of a colored light that corresponds to the location of the marker beacon station.

The Outer Marker (OM) station, located approximately 4.6 miles from the end of the runway, amplitude modulates the 75-Mhz carrier at 400 Hz. The OM station is identified by a 400-Hz audio tone consisting of dashes (approximately 1.5 seconds long), and by the annunciation of a "blue" indicator light on the cockpit instrument panel.

Located approximately 0.6 miles from the runway is the Middle Marker (MM) station, which amplitude-modulates its 75-MHz carrier at 1,300 Hz. The MM is identified by a 1,300 Hz audio tone consisting of alternate dots (approximately 0.5 second long) and dashes, and the annunciation of an "amber" indicator light in the cockpit.

LOCALIZER FREQUENCY (MHz)	PAIRED GLIDESLOPE FREQUENCY (MHz)	GLIDESLOPE FREQUENCY (MHz)	PAIRED LOCALIZER FREQUENCY (MHz)
108.10	334.70	329.15	108.95
108.15	334.55	329.30	108.90
108.30	334.10	329.45	110.55
108.35	333.95	329.60	110.50
108.50	329.90	329.75	108.55
108.55	329.75	329.90	108.50
108.70	330.50	330.05	110.75
108.75	330.35	330.20	110.70
108.90	329.30	330.35	108.75
108.95	329.15	330.50	108.70
109.10	331.40	330.65	110.95
109.15	331.25	330.80	110.90
109.30	332.00	330.95	111.95
109.35	331.85	331.10	111.90
109.50	332.60	331.25	109.15
109.55	332.45	331.40	109.10
109.70	333.20	331.55	111.15
109.75	333.05	331.70	111.10
109.90	333.80	331.85	109.35
109.95	333.65	332.00	109.30
110.10	334.40	332.15	111.35
110.15	334.25	332.30	111.30
110.30	335.00	332.45	109.55
110.35	334.85	332.60	109.50
110.50	329.60	332.75	111.55
110.55	329.45	332.90	111.50
110.70	330.20	333.05	109.75
110.75	330.05	333.20	109.70
110.90	330.80	333.35	111.75
110.95	330.65	333.50	111.70
111.10	331.70	333.65	109.95
111.15	331.55	333.80	109.90
111.30	332.30	333.95	108.35
111.35	332.15	334.10	108.30
111.50	332.90	334.25	110.15
111.55	332.75	334.40	110.10
111.70	333.50	334.55	108.15
111.75	333.35	334.70	108.10
111.90	331.10	334.85	110.35
111.95	330.95	335.00	110.30

Table 5-1. Localizer and glideslope frequency pairing.

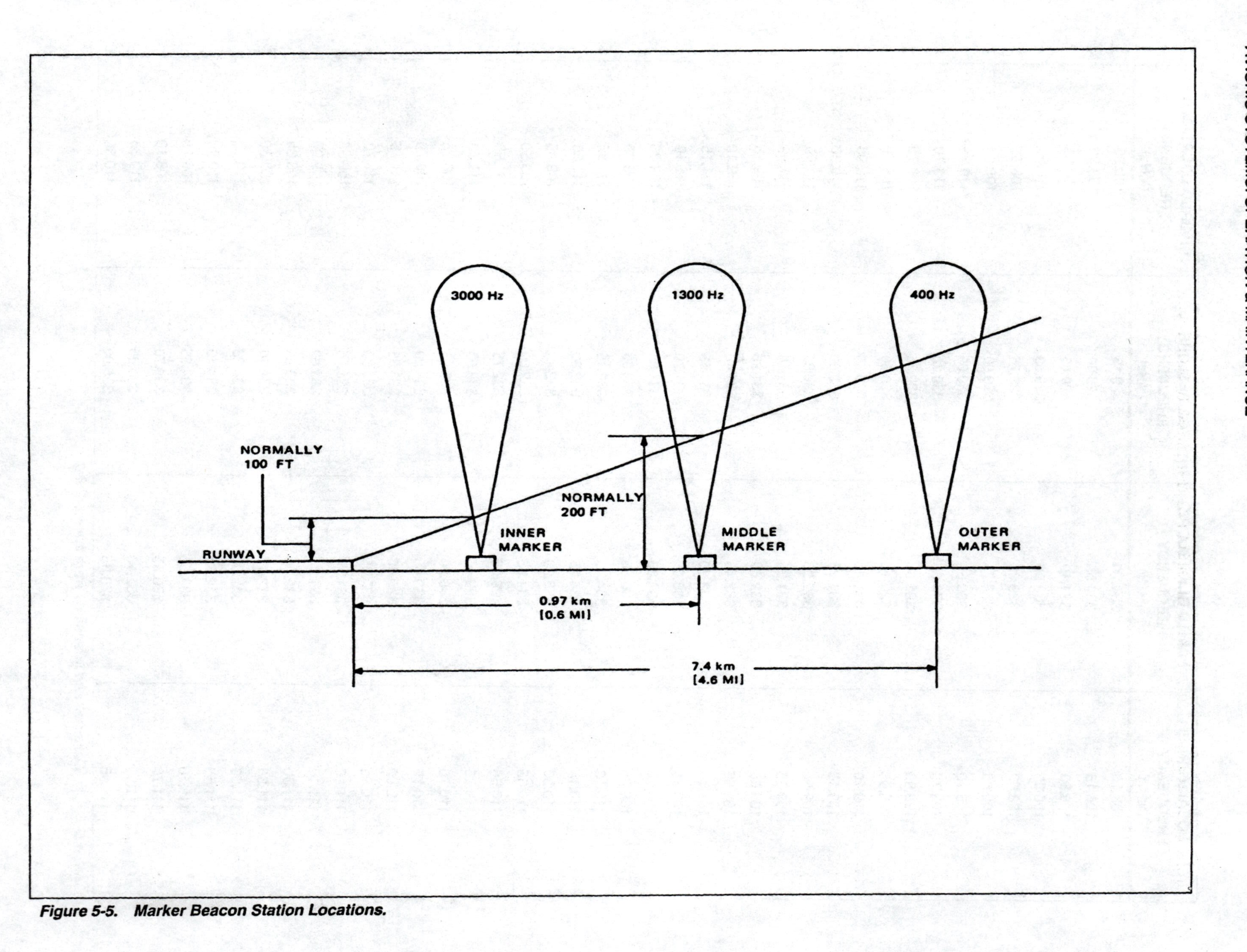

Figure 5-5. Marker Beacon Station Locations.

Aircraft certified for Category II operation are permitted to land if visual contact is established with the runway at a 100-foot decision height. Airports equipped with Category II instrument landing systems include Inner Markers (IM) located approximately 1/4 mile from the runway. The IM station amplitude-modulates its 75-MHz carrier at 3,000 Hz and is identified by a audio tone consisting of dots and the annunciation of a "white" indicator light.

The back-course marker, where installed, normally indicates the back-course final approach descent point. This marker is also modulated at 3,000 Hz and is identified with 75 to 95 two-dot combinations per minute which energizes the same "white" annunciator light as used for the IM. When flying a back-course approach, or outbound on a front-course approach, the pilot corrects by flying away from the left-right deviation bar instead of steering towards it as is normally done.

Compass locater nondirectional beacons (NDB), rated at 25-watts output and operating in the frequency range of 200 to 415 kHz, are installed at the outer and middle markers. The low to medium frequency ADF system utilizes the NDB to provide bearing information to the outer or middle marker in order for the pilot to intercept the ILS approach. Compass locator beacons are identified by a CW signal modulated with a 1,020-Hz tone that transmits a three-letter station identification code. The 1,020-Hz tone is keyed with the first two letters of the ILS identification on the outer locator and the last two letters on the middle locator.

MARKER BEACON CIRCUIT THEORY

Marker Beacon stations operate on a carrier frequency of 75 MHz and are identified with 400-Hz, 1,300-Hz or 3,000-Hz amplitude-modulated tones. They transmit a very narrow vertical beam which is approximately 2,400 feet wide by 4,200 feet long at 1,000 feet above the ground. Since the RF field strength is high within the pattern of the beam, an airborne MB receiver with a typical sensitivity of 2,500 microvolts would receive the signal for roughly 15 seconds if the aircraft is flying on the localizer course at a speed of 120 knots.

Airborne MB systems employ superheterodyne receivers with a 75-MHz crystal-controlled local oscillator. The intermediate frequency will be typically set at 6 kHz or greater to provide image rejection and AGC will be used to prevent saturation when the aircraft is passing directly over the station at a low altitude. A high/low sensitivity switch is sometimes provided in the cockpit to select an increase in receiver sensitivity to approximately 200 microvolts in the event that the pilot prefers to receive the marker signal over a wider range.

As shown in Figure 5-6, the audio output from the detector of the MB receiver is routed to three bandpass filters, which select the audio frequency signal that is present. If the aircraft is passing over the outer marker station, a 400-Hz amplitude-modulated RF signal will be detected by the MB receiver. The 400-Hz audio will be passed through its respective filter and amplified by the lamp driver to activate the "blue" OM light on the aircraft instrument panel.

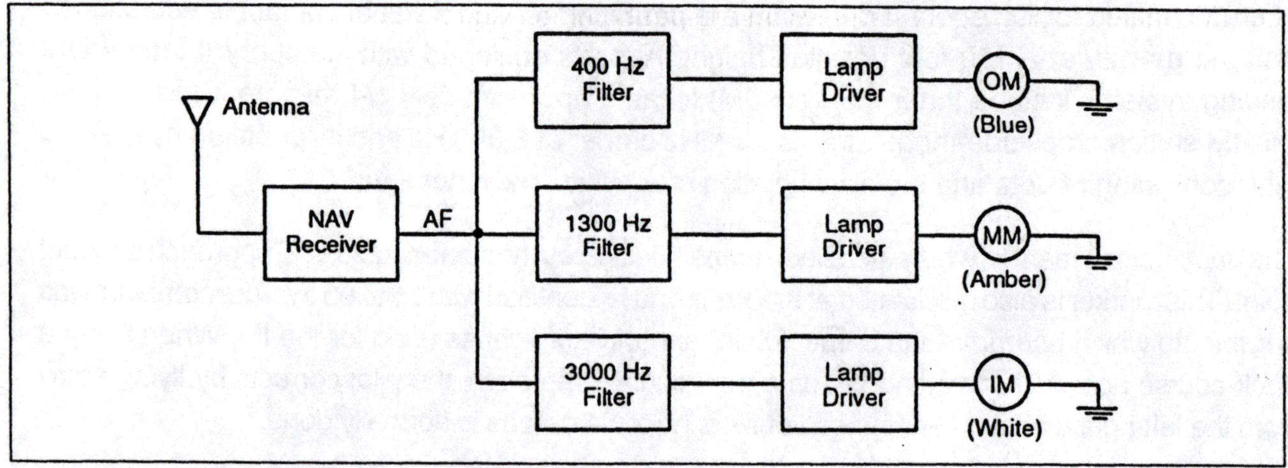

Figure 5-6. ***Typical Marker Beacon Lamp Circuit.***

As the aircraft descends along the ILS approach path, it passes above the middle marker station. At this point, the MB receiver detects and filters the 1,300-Hz audio to light the "amber" MM light. At the inner marker, a 3,000-Hz signal is received which activates the "white" IM indicator light.

OPERATION OF THE LOCALIZER PORTION OF THE S-TEC VIR-351 NAVIGATION RECEIVER

In the previous chapter, a detailed technical discussion was presented on the operation of a typical VOR/LOC receiver, such as the S-TEC VIR-351, with the exception of the localizer converter. A block diagram of the localizer converter, referred to in Figure 4-9, is shown in Figure 5-7.

Recall that the VIR-351 tunes a frequency range from 108.00 MHz to 117.95 MHz with 50-kHz spacing between channels. This spacing establishes 200 channels in which 40 are used for localizer guidance on ILS equipped runways. The VIR-351 navigation receiver processes the

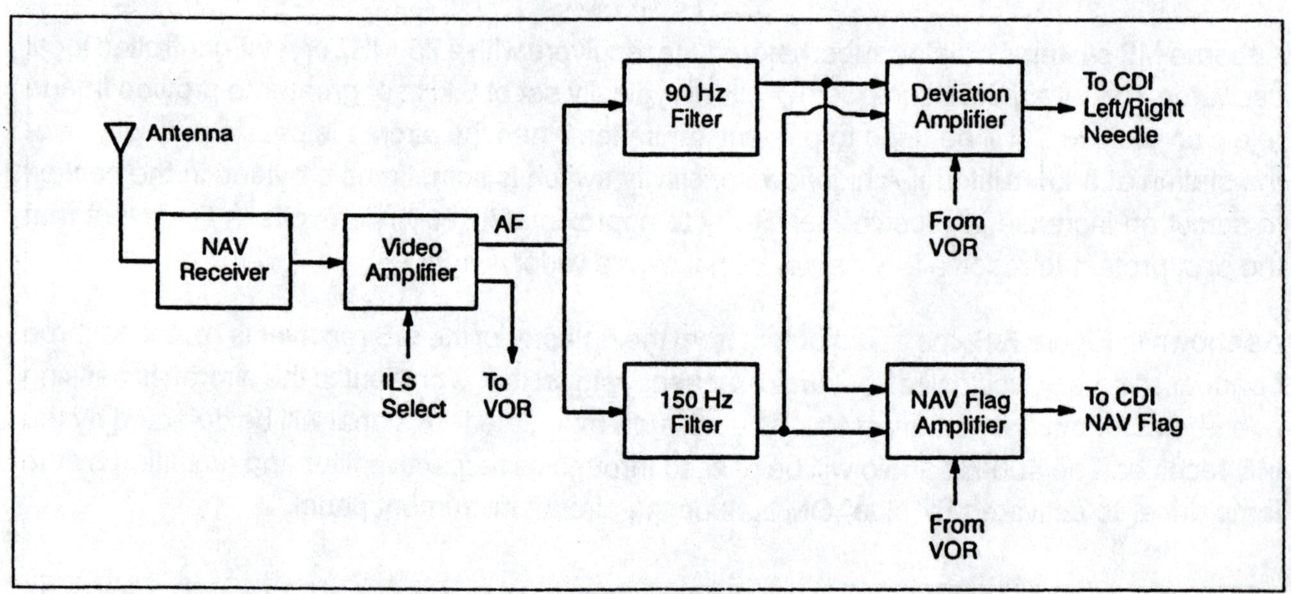

Figure 5-7. ***S-TEC VIR-351 Localizer Converter Block Diagram.***

VOR and localizer signals separately while using common video and deviation amplifiers. ILS logic, supplied by the front panel frequency selection switches, automatically controls the receiver mode of operation.

Selection of an ILS frequency automatically disables operation in the VOR mode. During operation in the ILS mode, the localizer converter receives the detected signal from the receiver section, amplifies the composite signal in the same video amplifier that is used for VOR operation, and finally isolates the 90-Hz and 150-Hz components of the input signal in two bandpass filters. The isolated 90-Hz and 150-Hz signals are then rectified and compared by a matched resistive differential network to determine the amplitude difference between the two signals. This difference data, which is directly proportional to the localizer course deviation, is applied to the deviation amplifier. Equal amounts of each signal result in a localizer left/right deviation bar that is centered on the CDI, while an unproportional relationship results in a deviation of the left/right bar from center.

S-TEC GLS-350 GLIDESLOPE RECEIVER OPERATION

The S-TEC GLS-350, a typical aircraft glideslope receiver, contains a double conversion, superheterodyne receiver that receives RF signals in the frequency range of 329.15 MHz to 335.0 MHz in 150-kHz increments. As shown in Figure 5-8, the RF input signal from the glideslope antenna is coupled through a preselector to the first mixer where it is combined with the synthesized first injection frequency.

The synthesizer is a single-loop digital frequency synthesizer that phase locks a voltage-controlled oscillator operating over the 269.84-MHz through 275.69-MHz frequency range to a single 2.4-MHz crystal oscillator that determines the overall frequency stability of the

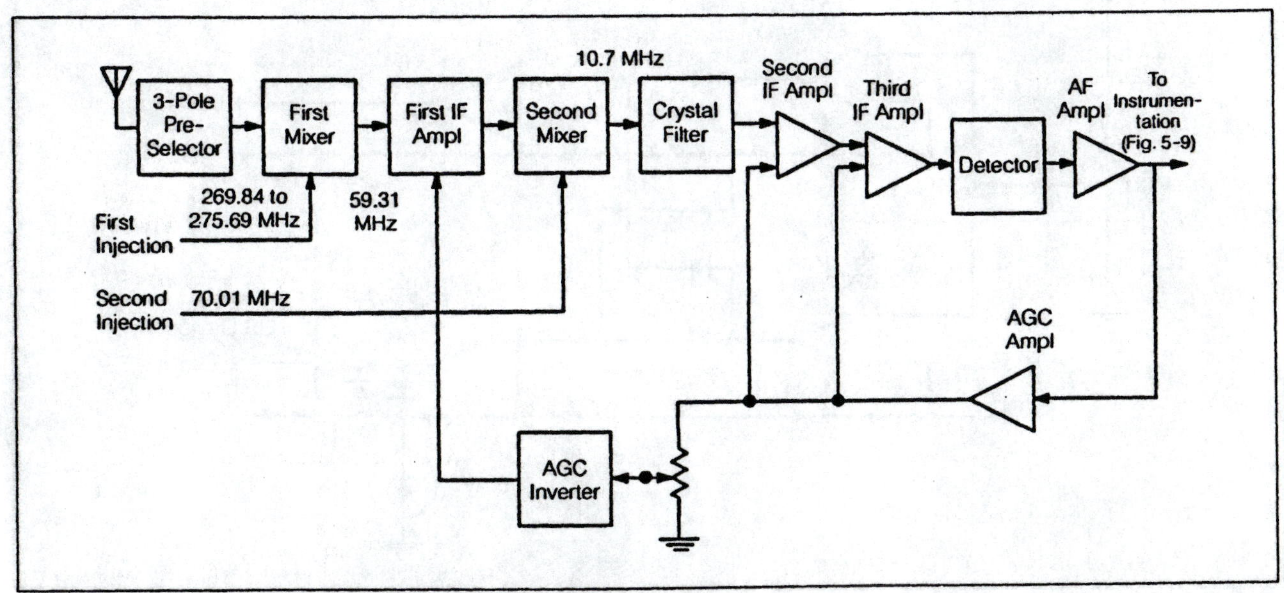

Figure 5-8. S-TEC GLS-350 Receiver Block Diagram.

oscillator. The synthesizer also contains a crystal oscillator operating at 70.01 MHz to provide an injection frequency to the second mixer.

The 59.31-MHz output of the first mixer is amplified and applied to the second mixer where it is combined with a second injection signal of 70.01 MHz supplied by the crystal oscillator in the synthesizer. The resulting 10.7-MHz intermediate frequency is coupled through a crystal filter to provide adjacent channel selectivity, and then to the second IF amplifier stages. AGC voltage controls the gain of all three IF amplifier stages.

The detector recovers the composite modulation signal and produces a DC voltage proportional to the input level. This DC voltage is amplified and compared to a fixed reference voltage by the AGC circuit. The detector output is also applied to the audio amplifier, which supplies the composite modulation signal to the instrumention circuitry that provides the up/down deviation and GS flag outputs.

The detected output signal is coupled to the instrumentation circuitry, shown in Figure 5-9, through course width adjustment (R131), and centering adjustment (R133), to two active bandpass filters (BPF) tuned to 90 Hz and 150 Hz. The outputs of these filters are detected and amplified to drive the deviation and warning flag indicators.

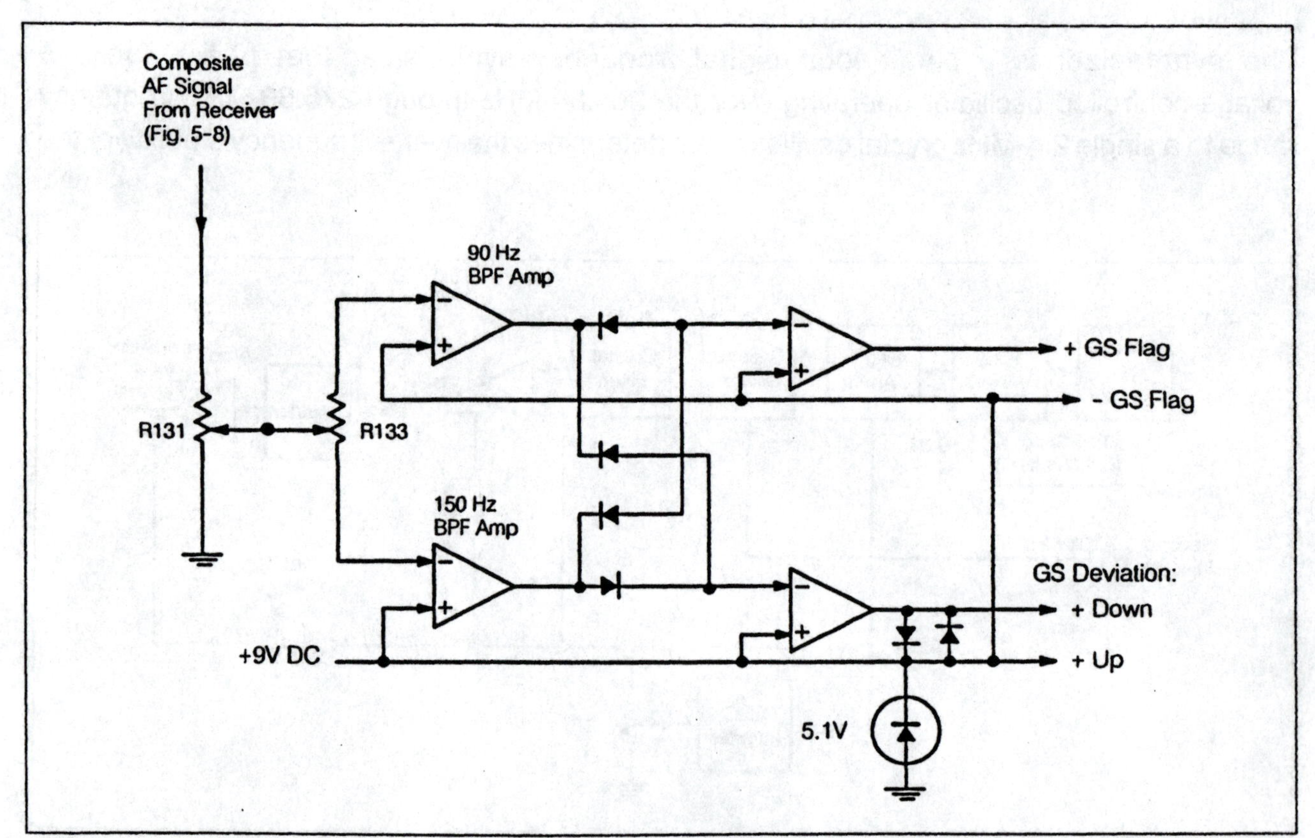

Figure 5-9. GLS350 Glideslope Instrumentation Block Diagram.

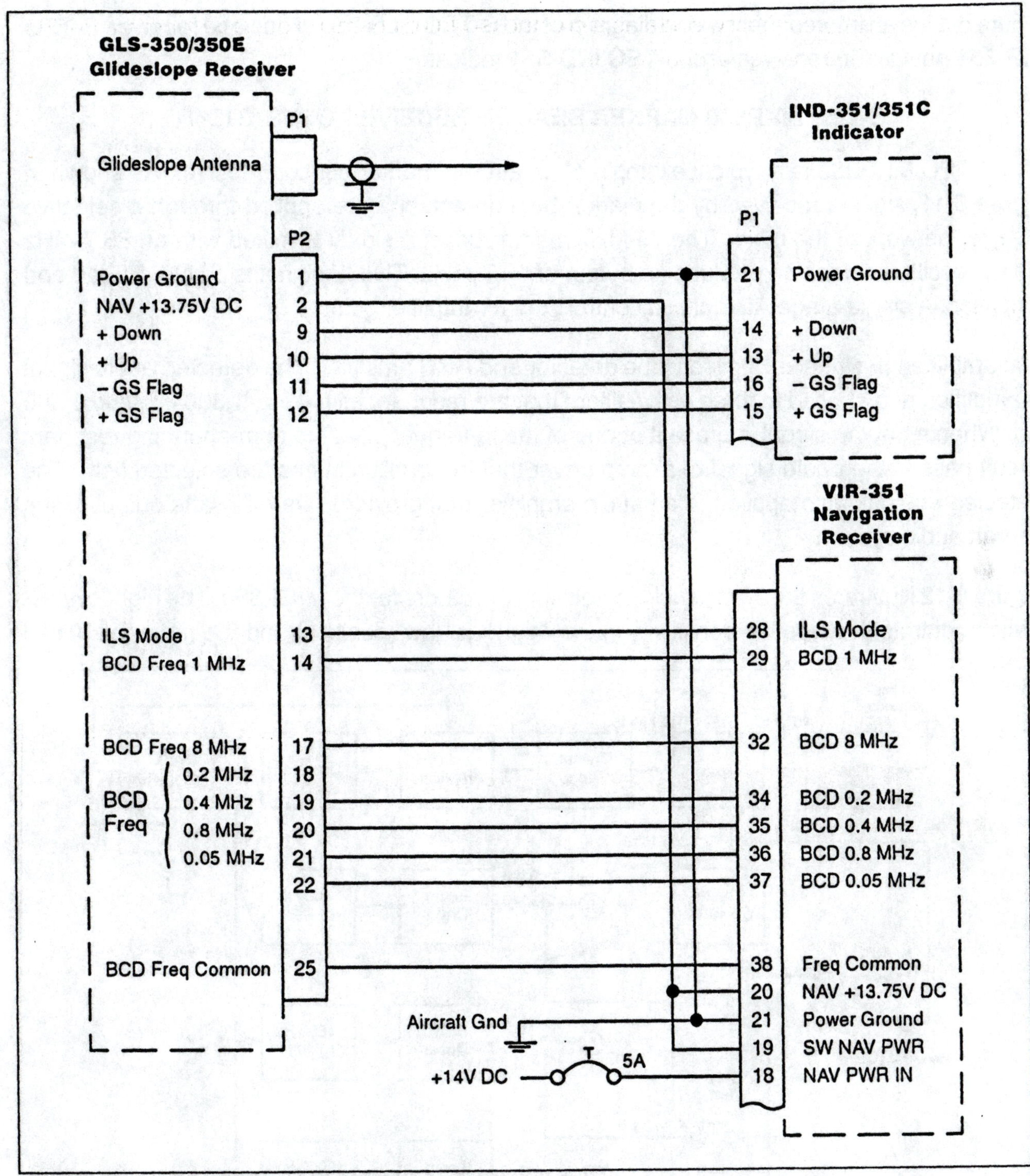

Figure 5-10. GLS-350 Interconnect Wiring Diagram.

Figure 5-10 is an interconnect wiring diagram of the S-TEC GLS-350 glideslope receiver, S-TEC VIR-351 navigation receiver, and S-TEC IND-351 indicator.

S-TEC MKR350 MARKER BEACON RECEIVER OPERATION

The S-TEC MKR350 is a typical example of an aircraft marker beacon receiver. As shown in Figure 5-11, signals received by the marker beacon antenna are applied through a selective LC filter network to the mixer. The 75-MHz marker beacon signal is mixed with an 85.7-MHz crystal oscillator signal to produce an output of 10.7-MHz. This IF signal is crystal filtered and amplified. A single-stage AGC circuit controls the IF amplifier gain.

The amplified IF signal is applied to the detector and AGC amplifier. The detected audio signal is amplified and applied to three active filters that are resonant at 400 Hz, 1,300 Hz, and 3,000 Hz. When an audio signal is present at one of these frequencies, the corresponding resonant circuit passes the audio signal to a lamp driver that in turn illuminates the selected light. The detected audio is also applied to an audio amplifier that provides five milliwatts output to the aircraft audio system.

Figure 5-12 illustrates the front panel controls and indicators for the MKR-350. The high/low/test switch controls the receiver sensitivity in the "high" or "low" position, and the momentary test

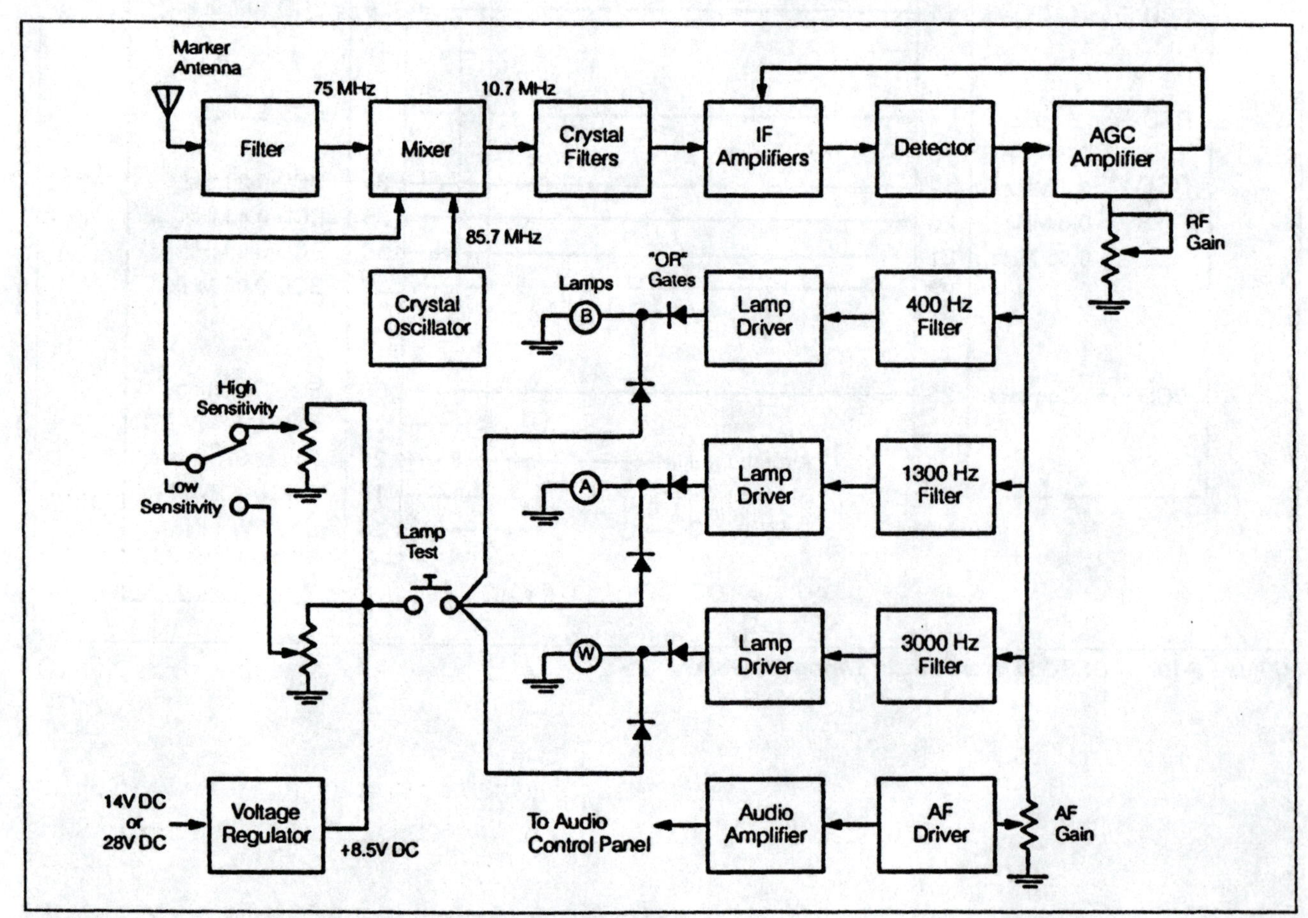

Figure 5-11. S-TEC MKR-350 Marker Beacon Block Diagram.

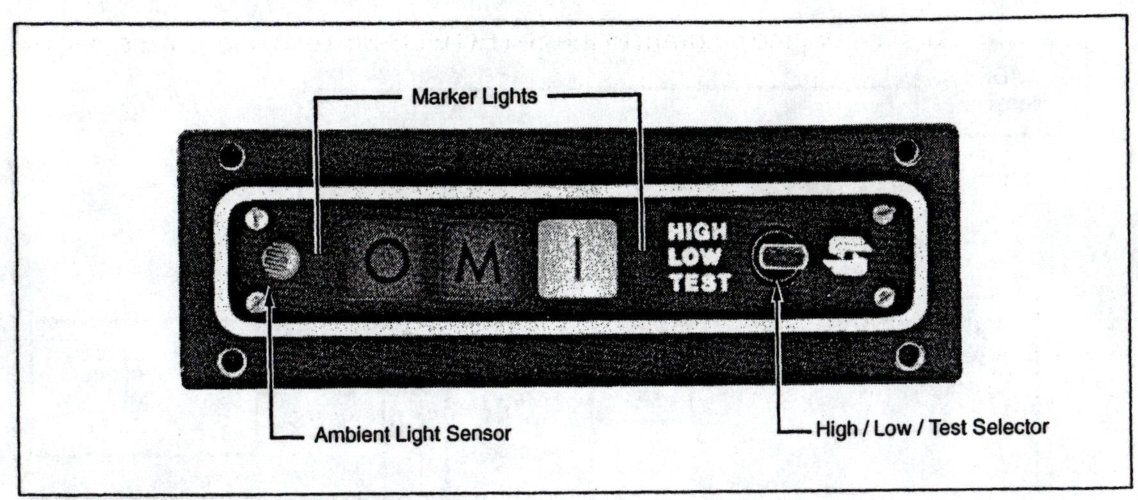

Figure 5-12. S-TEC MKR-350 MB Receiver.

position applies a test voltage to illuminate all three indicator lights. On the left side of the panel is a photocell which monitors the cockpit ambient light level and automatically adjusts the marker display brightness to an optimum level. In installations where the marker receiver is separate from the controls and displays and is remotely mounted in the radio rack, an additional switch is added to lower the marker lamp intensity during night operations by means of a series dropping resistor or zener diode in place of the ambient light sensor.

ROCKWELL COLLINS VIR-30 VOR/ILS NAVIGATION RECEIVER OPERATION

As previously mentioned, since the localizer and omnirange operate in the same frequency range, the instrumentation circuitry for both functions share a common VHF receiver and are packaged in one piece of equipment, such as in the operation of the S-TEC VIR-351. However, radio navigation equipment used on larger aircraft for corporate or commercial airline applications commonly incorporate the glideslope receiver and sometimes the marker receiver into an integrated package along with the VOR and localizer circuitry. A typical example of such an integrated navigation system is the Rockwell Collins VIR-30.

The VIR-30 radio navigation system supplies automatic VOR (for Radio Magnetic Indicator (RMI) operation) and manual VOR information; localizer and glideslope deviation outputs; high and low-level flag signals; to/from information; marker beacon lamp signals; and VOR, localizer, and marker beacon audio outputs. The system is compatible with standard 2-out-of-5 and ARINC 429 control units and provides a full complement of navigation outputs to interface with analog area navigation and flight control systems. The signal interface diagram for the VIR-30 navigation system is shown in Figure 5-13.

The VIR-30 navigation receiver provides sine/cosine VOR outputs to drive the Rockwell Collins RMI-30 Radio Magnetic Indicator. Magnetic compass information is combined with the VOR data within the RMI-30 to display continuous position along a VOR radial. If a three-wire synchro type (conventional) RMI is used in the system in place of the RMI-30, the VIR-30 receiver

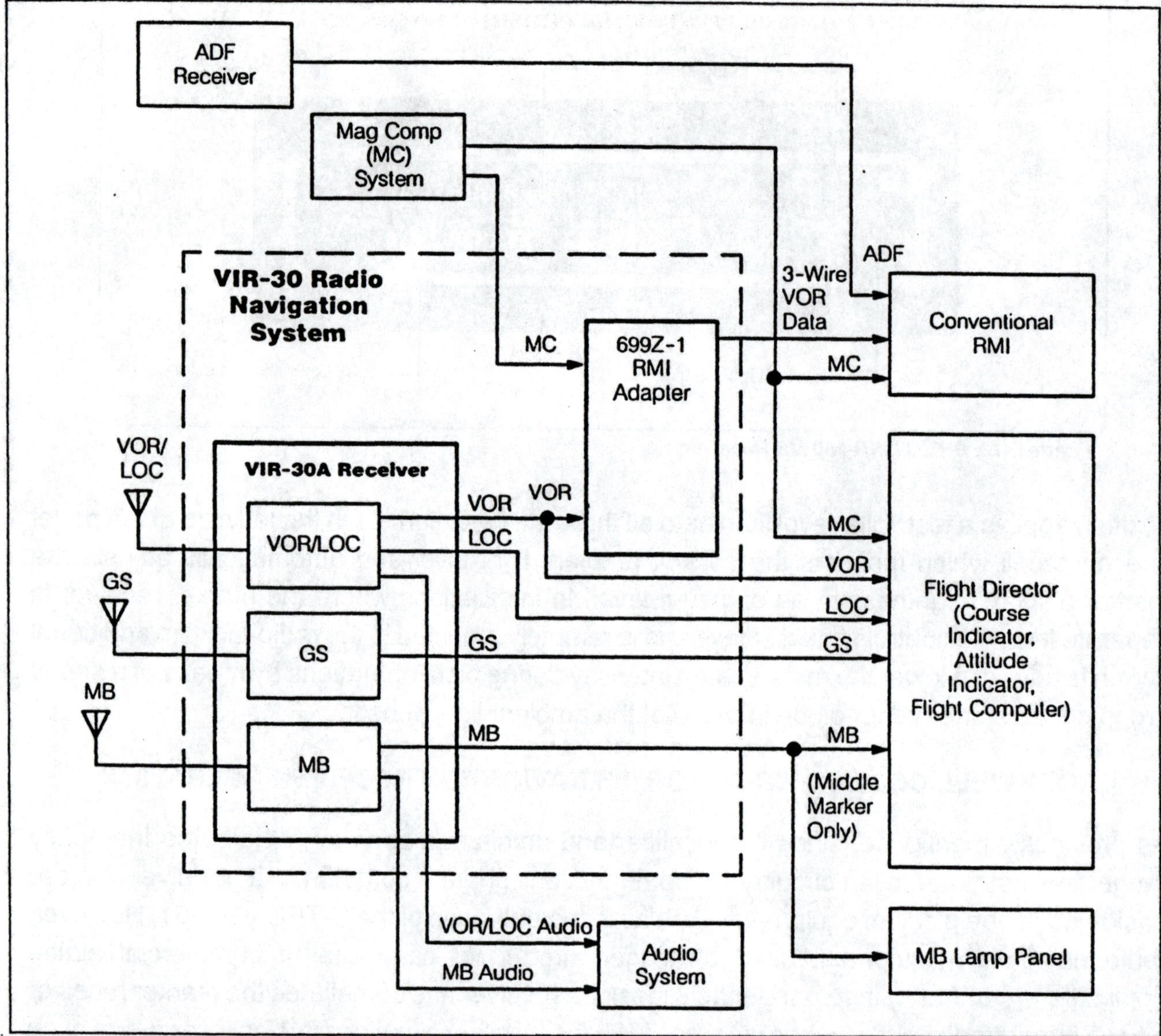

Figure 5-13. ***Rockwell Collins VIR-30 Navigation Receiver Signal Interface Diagram.***

sine/cosine output must be converted by the Rockwell Collins 699Z-1 RMI adapter to three-wire relative bearing data.

The VOR information on the RMI is called automatic VOR because no manual adjustments (such as OBS setting) are required to display usable navigational information. Thus, the RMI automatically displays both magnetic and relative bearing to a VOR (or ADF) station. As shown in Figure 5-14, the tail of the RMI bearing pointer, read against the rotating compass card, indicates the VOR radial that the aircraft is presently crossing or maintaining. The head of the RMI bearing pointer indicates bearing to the station. The displacement of the bearing pointer from the top index or lubber line (which represents the nose of the aircraft) measures the relative bearing from the aircraft heading to the VOR station.

Figure 5-14. Typical Radio Magnetic Indicator.
(Courtesy McDonnell Douglas Aerospace Corporation)

When an ILS frequency is selected, the RMI VOR bearing pointers park in a horizontal position (since they are not used during instrument landings). ILS data (localizer, glideslope, and marker beacon) is then received and processed to provide inputs to the course deviation indicator, flight director, and marker beacon lamp panel. The Flight Control Computer uses the LOC/GS information to provide steering cues on the Attitude Direction Indicator and guidance and control signals to the Autopilot. The middle marker signal controls the gain of the Flight Control Computer upon approach. The operation of the flight director will be presented in more detail in Chapter Twelve.

ROCKWELL COLLINS VIR-432 NAVIGATION RECEIVER

The Collins VIR-432 Navigation Receiver, which is the same form factor as the VHF-20/21/22 shown in Figure 2-12, is a fully digital VOR/LOC, glideslope and marker beacon receiver. The unit provides both CSDB and ARINC 429 interfaces. The VIR-432 can be tuned by the Collins CTL-32 CSDB NAV control, or an ARINC 429 control unit. Two CSDB and two ARINC output ports, standard VOR/LOC and MB audio outputs, and marker beacon lamp outputs are provided. Microprocessor technology allows a variety of interface and installation options, as well as full-time monitoring and self-diagnostic capabilities.

The VIR-432 uses digital techniques to read serial input data, compute the navigation situation, and generate serial data outputs. The heart of the computation circuits is an 80C51 instrumentation microprocessor. An 80C31 microprocessor controls the input/output function. A block diagram of the VIR-432 is shown in Figure 5-15.

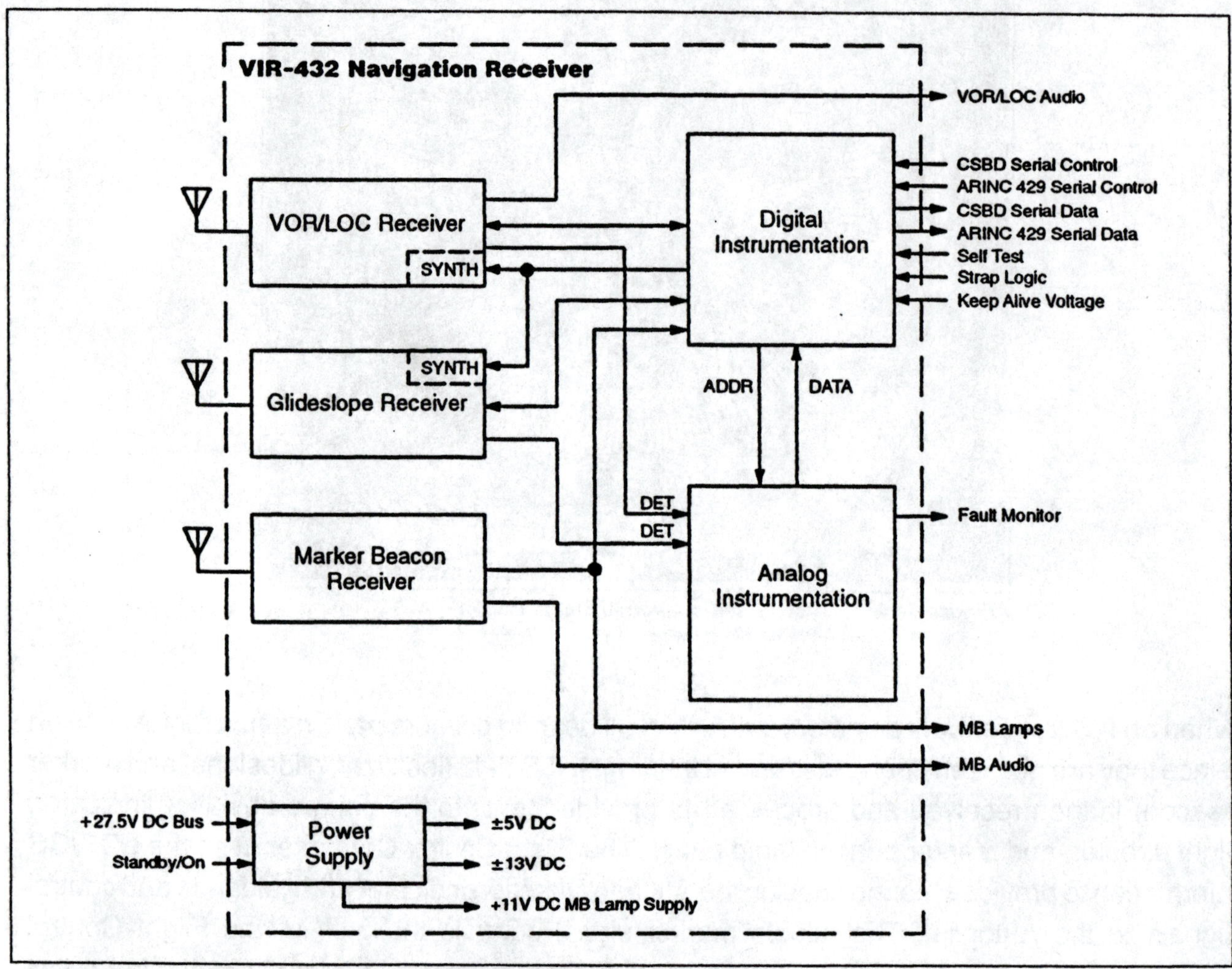

Figure 5-15. VIR-432 Navigation Receiver Block Diagram.

The VOR/LOC receiver is a dual-conversion receiver with IF frequencies of 20.05 MHz and 455 kHz. Frequency conversion is performed by two mixer circuits. The first is a dual-gate MOSFET mixer with synthesizer injection. The second is a dual-gate MOSFET mixer with injection from a 20.505-MHz crystal-controlled oscillator.

RF selectivity is determined by a four-pole, varactor-tuned preselector/filter network. Selectivity is determined by a 20.05-MHz crystal filter. The detected VOR/LOC signal is applied to AGC circuits, an audio amplifier, and to the analog instrumentation circuits before being applied to the digital instrumentation card.

The GS receiver is a dual-conversion receiver with IF frequencies of 60.95 MHz and 25.5 MHz. Frequency conversion is performed by two mixer circuits. The first is a JFET mixer with synthesizer injection. The second is a dual-gate MOSFET mixer with injection from a 35.45 MHz crystal-controlled oscillator.

RF selectivity is provided by a 25.5-MHz crystal filter. The band-limited signal is amplified by a two-stage IF amplifier and demodulated by a transistor detector. Detected DC voltage is applied to AGC circuits. The 90/150-Hz audio is applied to the analog instrumentation circuits and then to the digital instrumentation card for processing. Selectivity of the single-channel MB receiver is provided by a 75 MHz crystal filter. The band-limited signal is amplified by a two-stage RF amplifier and demodulated by a diode detector.

The detected signal is applied to an AGC circuit and audio amplifier and to a filter/lamp driver network. The filter/lamp driver network activates the proper 400-Hz, 1,300-Hz, or 3,000-Hz lamp drive output. MB lamp status is also applied to the digital instrumentation card.

The analog instrumentation processes all non-digital navigation signals into the proper format for analog-to-digital conversion. The processed detector outputs are applied to an analog-to-digital (A/D) converter for use by the 80C51 microprocessor.

The analog instrumentation includes the fault monitor output driver circuit. A logic signal is converted to a +27.5-volt DC, 200-mA annunciator output when no navigation failures are found. The output becomes an open circuit if a navigation failure is detected.

A microprocessor in the digital instrumentation receives and decodes both serial control data from ARINC 429 or CSDB sources. The parallel data received from strap option inputs select the microprocessor operating mode. The microprocessor then tunes the receivers to the frequency selected by the serial control.

GS and VOR/LOC signals from the receivers and monitor signals from other cards are multiplexed by the analog instrumentation. The multiplexed signal is then sampled by an A/D converter in the digital instrumentation card. The resulting data is formatted by a second microprocessor and transmitted serially to CSDB and ARINC 429 display units in the avionics system.

The pulse-width-modulated, switching-regulator-type power supply derives +5-volt DC and plus or minus 13-volt DC from the +27.5-volt DC input. The pulse-width modulator drives a push-pull network feeding three full-wave rectifiers. Rectifier outputs are filtered and regulated to supply the internally required voltages. A series pass regulator derives the marker beacon +11-volt DC source, and an inverter derives a -5-volt DC reference level.

AUTOMATED TEST EQUIPMENT

Modern avionics equipment, such as the Rockwell Collins VIR-432 previously discussed, use Automated Test Equipment (ATE) for bench testing, calibration, and troubleshooting. ATE has the advantage that minimal reference to the manufacturer's maintenance manuals is required.

The VIR-432 ATE consists of an Apple personel computer (PC), and Collins software to test the radio. A Collins Proline II Interface Card and ARINC 429 interface card are installed in the Apple PC. Step-by-step procedures are displayed on the computer monitor along with the test results, as shown in Figure 5-16. A printer option allows the avionics technician to retain a hard copy of the test results for reference.

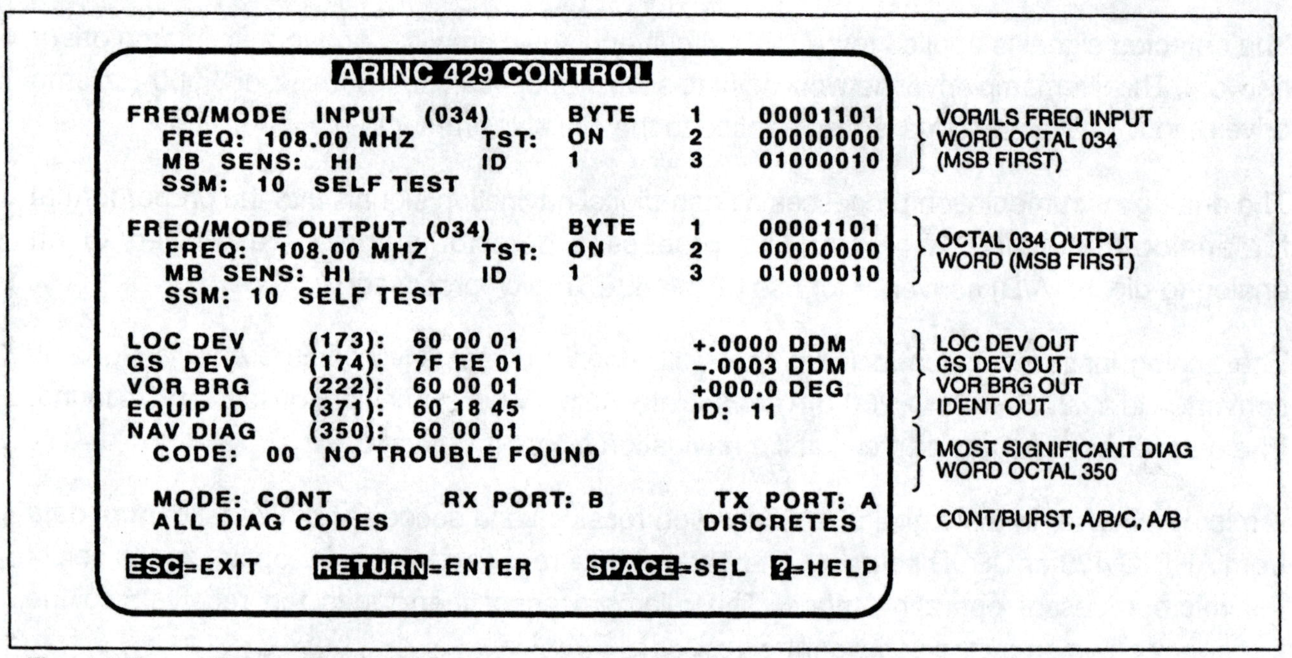

Figure 5-16. VIR-432 ATE Display Screen. (Courtesy Rockwell Collins Avionics)

The test programs are written to provide explicit operator instructions and to allow a variety of troubleshooting approaches. Some selected test operations are automatically performed by the software to lower the test time and enhance the technician's efficiency. A diagram of the VIR-432 ATE test set-up is presented in Figure 5-17.

MICROWAVE LANDING SYSTEMS (MLS)

As previously mentioned, ILS systems provide only one localizer and one glideslope path to the runway. Because of this, aircraft must line up on a single approach zone, which severely limits the flexibility of the air traffic controller in routing aircraft in and out of a given airport. Because of this limitation, Microwave Landing Systems were recently developed and, up until June of 1994, were being installed at major airports until the U.S. Department of Transportation abandoned MLS in favor of the newer differential GPS satellite navigation systems discussed in Chapter Eleven.

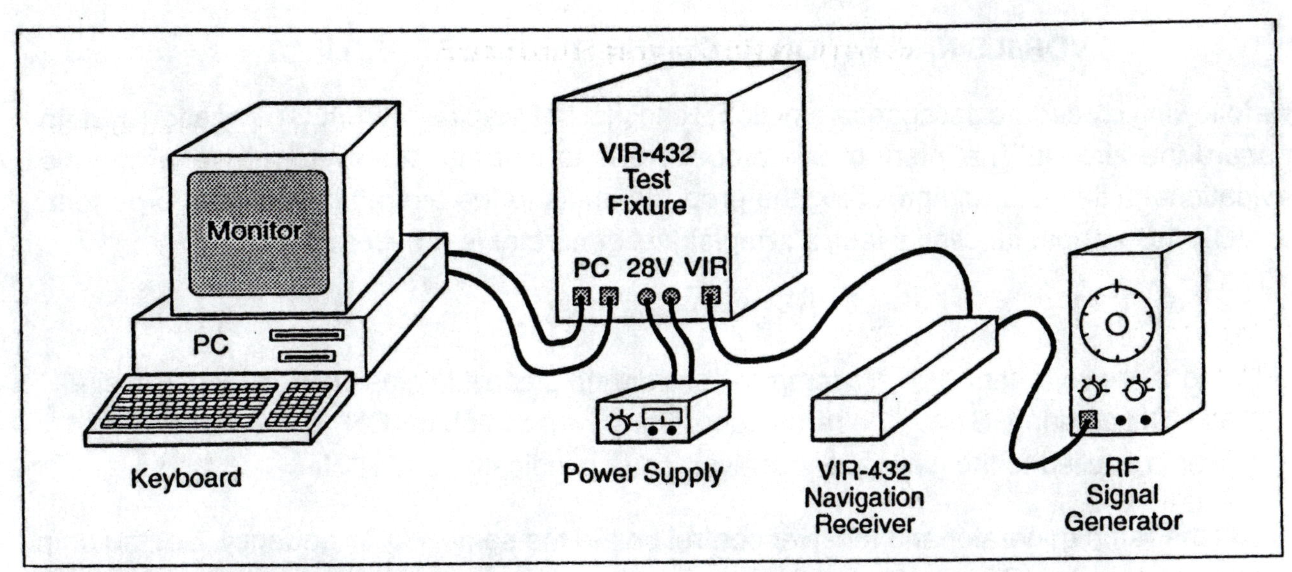

Figure 5-17. VIR-432 Test Set-Up Diagram.

MLS provides a multitude of azimuth (left/right) and elevation (up/down) paths to the runway, which results in reduced fuel costs and less noise over populated areas. In addition, instead of being limited to 40 ILS channels, the MLS operates on 200 channels, which allows for the installation of more MLS facilities in high-traffic areas without causing interference between facilities. There are currently about half a dozen airports equipped with MLS in the United States.

MLS operates in the frequency range between 5,031 MHz and 5,090 MHz with channels spaced every 300 kHz. It uses a time-referenced scanning beam, which allows the pilot to determine his or her position in the approach zone relative to the runway. The airborne MLS receiver detects and measures the time it receives the radiated pulse from the ground station as the microwave beam sweeps back and forth through the approach zone and provides the same left/right and up/down signals to the CDI as a conventional VOR/LOC navigation receiver. In addition, continuous distance information, in lieu of a few marker beacon distance indications, is provided by Precision Distance Measuring Equipment (DME), which operates in the same frequency range as conventional Distance Measuring Equipment.

CONCLUSION

This chapter discussed the operation and maintenance of the localizer, glideslope, and marker beacon systems, which are required for instrument landings. Microwave landing systems were only briefly discussed since they will be replaced with the newer differential GPS landing systems.

Chapter Four presented VOR navigation systems, which are used for in-route navigation. However, the VOR receiver only presents bearing information to or from a VOR ground station. Distance Measuring Equipment, which is presented in the next chapter, provides distance information to augment the utility of VOR navigation.

VOR/ILS NAVIGATION RECEIVER RAMP TEST PROJECT

The following procedure describes a functional checkout of a typical VOR/ILS navigation system onboard the aircraft. The intent of this procedure is to test that the system is receiving the navigational aid inputs and providing the proper outputs to the aircraft indicators. To perform the VOR/ILS system functional test, a ramp signal generator is required.

VOR System Test

1. Raise the whip antenna on the ramp generator and place it in close proximity of the aircraft VOR antenna. Switch the ramp generator power switch to "ON" and ensure that power is applied to the navigation receiver, course indicator, and RMI.

2. Set the ramp generator and receiver control unit to the same VOR frequency. Set the omni bearing selector (OBS) to 0°. Set the bearing control on the ramp generator to 0°.

3. Observe that the RMI points to 180 plus or minus 5°, course deviation bar is centered, NAV flag is out of view, FROM direction is indicated, and GS pointer is out of view.

4. Set the aircraft course selector to 10°, and observe that the course deviation bar moves to the left.

5. Set the bearing control on the ramp generator to 45°, 90°, 180°, 225°, 270°, and 315°, and verify that the RMI pointer indicates the reciprocal of the bearing selected. The deviation bar should exhibit movement left and right and the to/from flag will change.

6. Set the 30-Hz VAR control on the ramp generator to remove the 30-Hz variable signal from the VOR output, and observe that the NAV flag comes into view.

LOC/GS System Test

1. Raise the whip antenna on the ramp generator and place it in close proximity of the localizer and glideslope antennas. Switch the ramp generator power switch to "ON" and ensure that power is applied to the navigation receiver and course deviation indicator.

2. Set the ramp generator and receiver control unit to the same ILS frequency. Set the ramp generator deviation switches for an "ON-COURSE" output.

3. Observe that the course deviation bar and glideslope pointer are centered, NAV and GS flags are out-of-view, and VOR RMI pointer points to right wing.

4. Set the ramp generator glideslope deviation switch to "UP" and the localizer deviation switch to "LEFT". Observe that the glideslope pointer indicates full up deflection and the course deviation bar indicates full left deflection.

5. Set the glideslope deviation switch to "DOWN" and the localizer deviation switch to "RIGHT". Observe that the GS pointer indicates full down deflection and the course deviation bar indicates full right deflection.

6. Set the glideslope and localizer deviation switches to "ON-COURSE", and observe that the glideslope pointer and course deviation bar center.

7. Set the ramp generator glideslope "DELETE 90 Hz" and localizer "DELETE 150 Hz" switches to "ON". Observe that the glideslope pointer indicates fullscale up deflection, GS flag is in view, LOC deviation bar indicates fullscale right deflection, and the NAV flag is in view.

8. Set "DELETE 90 Hz" and "DELETE 150 Hz" switches to "OFF", and observe that the GS pointer and LOC deviation bar center and the GS and NAV flags go out-of-view.

9. Set the "LOC 1020-Hz" tone switch on the ramp generator to "ON", select NAV on the aircraft audio selector panel, and listen for an audible tone in the headset and loudspeaker.

Marker Beacon System Test

1. Apply power to the marker beacon receiver and depress the marker test switch. In the test mode, all three marker lights should be at maximum intensity.

2. Raise the whip antenna on the ramp generator and set it directly beneath the marker beacon antenna. Switch the ramp generator power switch to "ON".

3. Set the ramp generator for a 10,000 microvolt output with 90% modulation at 3,000 Hz. Set the marker beacon receiver sensitivity switch to the HIGH position and select MKR on the audio selector panel. A 3,000-Hz tone should be clearly audible in the aircraft audio system, and the white marker lamp should be brightly lighted.

4. Change the modulation frequency of the ramp generator to 1,300 Hz. A 1,300-Hz tone should be clearly audible, and the amber marker lamp should be brightly lighted.

5. Change the modulation frequency of the ramp generator to 400 Hz. A 400-Hz tone should be clearly audible, and the blue marker lamp should be brightly lighted.

6. Repeat the above procedure with the marker beacon receiver sensitivity switch in the LOW position.

VOR/ILS NAVIGATION RECEIVER LABORATORY PROJECT

The objective of this laboratory project is to become familiar with the proper procedure for settingup, testing, and calibrating a VOR/ILS navigation receiver. The equipment required consists of a VOR/LOC/GS receiver and CDI, a test panel to provide all the necessary interfaces and control functions, an RF signal generator, an audio oscillator, and a DVM. This procedure is very general in nature and therefore should be used only in conjunction with the applicable manufacturer's maintenance manual.

VOR Calibration

1. Connect the RF signal generator to the VOR antenna input and set the output frequency to the same VOR frequency selected on the control head. Adjust the VOR signal for a 0° radial with 30% modulation output and verify that the signal is being received. Adjust the receiver audio level control, if necessary.

2. Set the receiver reference and variable level controls to 4.5 volts RMS. Set the OBS to indicate 0° on the CDI and set the receiver VOR zero adjust control so that the left/right needle centers. Turn the OBS control until 180° is read on the course index and again set the zero adjust control to split the error difference between the two reciprocal readings.

3. Adjust the receiver VOR flag sensitivity at 375 to 385 microamps. Set the VOR course width sensitivity for 20° by turning the OBS to a reading of 10° off course and adjusting the course width sensitivity control to obtain a 150-microamp reading. Center the course width deviation with the phase balance control.

4. Test the VOR indicator error dispersement by adjusting the signal generator for radial outputs at every 30° and setting the OBS for zero needle deflection. Adjust the tracking control to obtain no greater than a 5° error at any radial and split the difference between reciprocal errors.

Localizer Calibration

1. Connect the RF signal generator to the localizer antenna input and set the output of the generator to correspond with the selected ILS frequency on the control head. Set the modulation level to 20% with both 90-Hz and 150-Hz signals present and equal in amplitude.

2. Adjust the LOC flag sensitivity control between 375 to 385 microamps. Adjust the LOC centering control to center the left/right needle on the CDI. Adjust the LOC deviation sensitivity control so that a 4-dB increase in either the 90-Hz or 150-Hz signals will cause a 90-microamp needle deflection.

Glideslope Calibration

1. Connect the RF signal generator to the GS antenna terminal and set the generator output … he selected LOC channel. Adjust the generator to 40% … d 150-Hz signals present and equal in amplitude.

2. … ontrol between 375 to 385 microamps. Set the GS centering … e up/down needle. Adjust the GS deviation sensitivity con- … either the 90-Hz or 150-Hz signals will cause a 78-micro-

REVIEW QUESTIONS

1. What does the term decision height mean?
2. What stations of the ILS facility provides the pilot with a vertical reference, horizontal directional reference, and distance reference along the approach path?
3. Which of these stations provide audio identification?
4. Describe the modulation components of the localizer and glideslope signals in respect to the ILS approach path.
5. When the course deviation indicator indicates a localizer full-right deflection and the glideslope pointer indicates a full-up deflection, what are the signal characteristics of the localizer and glideslope signals? What corrections must the pilot make to intersect the center of the ILS approach path?
6. What is the frequency range and spacing of the localizer and glideslope channels?
7. Define the term "course width." What is the course width for full-scale localizer and glideslope needle deflections?
8. What is the angle of the radiated glideslope signal?
9. On what frequency do marker beacon transmitters operate and how are the various marker stations identified?
10. What two marker beacon stations light the same marker light and which one of the three lights are we referring to?
11. What is the purpose of the compass locator beacon and how are NDBs identified?
12. How are localizer frequencies distinguished from VOR frequencies if they both operate in the same frequency range?
13. How does the navigation receiver know to select the localizer instrumentation circuit as opposed to the VOR instrumentation circuit when an ILS frequency is selected?
14. Draw a simplified block diagram to describe the operation of the localizer, glideslope, and marker beacon instrumentation circuits.
15. What is the purpose of the NAV flag on the CDI and what conditions cause it to come into view?
16. How is a glideslope channel selected to correspond with the proper localizer channel for a given ILS facility?

17. What is the typical receiver sensitivity for a marker receiver in the high and low sensitivity modes?

18. What is the function of an RMI adapter and when is it used?

19. What is the difference between manual VOR and automatic VOR? Describe the indications obtained on the RMI using a receiver equipped with automatic VOR.

20. Describe the advantages of MLS over conventional ILS.

Chapter 6

Distance Measuring Equipment

INTRODUCTION

The primary function of airborne Distance Measuring Equipment (DME) is to calculate the aircraft's distance to or from a selected enroute VOR navigation station or an approach ILS facility. The displayed DME distance and VOR bearing are used by the pilot to determine the aircraft's polar coordinates or distance from a fixed point (ground station) and direction from a fixed line (radial from magnetic north). In addition to providing nautical mile distance, most DME systems also compute and display the aircraft's velocity in respect to the ground station, known as ground-speed, and the time required for the aircraft to intercept the ground station.

The ground facilities that the airborne DME may use include VOR/DME, ILS/DME, TACAN and VORTAC. The VOR/DME facility is comprised of a VOR and DME station at the same location. The ILS/DME is a DME station located at the same site as an ILS station. TACAN (TACtical Air Navigation) stations are used by military aircraft for obtaining both bearing and distance information. Civil aircraft use only the distance measuring portion of the TACAN signal. Bearing information for civil aircraft is not provided by the UHF TACAN station since civil aircraft are equipped with VHF omnirange navigation receivers to obtain the radial bearing. A ground station equipped with both VOR and TACAN is known as a VORTAC station. Military aircraft obtain both bearing and distance information through the TACAN system, while civil aircraft use the VOR station for bearing information and the DME portion of the VORTAC for distance measurement.

The majority of airborne DME systems automatically tune their respective transmitter and receiver frequencies to the paired VOR/LOC channel (in the same manner that the glideslope is automatically tuned to its paired localizer frequency). The DME or TACAN station is identified by a 1350-Hz coded audio tone which is transmitted every 30 seconds.

An airborne DME transmits and receives pulse-modulated signals in the UHF frequency range of 960 MHz to 1215 MHz. The effective range of the airborne DME transceiver in nautical miles can be calculated by multiplying 1.23 by the square root of the aircraft's altitude in feet. For example, an aircraft at 30,000 feet would be able to receive a ground station 200 nautical miles away.

The principle of the DME operation is based on the airborne system transmitting paired pulses, known as interrogations, to the DME or TACAN ground station. The ground station receives the interrogation signals and replies (after a 50 microsecond delay) by transmitting paired pulses that are synchronous to the interrogation pulse pair back to the aircraft DME system on a different frequency. The time required for the round trip of this signal exchange is measured in the airborne DME transceiver and is translated into distance in nautical miles from the aircraft to the ground station. This distance is then displayed on the DME indicator.

Once the ground station reply signal is received, the airborne DME transceiver will initially go into a search mode, and will examine all signals received which have a regular time relation with respect to its own transmitted pulse pair. When the search circuit determines which received pulses are due to its own interrogations, it will lockon to them and go into a track mode, at which time, the slantrange distance will be displayed on the DME indicator.

DME NAVIGATION CONCEPTS

When an aircraft is flying at altitude, the direct distance to the station will be the slantrange or line-of-sight distance. The difference between the measured distance on the surface and the DME slantrange distance is called the slantrange error. The error will be maximum when the aircraft is directly over the ground facility, at which time, the DME will display altitude in nautical miles above the station. Slant-range error is minimum at low altitude and long range, and is negligible when the aircraft is one mile or more from the ground station for each 1,000 feet of altitude above the elevation of the station. Figure 6-1 illustrates the effect of altitude and range on slant-range error.

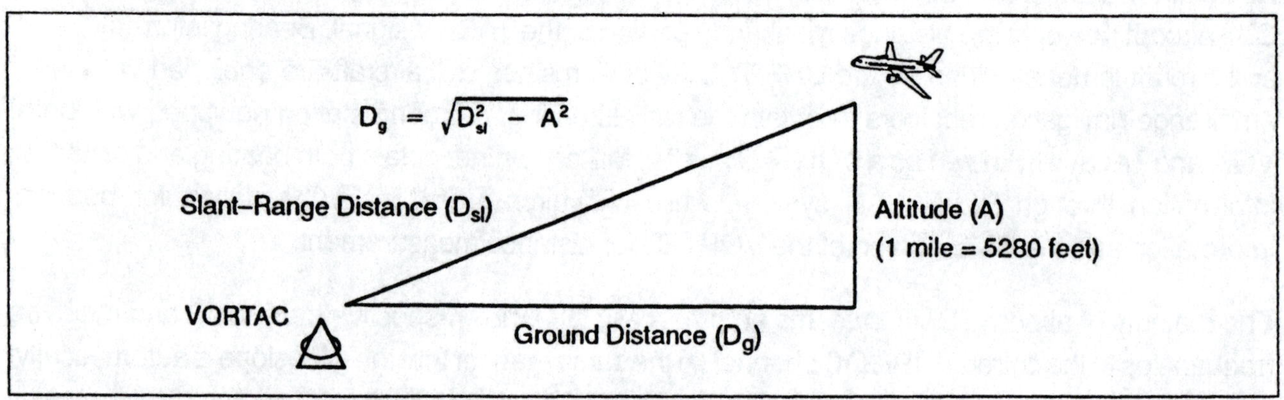

Figure 6-1. ***DME Slant-Range Distance.***

The airborne DME computes slantrange distance to or from the station as follows:

$$D = (T - 50\ \mu s) / 12.359$$

where:

D = slant-range distance in nautical miles

T = time in microseconds between transmission of the interrogating pulse pair and the reception of the corresponding reply pulse pair

50 ms = delay in DME ground station between reception of initial interrogation and transmission of a reply

12.359 ms = time required for RF energy to travel one nautical mile and return

Figure 6-2. DME System Operation.

The DME computes the aircraft ground-speed as the rate of change of the distance with respect to time. Since this computation is measured as a function of slantrange, the ground-speed would read zero when the aircraft is flying over the station or in a circle at a constant distance from the station. In both of these cases, the distance is not changing as would if the aircraft was flying directly to or from the DME station.

PRINCIPLES OF DME SYSTEM OPERATION

As shown in Figure 6-2, the operation of the airborne DME system is based on the transmission of paired pulses that are received and replied to by a ground station. The spacing between the pulse pairs transmitted by the airborne DME is either 12 or 36 microseconds apart, depending on the operating mode. The ground station reply also consists of paired pulses, spaced either 12 or 30 microseconds apart; however, the reply is transmitted on a different frequency. The time required for the round trip of this signal exchange is measured in the airborne DME transceiver, translated into nautical miles from the aircraft to the ground station, and displayed on the DME indicator.

A block diagram of the DME ground station and airborne DME is shown in Figure 6-3. The transmission exchange cycle begins when the airborne DME transceiver transmits pulse pairs on the receive frequency of the ground station; this may be any one of 200 channels ranging from 1,041 MHz through 1,150 MHz. Upon reception of the coded pulse pair interrogation, the ground station decodes the received signal and transmits a pulse pair reply (after a 50 microsecond delay) on a frequency offset by 63 MHz from the interrogation signal. The airborne DME receiver operates in the frequency range of 978 MHz through 1,213 MHz. The reception of the ground station pulse pair by the DME receiver concludes one complete DME cycle.

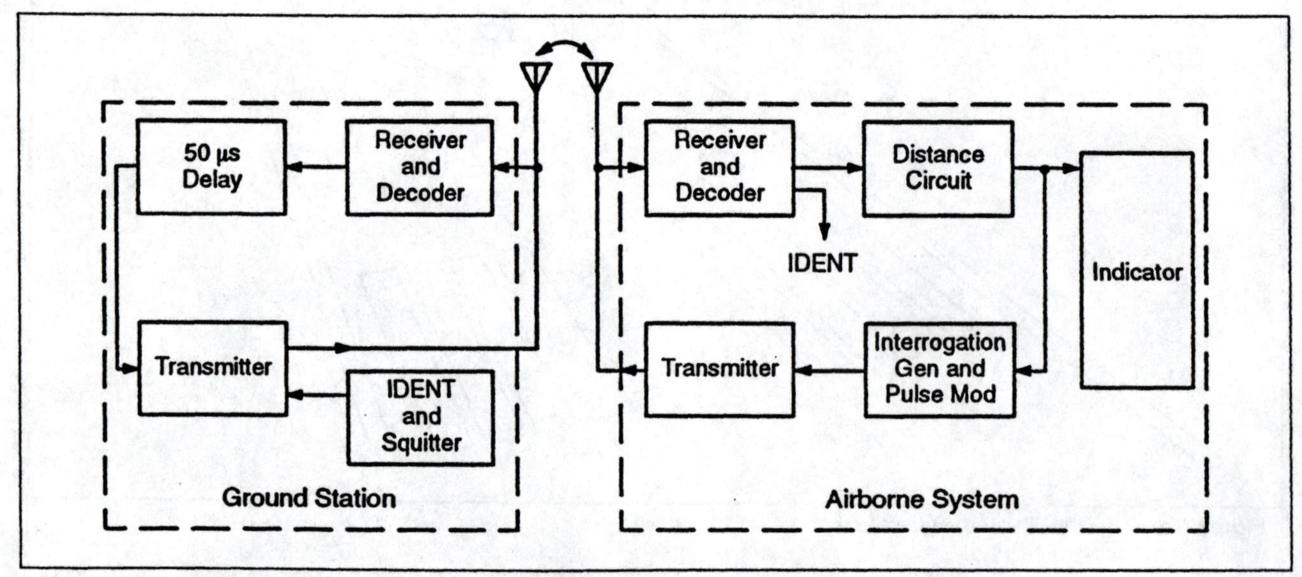

Figure 6-3. Typical DME System Block Diagram.

The 50 microsecond delay in the reply from the ground station is added to eliminate the possibility of uncoordinated operation when the aircraft and ground station are at close range. Without the delay, the airborne DME could still be transmitting its second pulse when the first pulse of the reply was received.

The DME ground station transmits a constant pulse repetition frequency (PRF) rate of 2,700 pulse pairs per second, consisting of replies to interrogations and random pulses, known as squitter. Squitter provides filler pulses between replies to interrogations to maintain the ground station transmitter at a constant duty cycle. As the number of interrogations increases, the squitter will be replaced with reply signals. In addition to transmitting squitter and reply pulse pairs, the ground station also transmits a 1,350-Hz coded audio identification signal every 30 seconds.

In actual operation, a given ground station will be interrogated simultaneously by a number of aircraft which are within range and tuned to the station's frequency. The ground station will then reply to all interrogations, and each aircraft will receive the sum total of replies to all aircraft. However, if the ground station receives more than 2,700 interrogations per second, it will reply only to the stronger interrogations rather than increase its PRF rate.

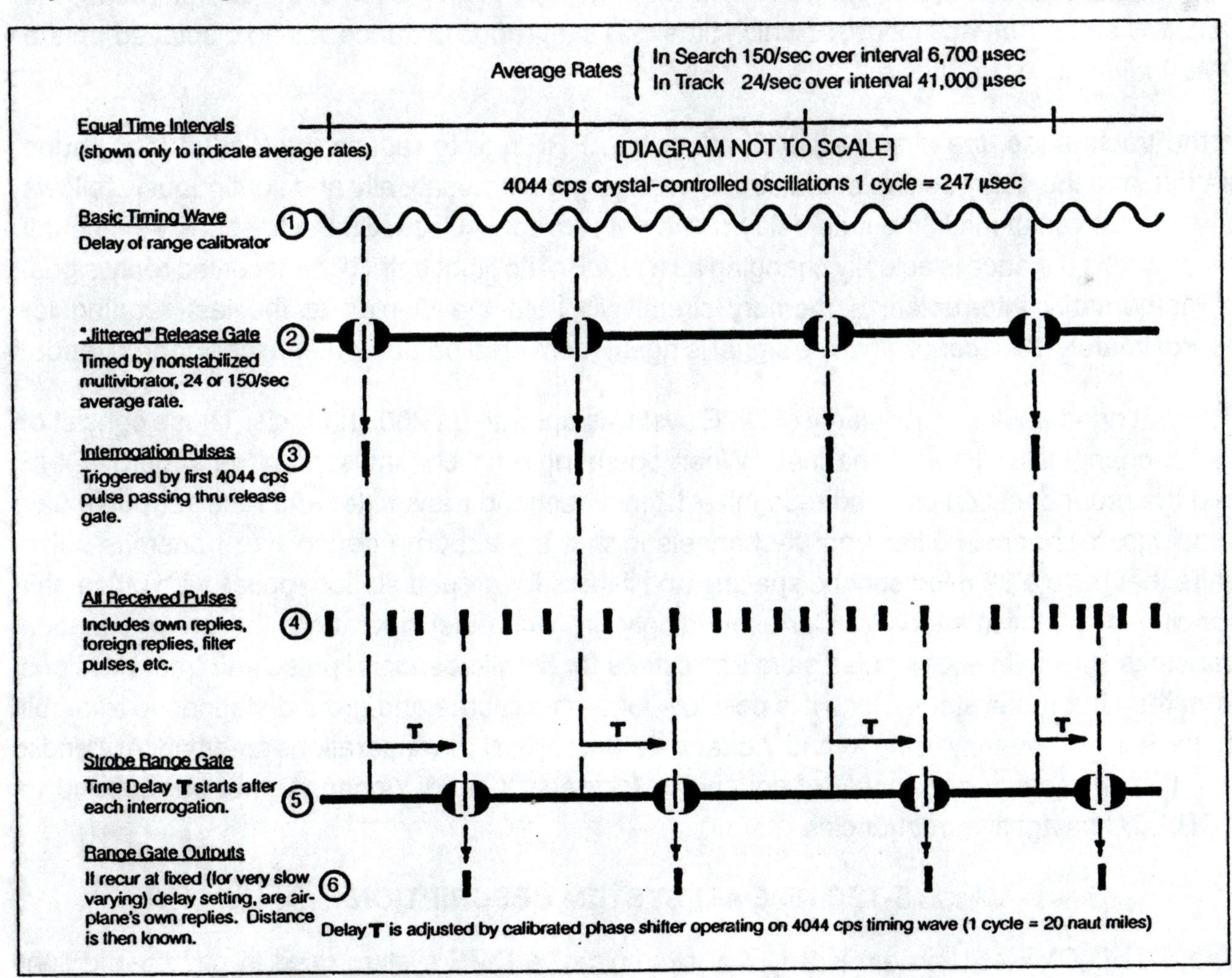

Figure 6-4. Pulse Sequence Process. *(Courtesy ITT Avionics)*

To prevent interference from replies to other DMEs, it is arranged that each DME's interrogation pulses occur at a rate which intentionally varies, within limits, in an irregular manner. This effect is caused by permitting a non-stabilized multivibrator circuit to exercise gross control over the interrogation rate. In order that the DME may distinguish replies to its own interrogations from squitter or replies to other DMEs, a "stroboscopic" search process is employed in the distance circuit. Stroboscopic refers to a technique wherein a particular set of recurring pulses is located at a point in time by matching to their periodicity an adjustable electronic time-gate. This process is illustrated in Figure 6-4.

The strobe locates the proper reply pulses by finding the fixed time delay, measured from its own previous interrogation pulse pair, at which a reply pulse is repeatedly received. The strobe progressively scans various time delay intervals by means of a sliding "time slot" or "range gate". It quickly tests each time slot position for the number of successive reply pulses received. If no replies or only sporadic replies are received, the strobe advances the range gate to test a slightly longer time delay interval. When, at some particular time delay interval, a sufficient number of recurrent pulses is detected, the strobe's search will be completed and the range gate will be lockedon to that particular time interval. Upon completion of the search mode, the DME will enter the track mode, at which time, the slantrange distance will be displayed on the DME indicator.

In the track mode, the airborne DME reduces its PRF rate to reduce the load at the ground station, and the delay setting of the strobe's range gate automatically and continuously follows any normal variations in the time delay of the proper reply pulses. Such variations will occur if the aircraft's distance is actually changing as a result of its flight path. If the received reply signal is momentarily interrupted, a memory circuit will hold the display at the last reading for approximately 10 seconds until the signal is again recovered before rentering the search mode.

As mentioned earlier, the majority of DME systems operate on 200 channels. These consist of 100 X-channels and 100 Y-channels. When operating on X-channels, both the airborne DME and the ground station use and recognize 12 microsecond transmitter and receiver pulse pair spacings. Y-channels differ from X-channels in that the airborne transceiver transmits pulse pairs that have a 36 microsecond spacing and listens for ground station replies with pulse pair spacing of 30 microseconds. Correspondingly, in Y-channel operation, the ground station transmits 30 microsecond pulse pairs and listens for 36 microsecond pulse pair transmissions from the airborne system. Since it is possible for both airborne and ground stations to transmit on the same frequency using X- and Y-channels, pulse pair time separations are used. Appendix B lists the transmit and receive frequencies for each X- and Y-channel that correspond to VOR/LOC navigation frequencies.

S-TEC DME-451 SYSTEM DESCRIPTION

The S-TEC DME-451 system is a typical example of a DME system used in private and light corporate aircraft. The DME-451 system consists of the TCR-451 transceiver, IND-451 indicator, and ANT-451 antenna. The IND-451 indicator, as shown in Figure 6-5, provides a continuous Light Emitting Diode (LED) readout of DME distance in nautical miles in the top display, while the bottom display is controlled by the display selector control. The information that may be selected for readout on the bottom display consists of any one of the following: ground-speed in knots, timetostation (TTS), elapsed time (ET), Greenwich mean time (GMT), and estimated time of arrival (ETA). TTS, ET, GMT, and ETA are displayed in minutes. The ET pushbutton starts, stops, and resets the elapsed time display each time the button is pressed.

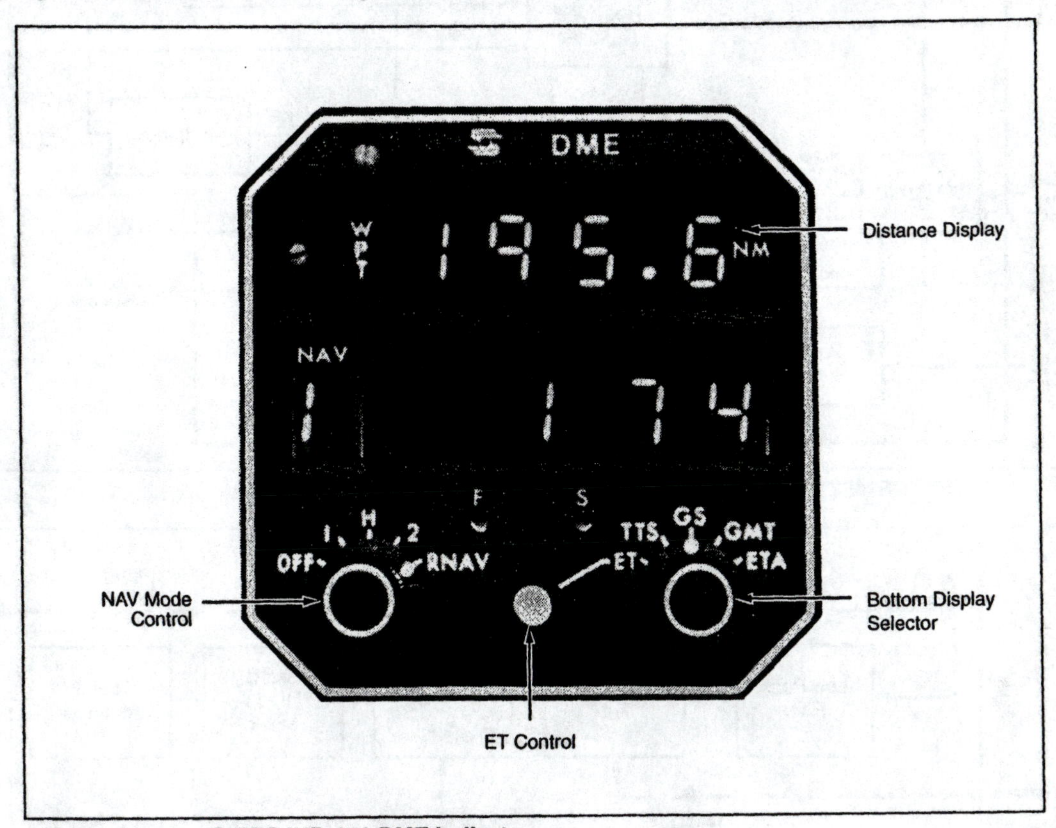

Figure 6-5. S-TEC IND-151 DME Indicator.

The NAV mode control on the IND-451 applies power to the system and provides selection of which VOR/LOC navigation receiver is to be used to control the frequency selection for the TCR-451 transceiver. The NAV mode control also provides a DME frequency hold (H) function to allow the navigation receivers to be tuned to an alternate frequency while the DME-451 system continues operation on its previous selected channel. The RNAV position on the mode control allows distance information to be displayed from the Area Navigation (RNAV) System. The operation of the RNAV system is discussed in Chapter Twelve.

A block diagram of the TCR-451 transceiver is illustrated in Figure 6-6. The 2outof5 frequency control information supplied by the VIR-351 navigation receiver or comparable VHF control head is applied to the TCR-451 transceiver for DME channel selection. Within the TCR-451, the 2 out of 5 control logic is decoded into DME channels that determine the transmit and receive frequencies necessary for compatible operation with the selected station.

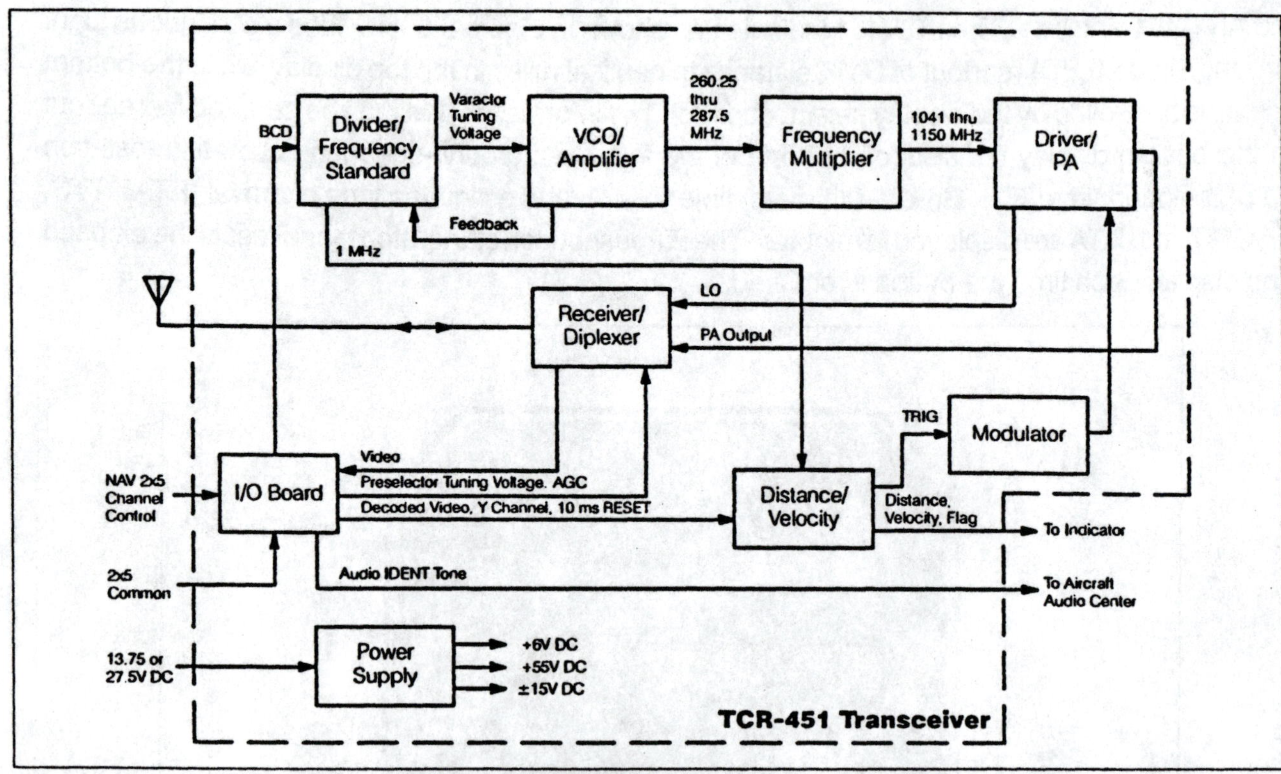

Figure 6-6. S-TEC TCR-451 Transceiver Block Diagram.

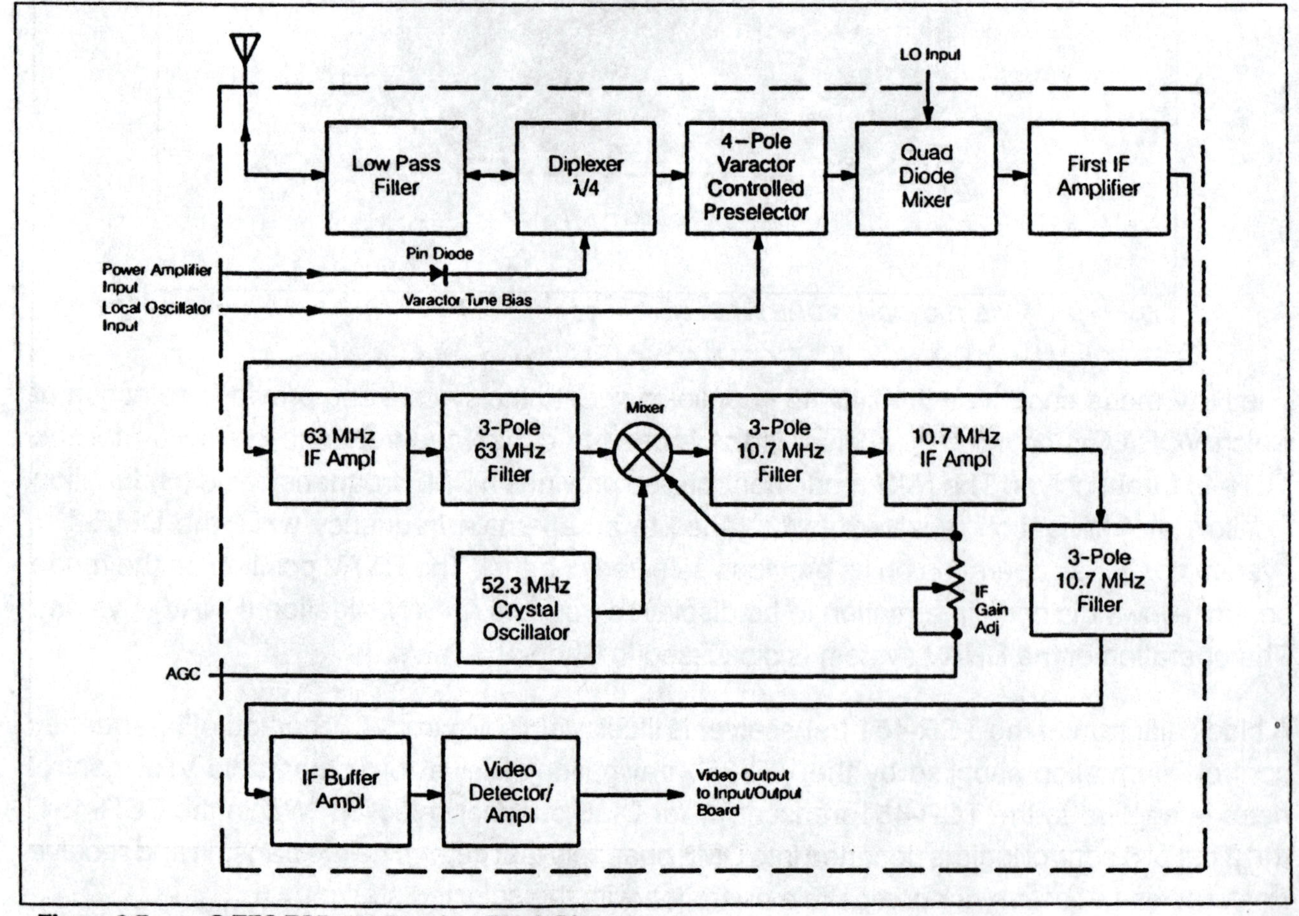

Figure 6-7. S-TEC TCR-451 Receiver Block Diagram.

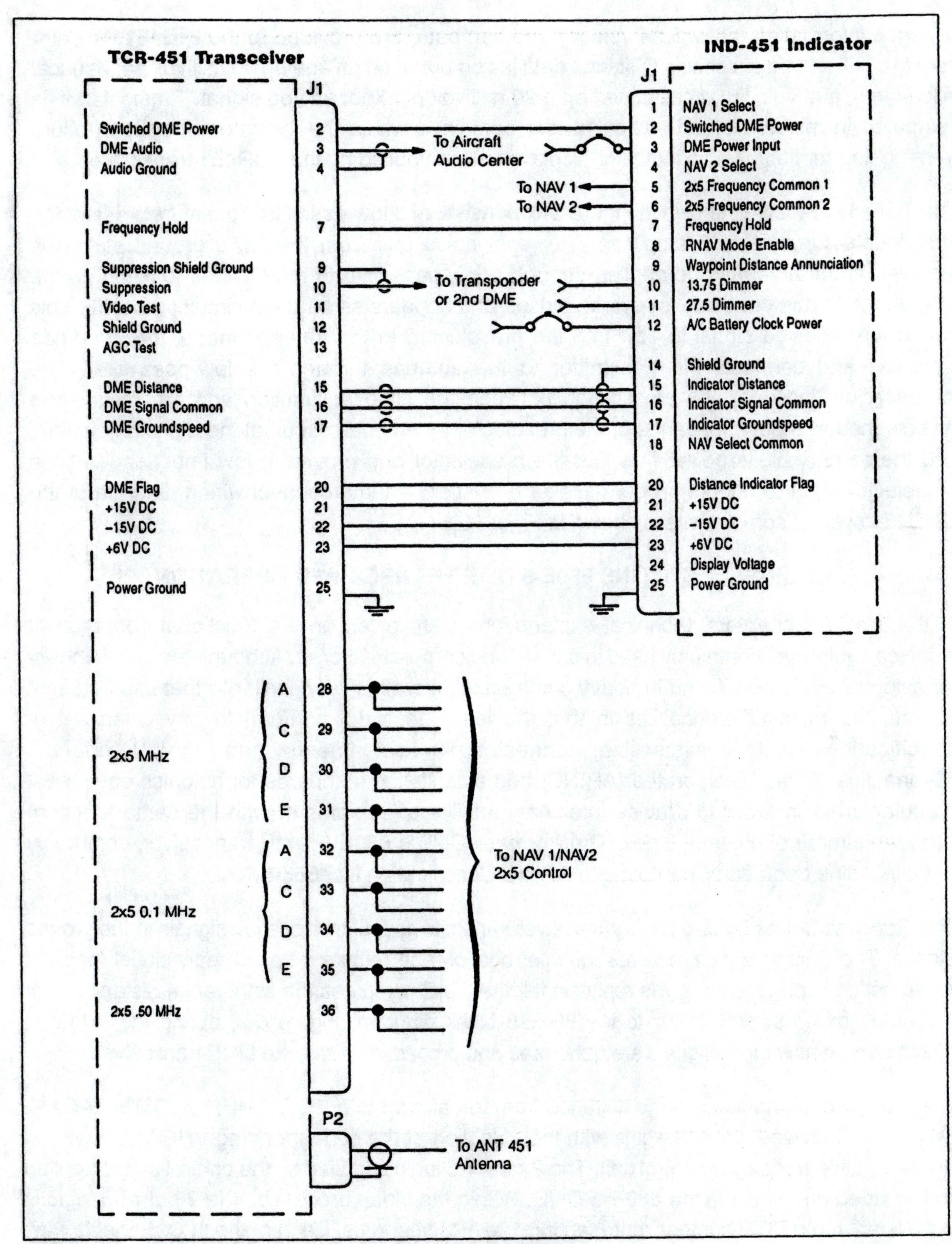

Figure 6-8. S-TEC DME-451 System Interconnect Wiring Diagram.

Distance information, as well as velocity and flag data, are provided to the IND451 indicator from the TCR-451 transceiver. Distance data is computed on an analog 40 millivolt per nautical mile signal, and velocity is calculated on a 20 millivolt per knot analog signal. Timetostation is computed in minutes by an analog divider contained within the indicator. A system failure warning flag that blanks the indicator display is also supplied by the TCR451 transceiver.

The TCR-451 receiver, shown in Figure 6-7, consists of a low-pass filter, diplexer, preselector, and double-conversion IF amplifier. The reply pulse pair from the DME ground station is received at the antenna and applied through a low-pass filter to the diplexer junction. In the receive mode, the pin diode is reverse-biased and appears as an open circuit to the RF. This forces the received signal to flow into the preselector. In the transmit mode, the pin diode conducts and connects the transmitter to the antenna through the low-pass filter. The transmission line quarter-wave section between the diplexer junction and the preselector reflects the low input impedance of the preselector as an open circuit at the diplexer junction, and therefore, inhibits power flow into the preselector and provide a low impedance at the preselector input to ensure proper diplexer operation. An interconnect wiring diagram of the DME451 system components is shown in Figure 6-8.

ROCKWELL COLLINS 860E-5 DME TRANSCEIVER OPERATION

In the previous chapters, technical explanations were given on the functional operation of avionics equipment commonly used in private and corporate aircraft. Although the circuit theory for avionics equipment found in heavy commercial aircraft is very similar to that used on light aircraft, the main difference lies in that the latter does not conform to any given set of specifications as to chassis size, connector pin assignments, and signal interfaces. Aeronautical Radio Incorporated (ARINC) has established standards for avionics equipment manufacturers in order to provide interchangeability of equipment (with the same function) between aircraft of different types. The Rockwell Collins 860E-5 DME transceiver, commonly used in airline operations, conforms to ARINC Specification No. 568-5.

The Rockwell Collins 860E-5 DME transceiver transmits coded interrogation signals to the ground station. The ground station receives the interrogation and returns a coded reply signal for each interrogation. Upon receiving the reply signal, the DME computes the slant-range distance to or from the ground station. Refer to the 860E-5 block diagram, Figure 6-9, during the following discussion on how these signals are generated and processed within the DME transceiver.

Measurement of the slant-range distance from the aircraft to a VOR/DME, ILS/DME, TACAN, or VORTAC ground station begins with the selection of the corresponding VHF frequency on the navigation frequency control unit. The 2-out-of-5 logic supplied by the control unit is applied to the video processor in the 860E-5 DME. Within the video processor, the 2-out-of-5 logic is converted into a BCD number representing 1 to 126 channels. Each of the 126 channels may have X or Y spacing, thus producing 252 available DME channels. The 860E-5 transmit frequency range is 1,025 MHz to 1,150 MHz and the receive frequency range is 962 MHz to

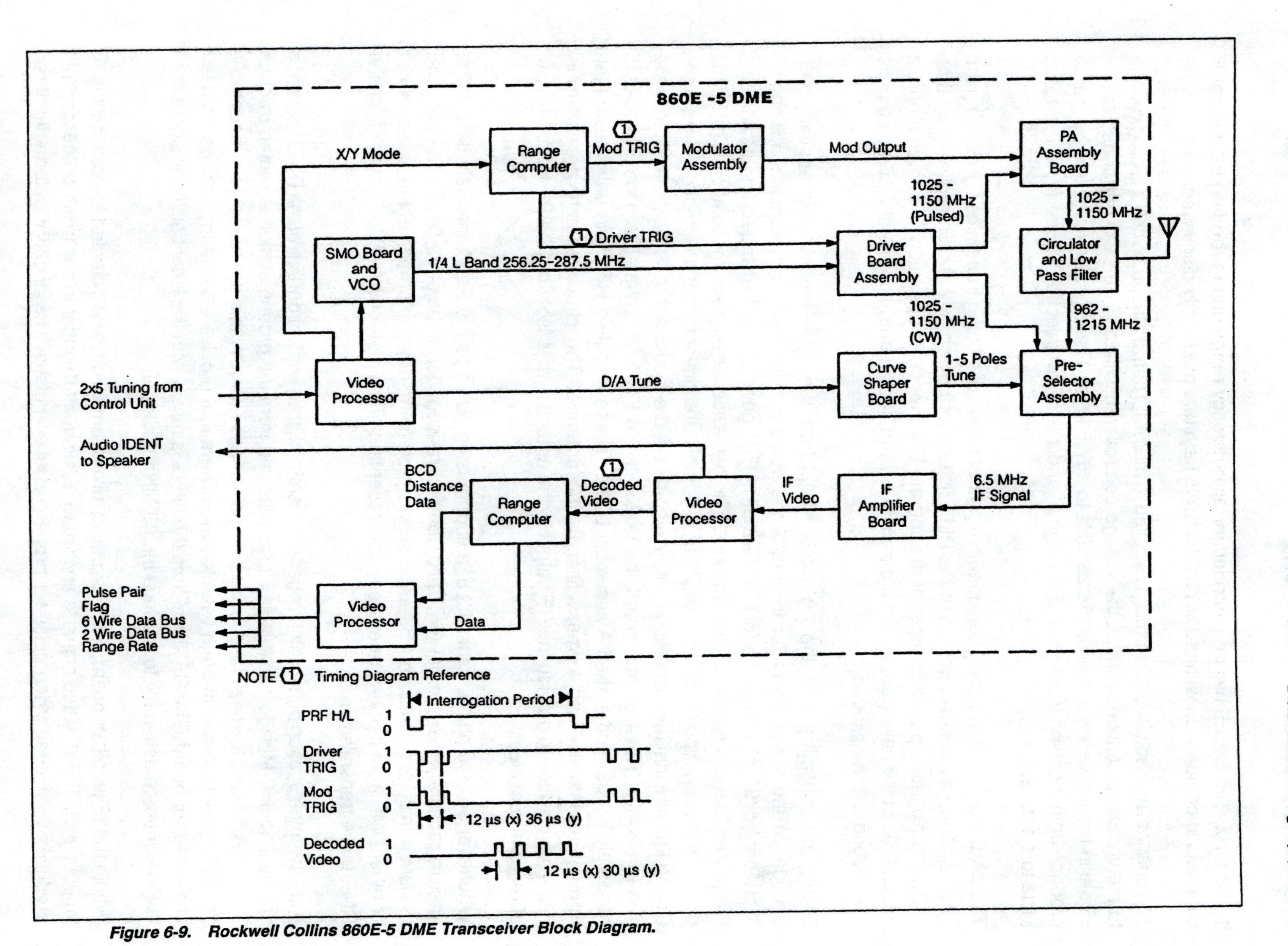

Figure 6-9. Rockwell Collins 860E-5 DME Transceiver Block Diagram.

1,213 MHz. The 860E5 has an extended lower frequency range, thus providing 52 additional channels in comparison with the S-TEC DME-451 system previously discussed.

The channel number, in BCD format, is applied to the SMO board that produces a tuning voltage to the varactors contained within the voltage controlled oscillator. The VCO generates an RF signal in the frequency range of 256.25 MHz to 287.5 MHz. This signal is multiplied by four to produce both a pulsed transmitter drive signal, and a receiver injection frequency in the 1,025 MHz to 1,150 MHz range.

The DME interrogation period begins with a pair of RF pulses being transmitted. Following the interrogation, the receiver portion of the DME listens for any ground station replies. The time of the interrogation period is dependent upon the DME mode of operation. In search mode, a nominal 90 pulse pairs per second are transmitted. In track mode, a nominal 22.5 pulse pairs per second are transmitted.

During the transmit phase, the X/Y mode signal from the video processor is applied to an encoder in the range computer to control the pulse pair spacing (12 microseconds for an X-channel, or 36 microseconds for a Y-channel). A PRF generator in the range computer initiates the pulses that are applied to the encoder. The two encoder outputs, modulation trigger and driver trigger, are applied to the modulator assembly and driver board assembly, respectively. During the time that the modulation output signal from the modulator assembly is applied to the power amplifier (PA) assembly board, the 1,025-MHz to 1,150-MHz signal from the driver board assembly is amplified in the PA assembly board. These amplified RF pulse pairs are routed through a circulator and lowpass filter to the antenna. The circulator acts an electronic microwave switch to provide the RF output to the antenna while isolating the RF from the input of receiver preselector.

Whenever a pulse pair is transmitted, a suppression pulse pair is simultaneously sent to the internal receiver, and external Air Traffic Control Transponder or the other DME (if a dual DME system is installed). These suppression pulses protect the receiver portion of the Transponder or other DME from damage due to possible reception of the high-powered RF energy from the DME pulse transmission.

After the interrogation pulse pair has been transmitted, the receiver portion of the DME becomes active. The 962-MHz to 1,213-MHz signal received from the ground station is routed through the circulator and lowpass filter to the preselector assembly. Within the preselector assembly, the RF signal is routed through varactortuned filters that receive their tuning voltage from the curve shaper board. The RF signal from the filters is mixed with the local oscillator signal from the driver board assembly to produce the 63-MHz IF signal.

Within the IF amplifier board, the IF signal is mixed, amplified, and detected to produce the IF video signal. The IF video signal is applied to a pulsepair decoder in the video processor that produces a decoded video pulse for properly spaced pulses. The decoder determines if the pulses have sufficient amplitude and are properly spaced for the channel selected (12 microseconds for an X-channel, or 30 microseconds for a Y-channel). The decoded video signal is applied to the range computer. If the IF video signal contains a 1,350-Hz audio identification code signal, it is amplified in the video processor and applied to the aircraft audio system.

In the range computer, the decoded video signal is applied to timing circuits that compute the slant-range distance to the ground station. The outputs from the range computer are routed through the video processor and made available to an external DME indicator for display. The distance data is available in either a pulse-pair format or by means of a digital data bus. The distance measurement from the pulse pair output is represented by the spacing between the pulses. In both the two-wire and six-wire data bus outputs, the distance measurement is represented by a 32-bit data word. The range rate output consists of a series of pulses, one pulse being outputted for each 0.01 nautical mile change in distance. The presence of the flag output indicates that a fault exists in the equipment. All data outputs are inhibited whenever a fault appears.

Rockwell Collins 860E5 Operating Modes

While the aircraft is on the ground, the DME system may be in the MANUAL STANDBY mode, as selected on the control head. In this mode, the DME transmitter is inhibited and the receiver is operative. If the ground station signal is received during the standby mode, the identification code will be audible but the digital distance indicator will display four dashes.

When the aircraft is airborne, the DME starts in the AUTOMATIC STANDBY mode. In this mode, the DME transmitter is inhibited and the receiver is operative. The DME will remain in this mode until the receiver determines that the antenna is receiving more than 450 squitter pulse pairs per second from a ground station. When this occurs, the DME switches to the SEARCH mode.

In the SEARCH mode, the DME interrogates the ground station by transmitting pulse pairs at a PRF of 90 pulse pairs per second. After each interrogation, the DME receiver searches the ground station signals for a reply pulse pair that is synchronous to the interrogation pulse pair. The receiver searches during the time a signal would be received from a ground station located between zero and 390 nautical miles away. The range computer counts the time from the interrogation pulse pair to the decoded reply pulse that it locates and stores this time.

Once the DME has located a decoded reply pulse, it waits until the next interrogation pulse pair is transmitted. It then counts out to the time at which the last decoded reply pulse was received and develops a range gate. Presence of a decoded reply pulse in the range gate means that, twice in a row, the DME has found a pulse located at the same time interval after the second interrogation pulse. When this occurs, the DME continues to develop a range gate at this same point in time for consecutive interrogation pulses. Location of seven decoded reply pulses in fifteen consecutive interrogation periods is the criterion necessary for the DME to switch to the PRE-TRACK mode.

Development of the range gate, at one point in time for fifteen consecutive interrogation periods, can be terminated. This termination will occur if the DME fails to find a decoded reply pulse during three consecutive interrogation periods, or if it loses the 7-out-of-15 decision. When a termination occurs, the DME begins to search outbound from the previous distance to 390 nautical miles, and then from zero to 390 nautical miles, until if finds another decoded reply pulse. When another decoded reply pulse is found, the range gate is then developed at that period of time. This process continues until the DME finds a point in time at which seven decoded reply pulses occur within fifteen consecutive interrogation periods. The DME will then switch to the PRE-TRACK mode.

In the PRE-TRACK mode, the DME determines the ground speed, or relative velocity of the aircraft with respect to the ground station. This is accomplished during the four-second PRE-TRACK mode by a velocity accumulator which fine-positions the range gate so that the reply pulses are centered within the range gate. The velocity accumulator determines both the direction of range gate movement, either inbound or outbound, and the slew rate of the range gate to track the reply pulses. During PRE-TRACK mode, the DME continues to interrogate the ground station at 90 pulse pairs per second and valid data is displayed on the indicator.

After the four-second PRE-TRACK mode, the DME switches to TRACK mode. During TRACK mode, the interrogation rate is decreased to 22.5 pulse pairs per second and the velocity accumulator and error detector continue to keep the reply pulses centered in the range gate. The criterion for maintaining track is that the DME continues to find at least seven synchronous decoded replies for every fifteen interrogation periods. If this criterion is not satisfied, the DME will go into MEMORY mode.

The nominal 11.4 second MEMORY mode is entered when a temporary or permanent loss of reply signal occurs. During MEMORY mode, the DME continues interrogations at the 22.5 pulse pairs per second rate, and distance is displayed as if the station were still being tracked. If the signal is re-acquired during MEMORY mode, the DME returns to TRACK mode. If the signal is lost for a length of time greater than 11.4 seconds, the DME reverts back to SEARCH mode.

DME NAVIGATION PROCEDURES

As previously mentioned, the polar coordinates of the aircraft may be determined by comparing the distance and bearing information available from the airborne DME and VOR systems. An approach procedure is also available that combines the use of VOR radials and DME information for interception of the localizer course. This type of "arc" approach provides the pilot with a smooth transition onto the approach path by eliminating much of the vectoring commonly used in interception.

Basically, when the DME arc approach is available, the pilot maintains a constant specified distance from the selected station by flying a circular pattern around it. When a predetermined VOR radial is intercepted, the pilot initiates an inbound turn to provide a smooth transition to

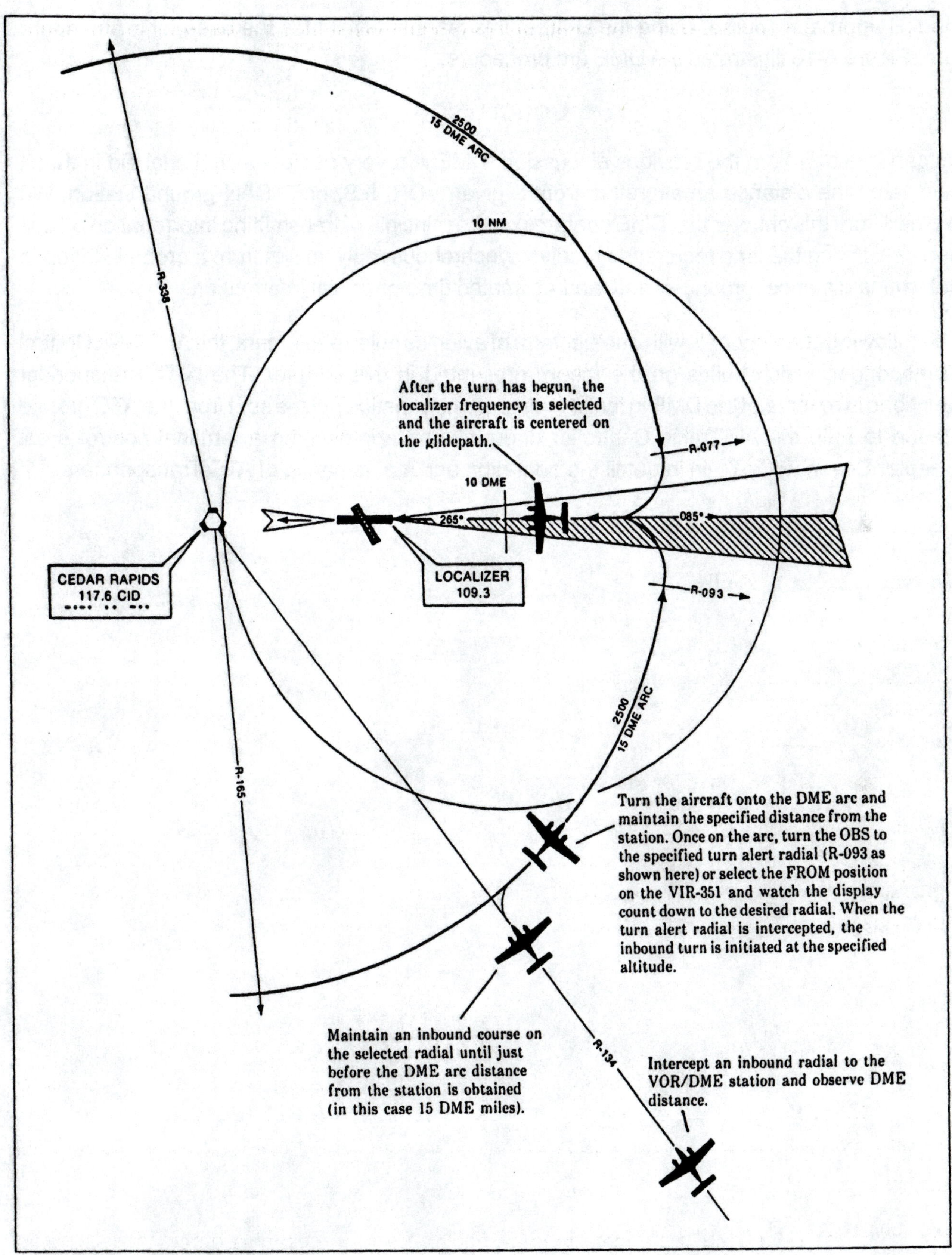
2500
15 DME ARC
10 NM
R-338
After the turn has begun, the localizer frequency is selected and the aircraft is centered on the glidepath.
R-077
10 DME
265°
085°
CEDAR RAPIDS
117.6 CID
LOCALIZER
109.3
R-093
2500
15 DME ARC
R-165
Turn the aircraft onto the DME arc and maintain the specified distance from the station. Once on the arc, turn the OBS to the specified turn alert radial (R-093 as shown here) or select the FROM position on the VIR-351 and watch the display count down to the desired radial. When the turn alert radial is intercepted, the inbound turn is initiated at the specified altitude.
Maintain an inbound course on the selected radial until just before the DME arc distance from the station is obtained (in this case 15 DME miles).
R-134
Intercept an inbound radial to the VOR/DME station and observe DME distance.

Figure 6-10. DME Arc Procedure.

the ILS approach course. Using the DME in this manner eliminates the need for a procedure turn. Figure 6-10 illustrates the DME arc procedure.

CONCLUSION

As can be seen from the previous discussion, DME is a very useful navigational aid in that it calculates the distance an aircraft is from a given VOR, ILS, or TACAN ground station. We learned from this chapter that DME operates on the principle of transmitting interrogation pulses and calculating the time required to receive synchronous reply pulses from a ground station to determine distance, ground-speed, and estimated time-of-arrival information.

The following chapter deals with another form of avionics pulse equipment, the Air Traffic Control Transponder, which builds on the theory presented in this chapter. The ATC Transponder functions in reverse of the DME in that it replies to interrogation pulses sent from the ATC ground station to help the Air Traffic Controller direct aircraft approaching a terminal control area. Chapter Seven will explain in detail the operation and maintenance of ATC Transponders.

REVIEW QUESTIONS

1. Define the primary and secondary functions of the airborne DME system.

2. Describe the various types of ground stations available for use by the airborne DME system.

3. If an aircraft is flying at an altitude of 30,000 feet and the DME indicator displays 10 miles to the ground station, what is the aircraft's actual distance from the station and what is the slant-range error?

4. In regard to Question No. 3, calculate the time required for the round trip signal exchange between the transmission of the interrogation from the airborne DME and when the reply is received from ground station.

5. What would be the DME indicator distance and groundspeed reading when an aircraft is flying at an altitude of 26,400 feet as it passes over a DME ground station?

6. If the airborne DME transceiver interrogates the ground station at a frequency of 1,041 MHz, on what frequency does the ground station send its reply signal?

7. What is the purpose of the 50 microsecond delay in sending the reply signal from the DME ground station?

8. Describe the three types of signals that are transmitted from the DME ground station.

9. How does the airborne DME and the DME ground station distinguish between X-channel and Y-channel operation?

10. What is the purpose of the DME hold function?

11. How is frequency selection of the DME transceiver accomplished?

12. What is the principal advantage of using avionics equipment that conforms to ARINC specifications?

13. What is the reason for reducing the airborne DME transmitter PRF rate in the track mode?

14. How does the airborne DME distinguish replies to its own interrogations from replies to other interrogations?

15. Why does the DME transceiver output a suppression signal to external L-band equipment?

16. Explain the functions of the range computer and video processor in the Rockwell Collins 860E-5 DME transceiver.

17. **Describe the six modes of operation available on the Collins 860E-5 DME transceiver?**

18. **How is the range gate developed in the Collins 860E5 DME transceiver?**

19. **What is the function of the velocity accumulator in the range computer of the Collins 860E5 DME transceiver?**

20. **Describe the DME arc approach procedure.**

Chapter 7

Radio Beacon Transponders

INTRODUCTION

The preceding chapters discussed the principles of airborne radio navigation, whereby the pilot can determine the aircraft's position by triangulation using two ADF or VOR stations, or by polar coordination using VOR bearing and DME distance information. Upon request, the pilot can relay the aircraft's position and altitude to the Air Traffic Control (ATC) center by means of the VHF communication system. Although this method is effective, it is not an optimum solution in high-traffic areas where the controller must be constantly informed of the exact position of all aircraft at all times within the controlled airspace. For this reason, the ATC center uses a ground-based radar surveillance system to automatically monitor the location of all aircraft within the control area without cluttering up the radio communication channels. With this information constantly displayed on the ATC radar scope, the controller is able to make timely decisions regarding handing over aircraft to the approach or departure control center, vectoring aircraft to avoid collision courses, maintaining safe altitude separation between aircraft, and locating and directing aircraft that are lost.

The ground-based ATC radar system consists of a primary surveillance radar (PSR) and secondary surveillance radar (SSR). The PSR locates and tracks aircraft within the control area by transmitting a beam of energy which is reflected from the aircraft and returned to the PSR antenna. The SSR transmits interrogation signals to the airborne radio beacon Transponder. Upon receiving the interrogation, the Transponder sends a coded reply signal back to the SSR system. Data received from the PSR and SSR are used in conjunction to develop the total air traffic situation display on the controller's radar scope. This enables the controller to identify Transponder-equipped aircraft in addition to determining the range and direction of all aircraft within the control area.

PRINCIPLES OF ATC RADAR SURVEILLANCE SYSTEM OPERATION

As shown in Figure 7-1, there are two types of radar systems installed at each ATC ground station. The first, called the Primary Surveillance Radar, operates on the principle of sending a narrow beam of energy, which is reflected from the aircraft under surveillance, and measuring its distance by noting the time lapse between the radar pulse transmission and its received echo. The second, called the Secondary Surveillance Radar, operates on the coded reply sent from the airborne radio beacon Transponder in response to an interrogation sent from the ground station. The PSR and SSR antennas are scan synchronized, and both radars are used in conjunction to develop the total air traffic situation display on a single CRT radar scope, called the Plan Position Indicator (PPI).

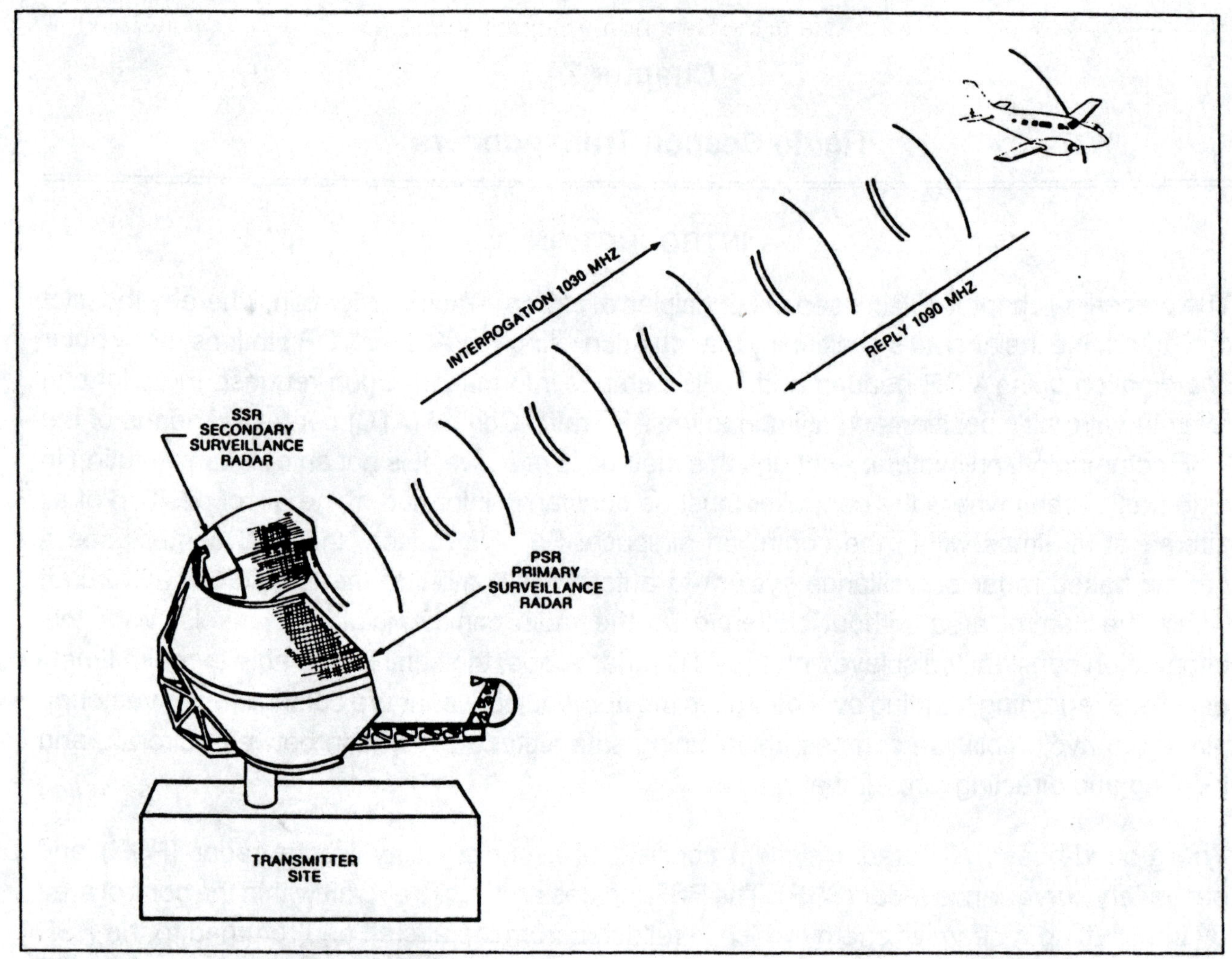

Figure 7-1. ***ATC PSR/SSR System.***

The PSR sends out radio waves in a very narrow beam. The ground antenna is made to rotate so that the position of the narrow beam of energy can be directed. When the directed beam strikes an object or target, some of it is reflected back to the radar antenna. This reflected signal is detected and processed to provide a display (indicated by a bright "pip") on the ATC PPI, which shows the location of the target (i.e., aircraft).

The PSR system works well in low traffic areas; however, as the air traffic increases in a given area, the PPI display becomes cluttered and specific targets may become difficult to distinguish from one another. Also, since the energy of a radiated RF signal is attenuated as the square of the distance it travels, the resulting weaker radar returns are accompanied by noise which tends to obscure the displayed target. Targets may also be lost due to ground clutter from terrain and precipitation unless a Moving Target Indicator (MTI) circuit is employed to detect and display only moving objects. Finally, the PSR has the distinct disadvantage in that the operator has no way of knowing the altitude of the aircraft unless the pilot reports it. All of the problems associated with the PSR system have been addressed with the introduction of Air Traffic Control Radio Beacon System (ATCRBS).

The ATCRBS incorporates the use of the Secondary Surveillance Radar in conjunction with the airborne radio beacon Transponder. The SSR was developed from the military Identification-Friend-or-Foe (IFF) system, in which an airborne radio beacon Transponder responds to ground radar interrogations on one frequency by transmitting coded replies on another frequency. The coded replies, displayed as short lines on the PPI, allow the controller to identify the various targets by having each one send back a different coded reply.

The desired code can be manually selected by the pilot on the Transponder control head in Mode "A" operation, or automatically set by an encoding altimeter or altitude digitizer for reporting the aircraft's altitude in Mode "C" operation. Since the reply signal from the airborne Transponder is stronger than the reflected PSR signal, it will reinforce the "pip" on the PPI to provide positive aircraft identification.

At the ATC radar ground station, received radar video and antenna azimuth information signals are relayed from the radar site to the air traffic control center, where the signals are processed and displayed on plan position indicators. Since radar coverage of each site includes a large area, several controllers are assigned to various segments of the area covered. Each controller's segment of the area is displayed on his respective PPI.

The PPI presents the operator with a maplike view of the space surrounding the area covered by the ATC radar antenna. Four dots appear on the PPI; one at the center, and one of each of the three 10-mile points out to the edge of the radar scope. These dots rotate, in sync with the rotation of the radar antenna, to display concentric circles that indicate range.

The incoming radar video signals are applied to a decoder control before being displayed. By adjusting the decoder to pass only a selected code, Transponders operating on the controller's code will appear as a short arc on the PPI, and as a bright (bloomed) arc when transmitting a

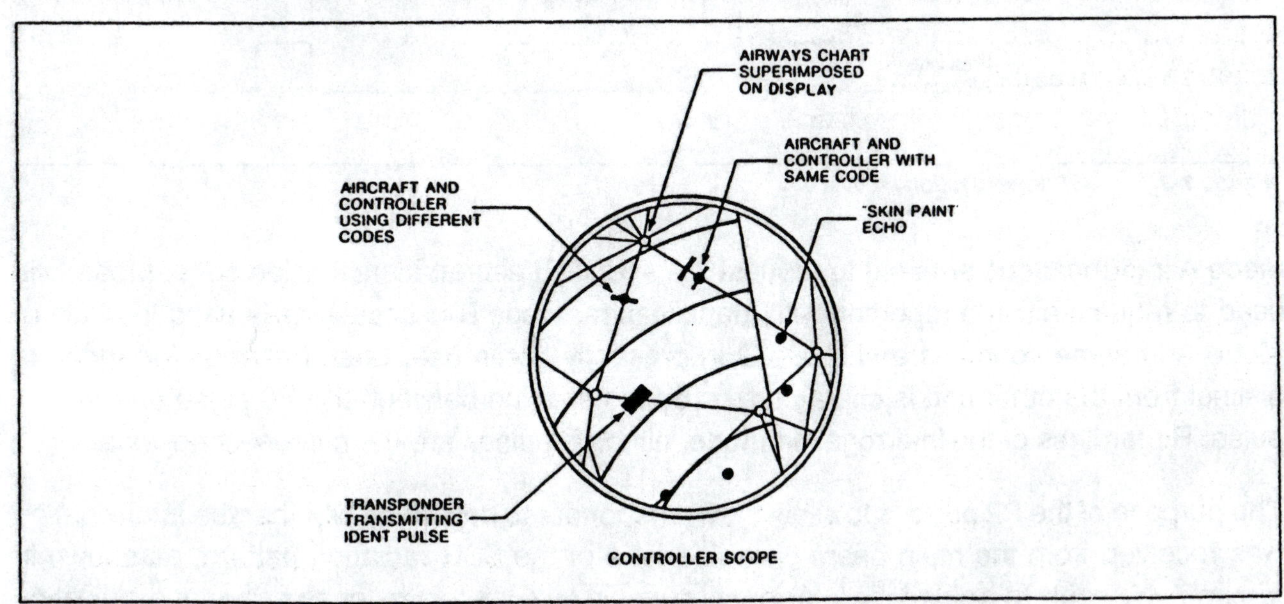

Figure 7-2. Typical PPI Display.

special position identification pulse. Replies from Transponders not transmitting the selected code will be filtered out. "Skin-paint" echoes detected by the primary surveillance radar will be displayed for all aircraft. An illustration of a typical PPI display format is shown in Figure 7-2.

PRINCIPLES OF RADIO BEACON TRANSPONDER OPERATION

As previously mentioned, the ATC radio beacon system incorporates the use of the groundbased SSR and the airborne radio beacon Transponder to determine the range and direction of aircraft responding to SSR interrogations. The following section will discuss the operation of the airborne Transponder in regard to receiving these interrogation signals and generating a coded reply signal to be transmitted back to the SSR ground station.

SSR Interrogation

An airborne Transponder transmits a reply signal on a frequency of 1,090 MHz in response to the SSR interrogation which is transmitted on a frequency of 1,030 MHz. Currently, in the United States, there are two types of SSR interrogations, Mode "A" and Mode "C", that may be transmitted by the ATCRBS ground station. The signal characteristics of the Mode A and Mode C interrogations are shown in Figure 7-3.

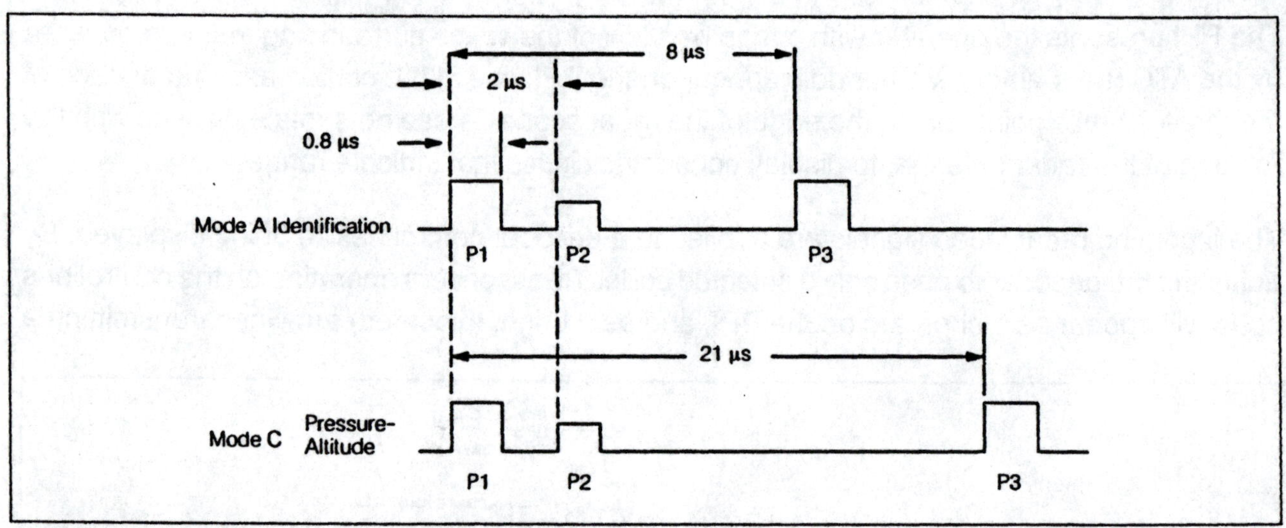

Figure 7-3. SSR Interrogation Modes.

Mode A interrogations are sent to request the specified aircraft identification code. Mode C is used to request altitude reporting with identification. Mode B is occasionally used in place of Mode A in some countries and Mode D in presently not in use. Each interrogation mode is distinct from the other and is characterized by the spacing between the P3 pulse and the P1 pulse. Regardless of the interrogation mode, all three pulses are 0.8 microsecond wide.

The purpose of the P2 pulse is to allow the Transponder to determine whether the interrogation was received from the main beam or a side lobe of the SSR radiation pattern, as shown in Figure 7-4. A reply to a side-lobe interrogation would give the controller an erroneous indication

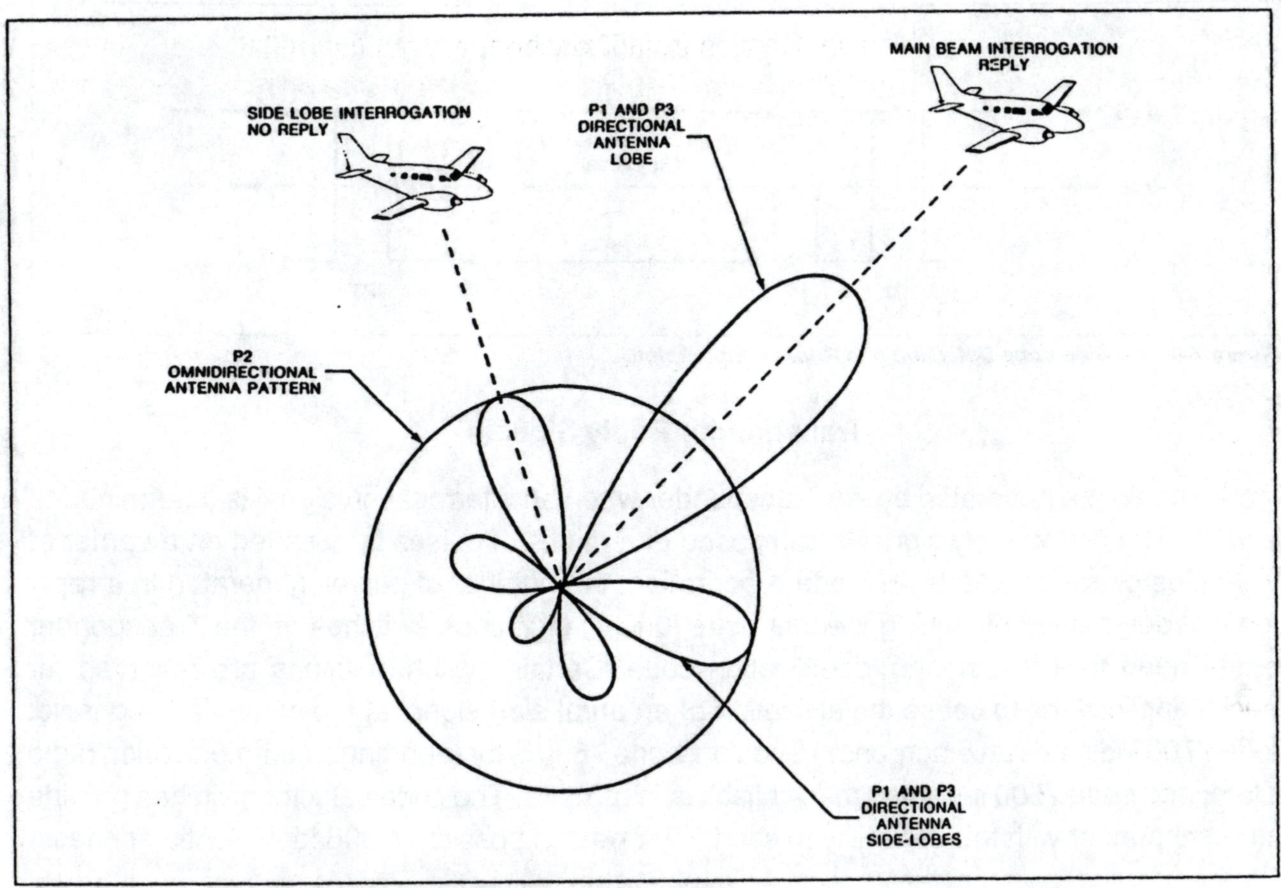

Figure 7-4. Propagation Pattern of SSR Interrogation Signal.

of the aircraft's position. For this reason, Side-Lobe Suppression (SLS) is used to inhibit the Transponder's reply in response to a side-lobe interrogation.

The three-pulse SLS interrogation method uses a directional radar antenna that transmits a pair of pulses referred to as PI and P3 pulses. As previously mentioned, the time spacing between these pulses determines the mode of operation. Two microseconds after the P1 pulse is transmitted from the directional antenna, the second pulse, P2, is transmitted from an omnidirectional antenna. The P2 pulse is used as a reference pulse for SLS determination. The signal strength of the omni-directional P2 pulse is just sufficient to provide coverage over the area that side-lobe propagation presents a problem.

Side-lobe interrogation is detected by the airborne Transponder SLS circuitry by comparing the amplitude of the P2 pulse in relation to the P1 pulse. When the omnidirectional P2 pulse is equal to or greater than the directional P1 pulse, no reply will be generated. Identification of the side-lobe interrogation is established before the P3 pulse is received. Therefore, the Transponder will be inhibited for a period lasting 35 microseconds, regardless of the interrogation mode. A valid main-lobe interrogation is recognized when the P1 pulse is at least nine dB larger than the P2 pulse, as shown in Figure 7-5.

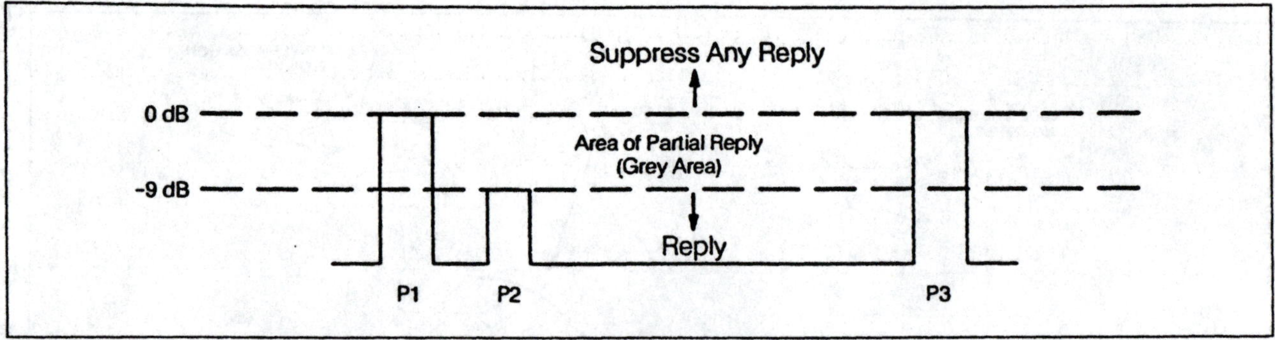

Figure 7-5. Side-Lobe Detection and Reply Suppression.

Transponder Reply Signals

Reply signals are generated by the Transponder when an interrogation signal is determined to be valid. The coded reply signal is composed of a series of pulses transmitted on a carrier of 1,090 plus or minus 3 MHz. In Mode A operation, the number of pulses generated in a reply signal is determined by setting the four octal (0 to 7) digit code switches on the Transponder control head to the assigned identification code. Certain switch positions are reserved for special applications to cause the activation of an aural alert signal at the controller's console: Code 7700 indicates an emergency condition, code 7600 is for reporting a communication radio failure, and code 7500 indicates that a hijack is in process. The code selector switches provide the Transponder with the capability to send any one of a possible 4,096 identification codes.

The Transponder replies to Mode C interrogations by generating pulses in the reply signal that corresponds with the aircraft's altitude. The received altitude information is then displayed directly on the controller's PPI. This information is not selected by the code switches on the control head, but is obtained directly from an encoding altimeter or altitude digitizer. These devices commonly use an optical encoder which is driven from an aneroid mechanism that is sensitive to variations in altitude. The encoder outputs a 10-bit parallel data code to the Transponder for the generation of Mode C replies.

The coded reply signal consists of various arrangements of code pulses within the boundaries formed by the two framing pulses, F1 and F2. Regardless of the mode of operation, these framing pulses are always present in the coded reply signal and are spaced 20.3 microseconds apart.

The reply code is divided into four pulse groups labeled A, B, C, and D. Each group contains three pulses that are assigned subscripts that indicate the binary weight of each. For example, the pulse arrangement in Figure 7-6 is assigned a reply code of 1324. The first digit (1) consists of the A1 pulse (=1), the second digit (3) consists of the B1 + B2 pulses (=3), the third digit (2) consists of the C2 pulse (=2), and the fourth digit (4) consists only of the D4 pulse (=4). The assigned reply code 0000 would cause no pulses to appear between the framing pulses, and code 7777 would result in all 12 pulses to be present between Fl and F2.

The Special Position Identification Pulse (SPIP), initiated upon request of the controller, is generated by momentarily depressing the IDENT button located on the Transponder control head. The SPIP causes a special effect on the controller's PPI that aids in determining the aircraft's position. This pulse occurs 4.35 microseconds after the last framing pulse (F2) and is transmitted with each Mode A reply for 15 to 20 seconds after releasing the IDENT button.

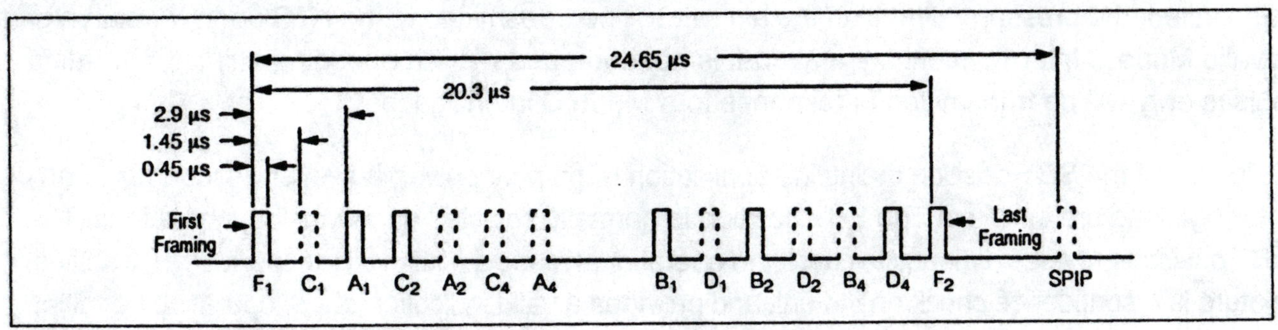

Figure 7-6. Transponder Reply Code Pulses.

S-TEC TDR950 TRANSPONDER OPERATION

The S-TEC TDR950, shown in Figure 7-7, is a typical example of an airborne radio beacon Transponder. The controls and displays available on the TDR-950 consists of a function selector (off, standby, on, altitude, or test), reply lamp/IDENT button/lamp dimmer, and code selection switches.

The function selector controls the application of primary power to the unit and the mode of operation. There is a 20-second delay from the time primary power is applied until the operation of the Transponder is obtained. This 20-second delay allows the transmitter tube to warm up and stabilize. Selection of the ON or ALT position enables the Transponder, causing it to become an active part of the ATCRBS. Reception of a valid interrogation in either of these two switch positions initiates the generation of a reply pulse train.

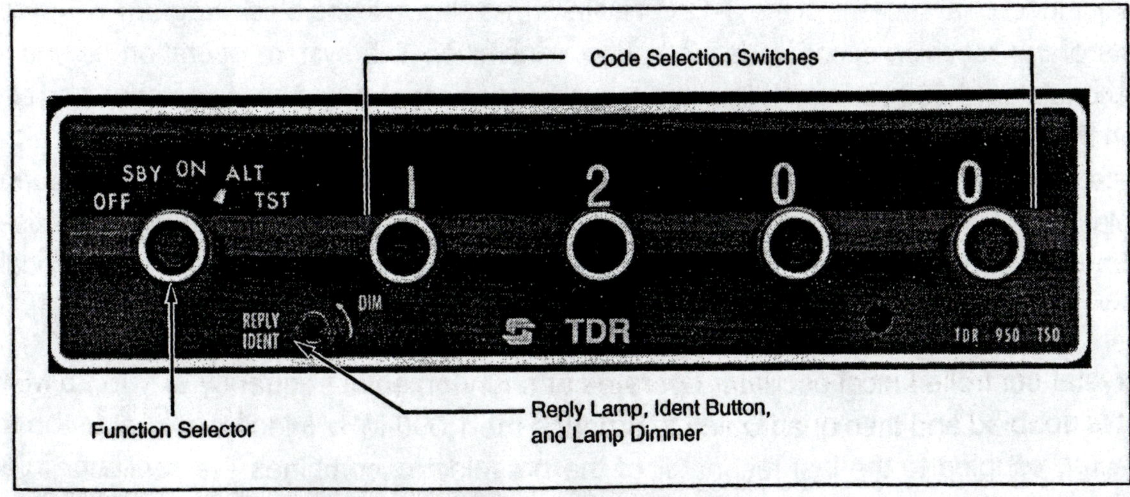

Figure 7-7. S-TEC TDR-950 Transponder. (Courtesy of S-TEC Corporation)

In the ON position, the configuration of the reply pulse train is determined by the front panel code selector switches and the mode of the SSR interrogation signal. Mode A interrogations initiate the generation of the complete code selected pulse train; Mode C interrogation (in the ON position) results in the transmission of F1 and F2 framing pulses only.

When the Transponder is used in conjunction with an encoding altimeter and the ALT position is selected, the pressure altitude of the aircraft will be transmitted to the ATC upon reception of a valid Mode C interrogation. If the aircraft is not equipped with an encoding altimeter, framing pulses only will be transmitted in response to a Mode C interrogation.

Selection of the SBY position maintains application of primary power; however, responses to any interrogations are inhibited. The SBY position is normally selected during taxiing operations. The TST position provides a means to check the operational readiness of the Transponder. The selftest feature is a confidence check on the unit and provides a valid indication of the operating condition by causing the REPLY lamp to illuminate. When the IDENT button is depressed, an additional SPIP is added to the normal reply pulses resulting in a unique identification pattern display on the ATC PPI. This SPIP is transmitted for approximately 20 seconds longer than the time during which the IDENT button is depressed. The IDENT button is depressed only when the ATC requests the aircraft to "squawk ident". The reply lamp serves as an IDENT transmission monitor by remaining steadily on for the period of the SPIP transmission.

The REPLY lamp will light whenever a response is made to a valid interrogation, the IDENT button is depressed, or the TST position is selected. When responses are made to interrogations, the REPLY lamp will flash on and off. Depressing the IDENT button or selection of the TST position places the reply lamp steadily on. Dimming of the REPLY lamp is accomplished by turning the IDENT button. This allows the pilot to maintain optimum REPLY lamp brightness for any cockpit ambient lighting conditions.

S-TEC TDR-950 TRANSPONDER CIRCUIT THEORY

The major functional sections of the S-TEC TDR-950, as shown in the block diagram in Figure 7-8, consists of the receiver, encoder, decoder, and transmitter. The system operation begins when pulse-modulated, 1,030-Hz signals from the ground station are received by the antenna and applied through the lowpass filter and diplexer to a three-pole preselector that routes the received signal to the receiver. The preselector passes the interrogation signal and rejects other frequencies, particularly the Transponder transmit frequency (1,090 MHz) and the band of frequencies used for DME transmission. The diplexer allows one antenna to be used for both transmitting and receiving. The low-pass filter attenuates harmonics of the Transponder reply transmission.

The crystal-controlled local oscillator operates at a fundamental frequency of 136.25 MHz. Its output is doubled and then quadrupled to produce the 1,090-MHz injection signal to the mixer. The mixer, coupled to the last resonator of the preselector, combines the oscillator injection frequency with the received signal to produce an IF output of 60 MHz. The IF amplifier provides

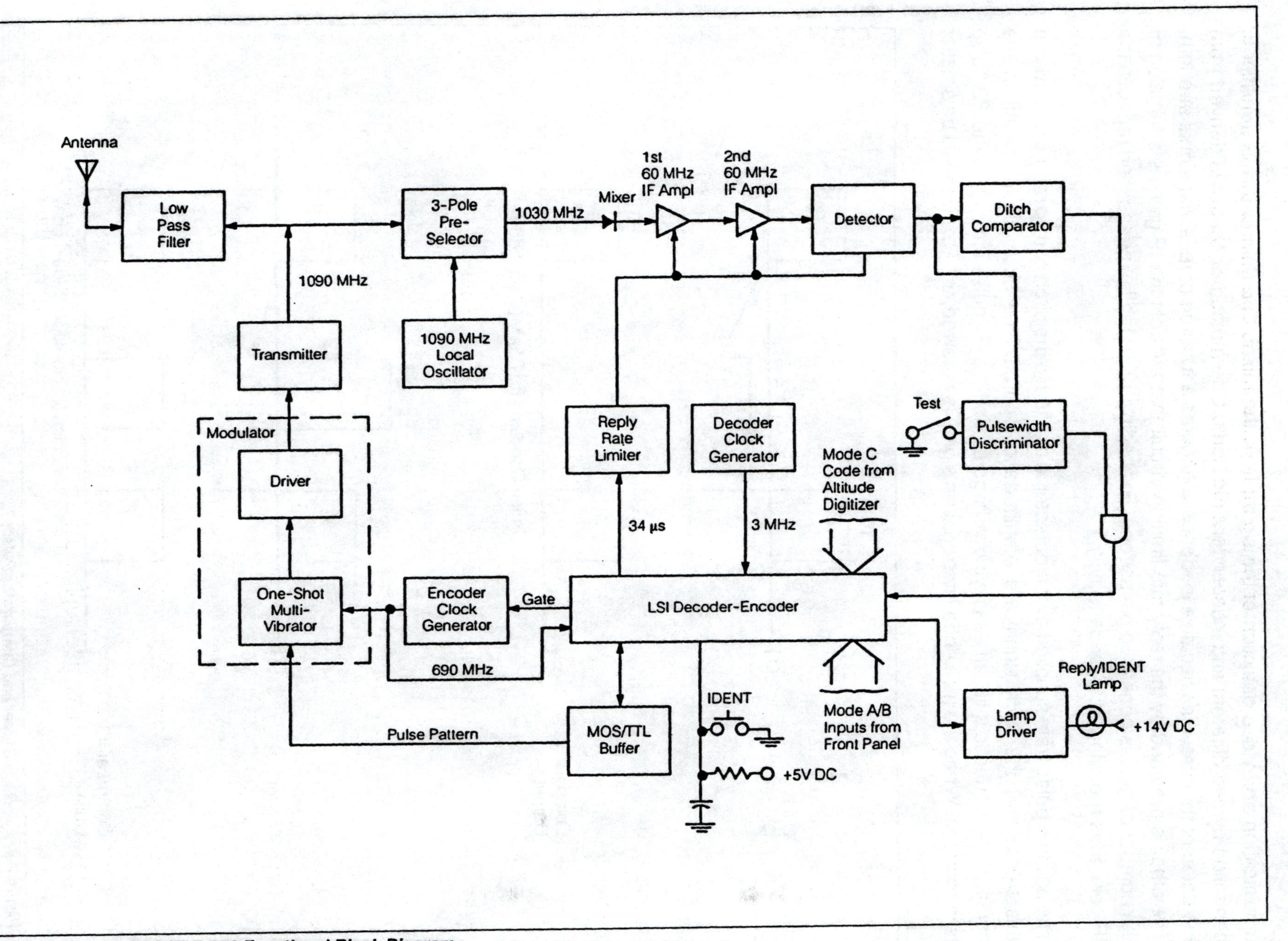

Figure 7-8. S-TEC TDR-950 Functional Block Diagram.

amplification and video detection of the signal from the mixer. The gain of the IF amplifier is controlled by two different and independent methods. One method of control is derived from the detector that provides negative feedback voltage as a function of received signal strength. The other is provided by the reply rate limiting circuit that controls the IF gain as a function of the number of replies generated per second. Exceeding 1,080 replies per second reduces the IF gain, thus reducing receiver sensitivity.

The positive pulse train supplied by the detector is applied to the ditch comparator circuit where this signal and the ditch signal are amplitude compared. The ditch digger circuit in the comparator produces a linearly decaying voltage behind the P1 pulse into which the P2 pulse is introduced. When the main-lobe interrogation signal is received, the P2 pulse will be smaller

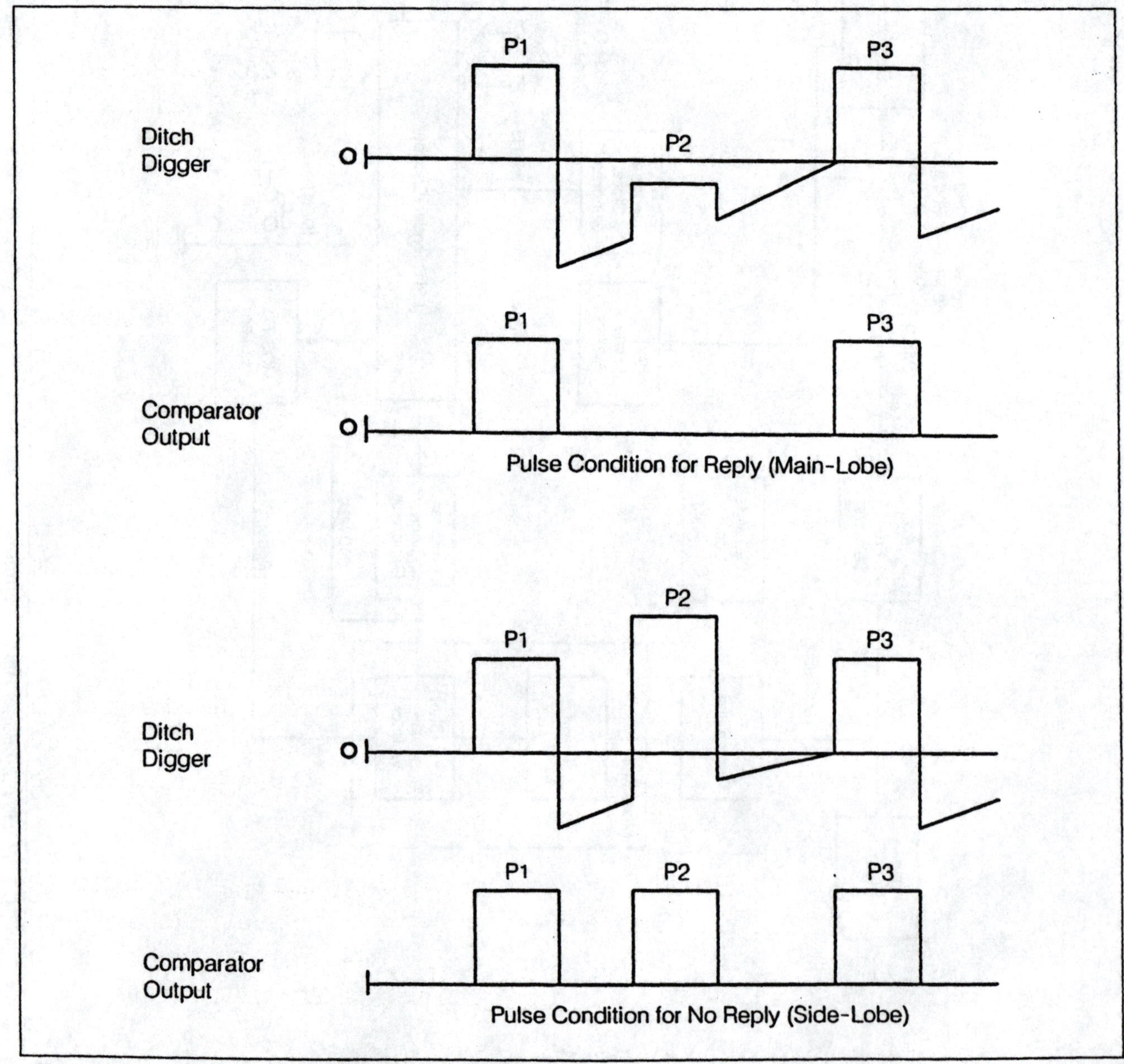

Figure 7-9. Ditch Digger and Comparator Pulses.

than the P1 pulse and will not appear in the comparator output. When interrogated by a side lobe, P2 will appear larger than PI and will be present in the output, as shown in Figure 7-9.

Whether an output of the comparator reaches the large-scale integration (LSI) decoder/encoder is also a function of the pulse width discriminator. The pulse width discriminator monitors the input to the ditch comparator and prevents an output from occurring for pulses less than 0.3 microseconds wide. Pulses 0.4 microsecond or wider will produce an output.

The LSI device performs the functions of both decoder and encoder. The decoder function determines validity and mode of the interrogation signal. If the interrogation pulse spacing identifies a mode, the LSI device will begin generating the encoder pulse train. If the pulse spacing does not conform to a particular mode, no further processing will occur within the LSI device.

Sidelobe interrogation is also recognized by the LSI device by detection of the P2 pulse. The P2 pulse will not be present at the input to the LSI device when interrogated by the main lobe; suppression of P2 under these conditions is the function of the ditch circuit. When a valid interrogation is received, the LSI device provides an output which initiates the generation of the reply pulse train. The reply code sequence is determined by the interrogation mode and the code set on the front panel controls or by the altitude digitizer outputs.

A decoder clock generator and an encoder clock generator are required to facilitate LSI device operation, as are inputs from the code selection signals. The decoder clock generator is crystal controlled and operates at 3.0 MHz. The encoder clock generator operates at 690 kHz and is actuated by the LSI device. The clock pulses from both generators are applied to shift registers within the LSI device that perform the decoding and encoding functions.

When an interrogation signal is determined valid, the LSI device enables the encoder clock generator that applies timing pulses to a one-shot multivibrator contained in the modulator section. The multivibrator drives the modulator that pulses the transmitter. The transmitter will be energized whenever a modulation pulse occurs. Reply pulse spacing is controlled by the encoder clock generator. The reply pulse pattern is controlled by the LSI encoder.

When triggered by the encoder clock generator and enabled by the encoder, the modulator will produce pulses required to drive the transmitter. When driven by the modulator, the transmitter will produce an RF reply pulse train at 1,090 MHz that is sent through the low-pass filter to the antenna. The transmitter is a cathode-pulsed, single-tube L-band oscillator.

For each reply, the LSI produces a 34-microsecond, positive-going pulse that is applied to the reply rate limiter. The reply rate limiter averages these pulses and causes a decrease in IF gain when the rate of reply exceeds 1,080 pulses per second. The 34-microsecond reply pulse is also processed and used to activate the lamp driver which turns on the reply lamp.

The SPIP (ident pulse) is initiated by depressing the IDENT button on the front panel which places the IDENT input to the decoder/encoder low (logic 0) for a 22-second duration determined by the time constant of a resistor across a capacitor forming a RC network. During the 22 seconds the ident pulse is being sent with replies, and the reply lamp on the front panel will remain steadily illuminated.

Selecting the test function disables the pulse width discriminator circuit, which establishes a clear path for noise present at the output of the ditch comparator to the LSI device. Distributed randomly throughout the noise present at the LSI input are pulses which will be interpreted by the LSI device as valid interrogation pulses. These pulses will be decoded, which in turn causes 34-microsecond reply pulses to be generated. When the Transponder is operational, the reply lamp will be steadily illuminated as long as the function selector is held in the spring-loaded test position.

The +14-volt DC primary power input is applied to a series regulator that develops +10-volt DC. The +5-volt DC is zener regulated from the +10-volt DC potential. The regulator also supplies the +10-volt DC to a DC-to-DC converter that derives 6.3-volt AC for the transmitter tube heater, 1350-volt DC for the transmitter anode, and -12-volt DC for use throughout the TDR-950 Transponder.

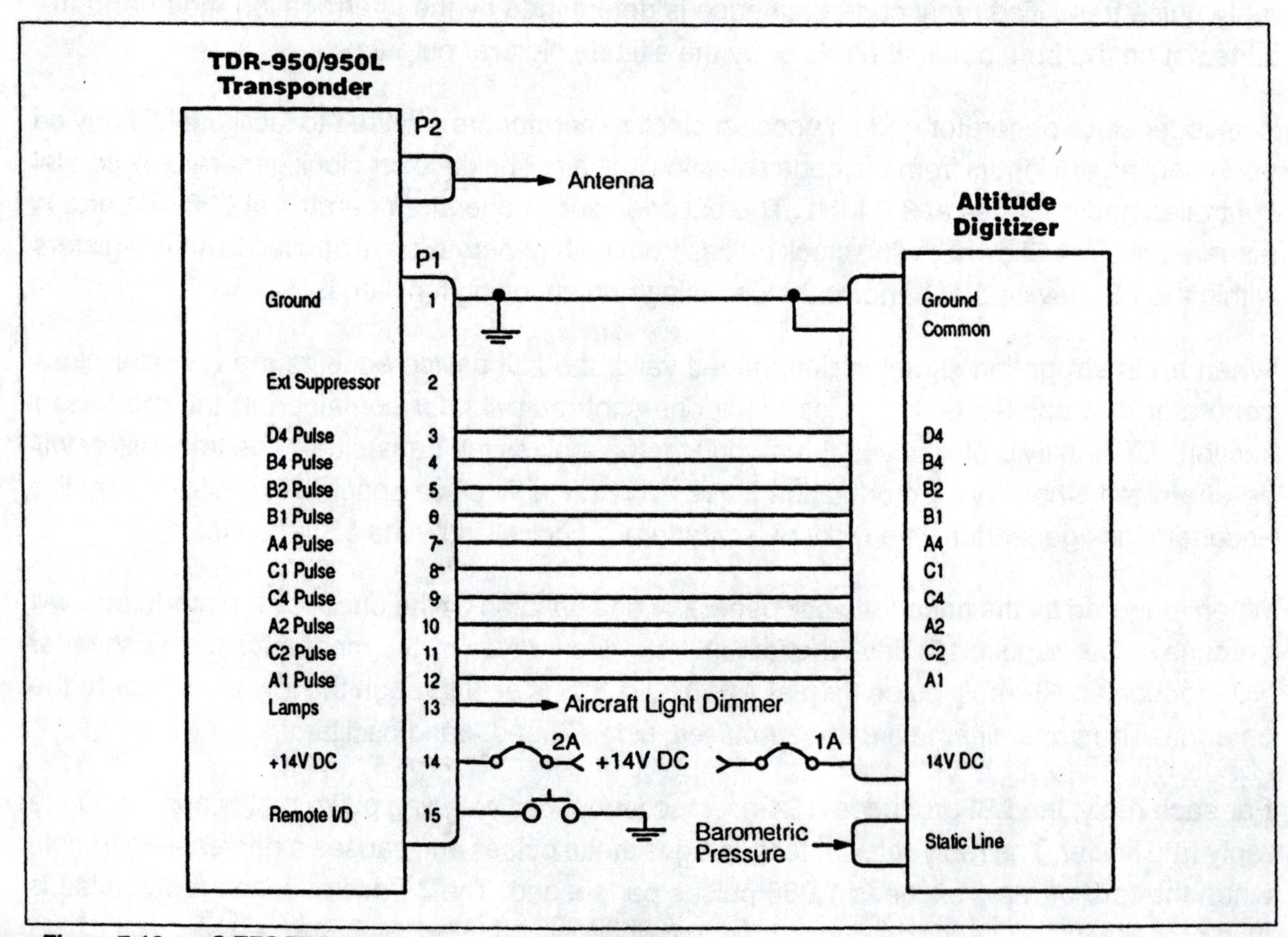

Figure 7-10. S-TEC TDR-950 Interconnect Wiring Diagram.

Figure 7-10 is an external interconnect wiring diagram of the TDR-950 Transponder and an atitude digitizer (or encoding altimeter) to provide altitude information for Mode C operation.

BENDIX/KING KT-76A/78A TRANSPONDER OPERATION

The King KT-76A/-78A is another example of a typical airborne Transponder. It is similar in operation to the S-TEC TDR-950. As shown in Figure 7-11, interrogation pulses from the SSR are received at the Transponder and are directed through the low-pass filter and duplexer to the 1,030-MHz bandpass filter and mixer. Here it is mixed with the local oscillator output and inserted in the IF amplifier. The LO frequency of 970 MHz, when mixed with the 1,030-MHz received signal, gives an IF frequency of 60 MHz.

The IF amplifier takes a signal input over a dynamic range of 50 dB and compresses the video output to a dynamic range of 15 dB. This is accomplished by successive detection of the amplified IF signal. A second input to the IF amplifier is the output from the Automatic Overload Control (AOC). This input line reduces the gain of the IF in an overload condition, so that only a specific portion of the interrogation received causes a reply. The detected video output from the IF is amplified in the video amplifier. The noise suppression circuit rejects any pulse with a width less than 0.3 microseconds.

The ditch digger and video switch circuits function together in comparing the amplitude of P1 and P2 to determine if the interrogation is from the main beam of the radar or a side lobe. The ditch digger produces a linearly decaying voltage behind P1 into which P2 falls. The characteristics of the ditch are such that if P2 is greater than or equal to P1, then the video switch is triggered by P2, but if P2 is less than P1, the video switch is not triggered by P2. The output of the video switch is a narrow pulse of about 150 microseconds. The output of the one-shot is a pulse of about 1 microsecond for input to the LSI chip.

The King Transponder LSI chip performs many of the logic and timing functions of the unit. These include the following:

1) The decoding of P1 to P3 pulse spacing for either a Mode A (8 microseconds) or Mode C (21 microseconds) reply.

2) Suppression of reply if P2 pulse is present.

3) The encoding of the serial pulse train output for either mode.

4) Insertion of the SPIP pulse in the serial output. When the IDENT button is pushed, the SPIP pulse is included in all replies for an internally timed period of 23 seconds.

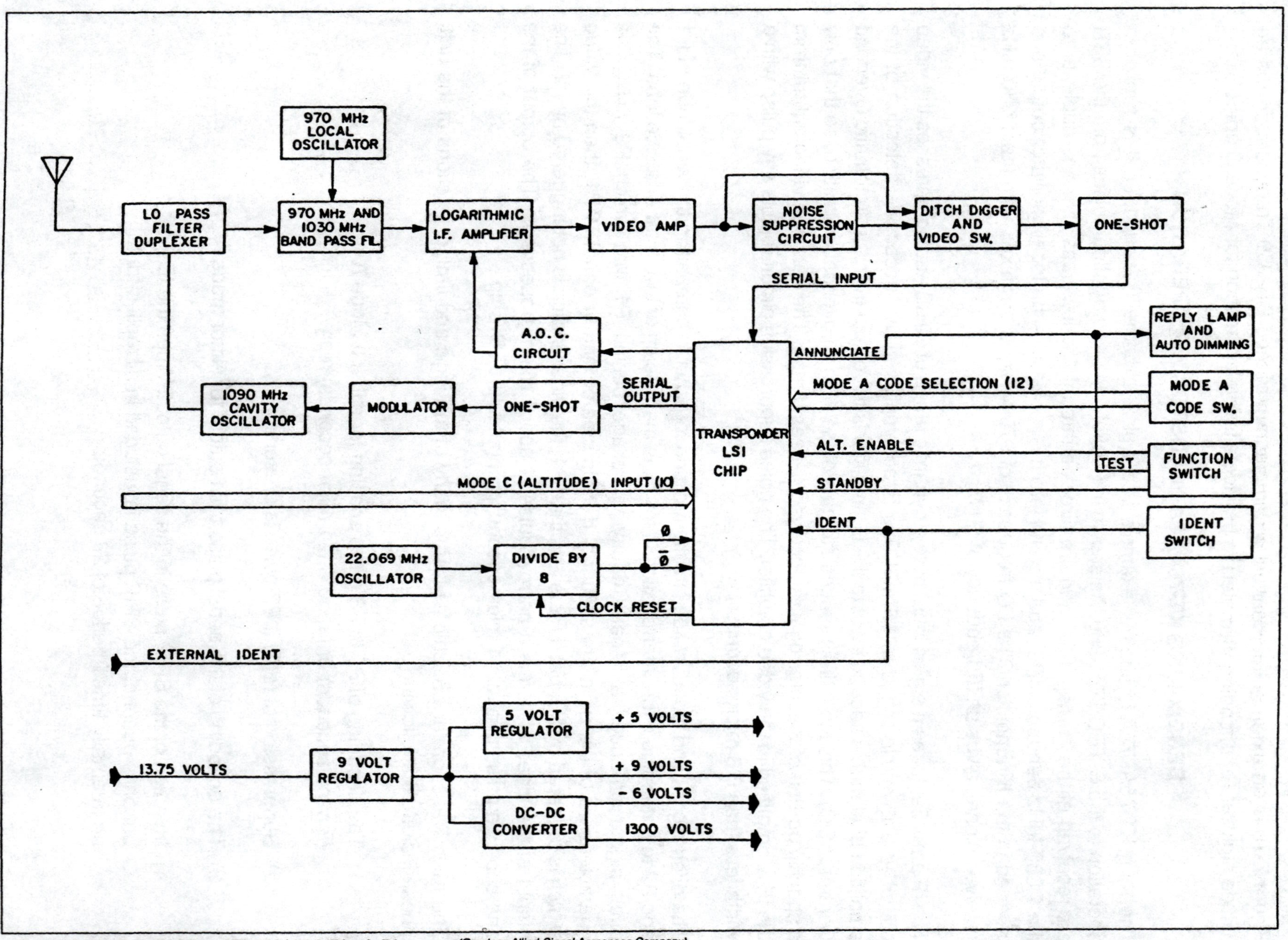

Figure 7-11. Bendix/King KT-76A/-78A Block Diagram. *(Courtesy Allied-Signal Aerospace Company)*

5) When power is first applied, all replies are suppressed for a period of 47 seconds. This turn-on delay also occurs whenever power is interrupted for more than 5 microseconds.

6) The annunciate output drives the REPLY light. This output goes high for 760 milliseconds whenever the Transponder sends a reply. Additionally, it goes high continuously for the 23-second period following a low on the ident input.

7) Mode C replies consist of framing pulses only when ALT ENABLE is low.

8) When the STBY pin is low, the LSI chip ignores the serial input line.

The LSI chip timing is derived from a 22.069-MHz crystal oscillator. This frequency is divided by eight to provide a two-phase clock at 2.758625 MHz. This divider is reset following an interrogation to synchronize the reply with the interrogation. The external suppression output of the LSI chip to the AOC circuit is a pulse 35 microseconds wide during each reply. These pulses are averaged and used to sense an overload condition. The receiver is desensitized to limit the reply rate to approximately 1,200 replies per second.

The Reply Lamp Driver switches the reply lamp on and off for each time the ANNUNCIATE line from the LSI chip goes high. A cockpit light level sensor automatically dims the reply lamp for lower cockpit lighting levels. Additionally, placing the function switch in the TEST position turns on the REPLY light.

The modulator takes the coded serial data and switches the transmitter cavity on and off accordingly. The oscillator RF output passes through the low-pass filter to the antenna where the power is radiated. The purpose of the low-pass filter is to attenuate higher harmonics of the 1,090-MHz transmission.

TRAFFIC ALERT AND COLLISION AVOIDANCE SYSTEMS (TCAS)

Traffic Alert and Collision Avoidance System were recently mandated by the FAA for commercial transport aircraft and are now in widespread use. There are two types of TCAS available, TCAS-I for 10 to 30 passenger aircraft, and TCAS-II for larger transports. TCAS-I displays the approximate bearing and relative altitude of all aircraft (typically out to 40 nautical miles), and indicates which are potential threats, or traffic advisories (TA), by the color used on the EFIS or multi-function display. In addition, TCAS-II computes and displays the optimum flight path strategy, or resolution advisory (RA), for the pilot to avoid a collision, whereas TCAS-I does not. If both aircraft are equipped with TCAS-II, the first to determine its RA transmits its intended actions to the other via Mode-S Transponders that are used with the TCAS-II systems. TCAS-II systems, although much more capable than TCAS-I, sell for approximately $125,000 compared to only $50,000 for a TCAS-I system. Following is a brief description of a typical TCAS-II system.

ROCKWELL COLLINS TCAS-II WITH MODE-S TRANSPONDER SYSTEM

TCAS-II operates as an airborne surveillance radar system interrogating the ATC Transponders of nearby aircraft. Range, relative bearing, and altitude (if Mode S/C-equipped) of the nearby aircraft are measured by the TCAS-II system. The system provides both aural and visual warnings if the TCAS-II computer predicts a potential collision. The TCAS-II system provides no indication of traffic conflicts with aircraft not equipped with transponders. TCAS-II computes two types of advisories, Resolution Advisory (RA) and Traffic Advisory (TA). An RA is the form of a commanded vertical maneuver to achieve safe separation of the aircraft involved. A TA indicates the relative position of aircraft in close proximity to assist the flight crew in visual acquisition.

The FAA has recently issued an Airworthiness Directive (AD) that changes the TCAS logic to prevent displacements against legally separated aircraft. TCAS systems that do not incorporate this change will continue to experience unnecessary resolution advisories when two aircraft are on a final approach to parallel runways. The AD note requires that the RA logic calculate the range (in feet) divided by range rate (in feet per second) to determine if the two aircraft are indeed on a collision course. If the result is more than 20 seconds but less than or equal to 35 seconds, the TCAS will output a TA; however, if the result is 20 or less seconds, the system will output an RA.

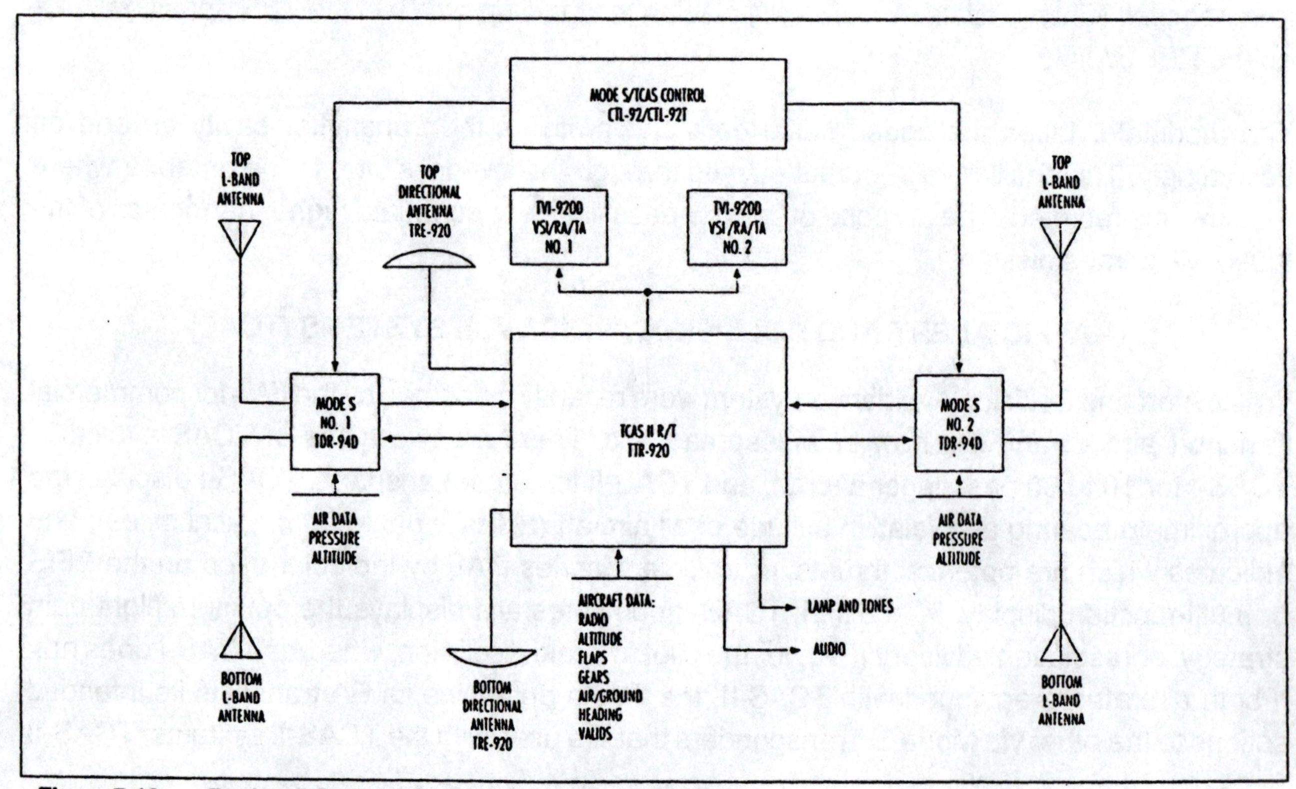

Figure 7-12. Rockwell Collins TCAS-II System Block Diagram.

A TCAS system block diagram is shown in Figure 7-12. The TTR-920 Receiver/Transmitter unit is mounted in the equipment rack. Two TRE-920 directional antennas are mounted on the aircraft, one top-mounted and one bottom-mounted. The directional antenna allows the TCAS-II system to determine the bearing of other transponder-equipped aircraft within, but not limited to, a 14-nautical mile radius.

Interfaces for two Mode-S Transponders are provided, although only one is used at a time. The second Mode-S is used for system redundancy. The Mode-S Transponders require antenna diversity when used in TCAS-II installations. This provides coordination of maneuvers to avoid a collision between two TCAS-II-equipped aircraft through the data link feature of the Mode-S Transponders. The top and bottom Mode-S antennas help ensure that the exchange of coordination data is not "shadowed" by the airframes of the two aircraft, whether the first aircraft is above or below the second aircraft.

The control of the TCAS-II system is accomplished through the Mode-S Transponder. The control function not only provides control of two Mode-S Transponders, but provides manual selection of required TCAS modes, as well as control of display modes and range where available.

The TDR-94D Mode-S Transponder operates in Mode-A, Mode-C and Mode-S. Whether the Transponder replies in Mode A/C or Mode-S is determined by how it is interrogated by the ground radar station. When interrogated in Mode A/C, the response code is the code set on the control head or provided by the altitude digitizer. The Mode-S code is set by the installer by installing jumpers on the rear connector (straps) on the 24 strapping pins. The strapping is based on the aircraft's "tail" number.

There are a variety of methods of displaying traffic information and RA information in the cockpit. There is a set of basic requirements defined in FAA Advisory Circular 20-131 that must be adhered to when implementing these cockpit functions.

There are two basic requirements:

1) Both pilot and copilot must have visual access to a traffic display. This may be implemented through a centralized traffic display or individual traffic displays (pilot and copilot).

2) Both pilot and copilot must have dedicated RA displays, as shown later in Chapter Ten.

CONCLUSION

In the previous chapter, we saw how the DME interrogates a ground station and uses reply pulses to calculate distance from the station. Whereas, the Transponder responds to ATC ground station interrogations with reply pulses to aid the controller in identifying and directing air traffic. We learned that the ground-based ATC radar consists of both a primary and secondary surveillance radar, and that the SSR is used to transmit interrogations, and receive replies while the PSR transmits a beam of energy that is reflected back and processed to provide a pip on the PPI.

Corporate and commercial air transport aircraft also carry onboard a primary surveillance radar system designed to locate adverse weather conditions and for mapping the terrain. The operation and maintenance of the Airborne Weather Radar System is discussed in the following chapter.

REVIEW QUESTIONS

1. What is the purpose of the ATC radar surveillance system?
2. Compare the differences in operation between the primary and secondary surveillance radar systems.
3. What are the advantages and disadvantages of the PSR?
4. Define the significance of the various symbols displayed on the controller's PPI.
5. What is the frequency of the interrogation and reply signals received and transmitted by the ATCRBS Transponder?
6. What information does the ATC receive from the Transponder when interrogating in Modes A and C?
7. How does the Transponder differentiate between Mode A and Mode C interrogations?
8. What is the purpose of the P2 interrogation pulse?
9. Describe the operation of the SLS interrogation method at the ground station and the interpretation of these pulses in the airborne Transponder.
10. Explain the operation of the device used for encoding Mode C replies.
11. What is the spacing between the reply framing pulses?
12. What reply pulses would be sent in response to a Mode A interrogation if the code selector switches are set to 3476?
13. What is the purpose of the SPIP and what action is required to initiate it?
14. Where does the ident pulse appear in the pulse train and what is its duration?
15. What reply pulses are sent if the Transponder control is selected to receive Mode A interrogations and instead it receives Mode C interrogations, or if it is selected to Mode C and does not have an altitude encoder input?
16. What happens within a ATC Transponder when the test function is selected?
17. When does the reply lamp illuminate steady and when does it blink on and off?
18. Describe the two methods of IF gain control in the S-TEC TDR-950 Transponder.
19. What is the function of the ditch digger and comparator circuits in the TDR-950?
20. Define the two types of advisories provided by the TCAS-II system. Which one of these advisories is not available to TCAS-I users?

REVIEW QUESTIONS

1. [illegible]
2. [illegible] system?
3. [illegible]
4. [illegible]
5. [illegible]
6. [illegible]
7. How does the [illegible] between [illegible] interrogations?
8. What is the [illegible]
9. [illegible]
10. Explain [illegible]
11. What is the [illegible]
12. [illegible]
13. What is the [illegible]
14. [illegible]
15. What [illegible]
16. [illegible]
17. [illegible]
18. Describe the [illegible]
19. [illegible]
20. [illegible]

Chapter 8

Weather Radar Systems

INTRODUCTION

As discussed in the previous chapter, the ground-based ATC radar surveillance system provides the ATC center with information concerning the location of aircraft within the control area. Radar systems are also commonly found onboard aircraft to warn the pilot of impending storm conditions along the flight path. A secondary function of the airborne radar is ground-mapping, in which the system is used to "map" the terrain beneath the aircraft for navigation purposes. More recently, Weather Radar Systems have been adapted to provide turbulence/windshear detection as well. Windshear is a suddent shift in wind speed and direction due to a rapid downward rush of cooled air. A pilot flying through windshear at low altitude first would experience a strong headwind and increased lift, then a tailwind and a sharp drop in lift. Airborne radar systems used on civil aircraft are termed Weather Radars since their primary function is weather detection and avoidance.

Both ground and airborne radar systems operate on the echo principle in which high-energy RF pulses are directed in a beam toward a reflecting target. The time interval required between the transmitted pulse and the received echo is computed and displayed in terms of distance and relative position of the target. Since most forms of turbulence are accompanied by precipitation which reflects radio energy, the weather radar can "map" the precipitation intensity pattern on the cockpit CRT radar indicator to allow the pilot to cross storm fronts at points of least turbulence, as shown in Figure 8-1. By tilting the radar antenna scan down, the indicator will outline principal topographical features, such as cities, lakes, rivers and coastlines, and will provide terrain contour display in mountainous areas. Airborne radar systems are designed only for weather detection and ground mapping functions, and therefore, should not be used for proximity warning of other aircraft or for collision avoidance.

WEATHER RADAR (WX) SYSTEM DESCRIPTION

A typical weather radar system consists of an antenna, receivertransmitter (RT), and control unit/indicator. Most weather radar systems operate in either the 5,200-MHz to 5,900-MHz (C band) or 8,500-MHz to 10,000-MHz (X band) frequency range. RF energy is coupled from the RT to the antenna by means of a hollow rectangular pipe called a waveguide.

The radar antenna is usually mounted in the nose of the aircraft and is protected by a fiberglass or plastic fairing, known as a radome. The rotatable radiating element of the antenna may consist of either a lead element with a parabolic reflector, or a flat-plate phased array which is electronically scanned to produce a highly-focused beam. Usually, the antenna scan pattern is

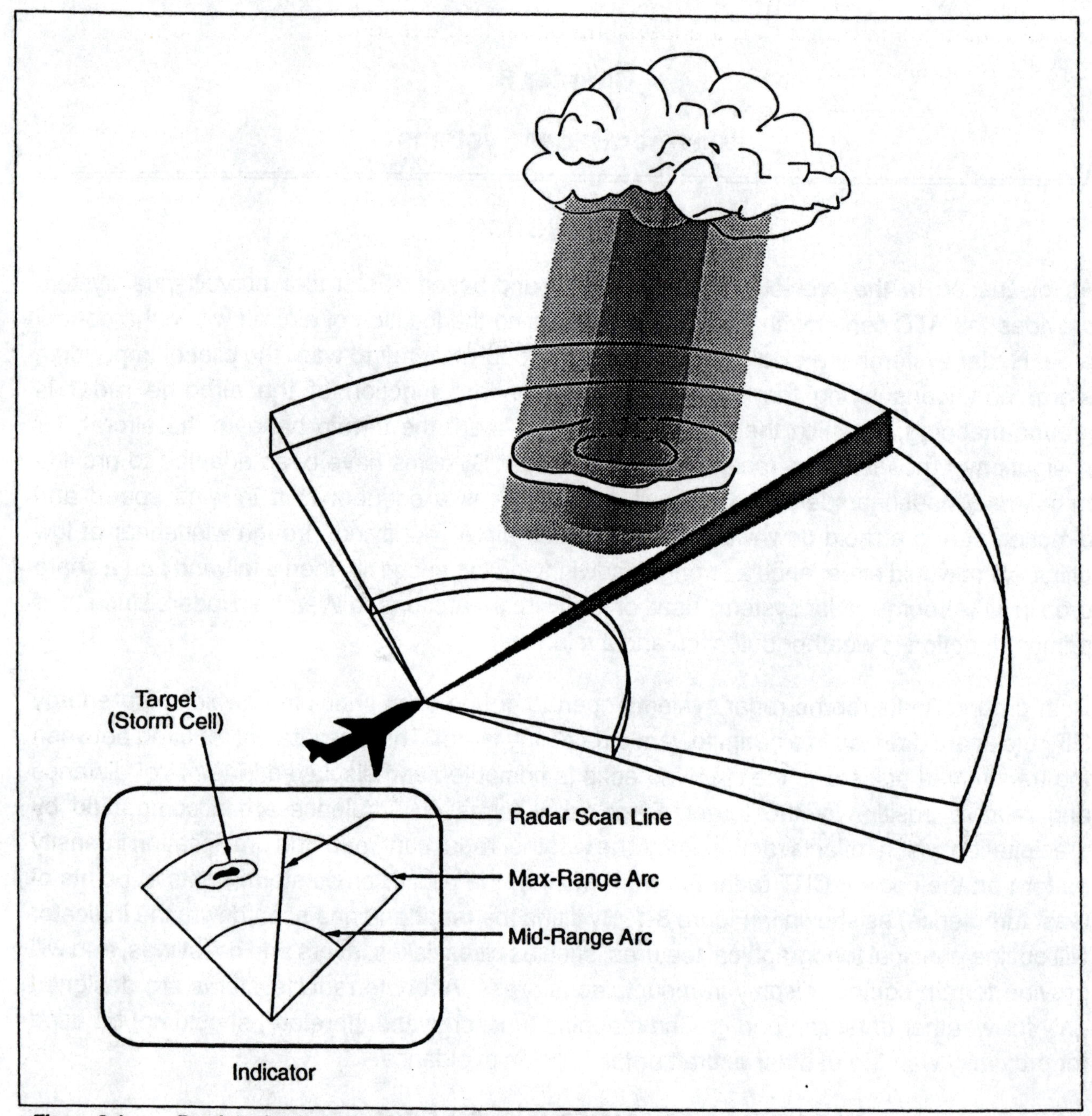

Figure 8-1. Precipitation Presentation on Weather Radar CRT.

corrected or stabilized by signal inputs received from the vertical gyroscope (VG) reference system to enable it to scan continuously in the same horizontal plane regardless of the pitch or roll of the aircraft. The operation of the VG is discussed in Chapter Eleven.

The cockpit control unit, which usually is located directly on the CRT indicator, provides mode selection for weather avoidance or ground mapping functions, target range selection (usually up to 300 nautical miles), antenna elevation tilt control, receiver sensitivity gain, and CRT brightness controls. A contour mode is also provided so that the most intense portion of the display, such as the center of a thunderstorm or peak of a mountain, will show up as a dark spot surrounded by light areas.

If the indicator has a color CRT, the intensity level, which corresponds to the amount of reflected energy received, will be identified with a color code. Also, by incorporating digital processing circuitry in the color indicator, enroute navigational information may be presented to provide a moving-map display of navigational routes that can be superimposed upon weather and terrain mapping information.

ANALOG VERSUS DIGITAL RADAR SYSTEMS

The first airborne radar system was developed prior to World War II for military applications to detect enemy aircraft and surface vessels. Later, airborne radar principles were applied to civil aviation for the purpose of detecting adverse weather conditions. The original weather radar system was based on an analog design concept. However, with advances in solid-state microprocessor technology, the digital radar system has emerged which offers several distinct advantages over the analog design.

The basic operating principles of the analog and digital design differ primarily in the manner in which the information is processed and displayed. In both concepts, the transmitter generates high-energy RF pulses that are radiated by the antenna. The moisture (rain drops and ice) present in clouds reflects some of this energy back to the antenna. It is the function of the radar receiver to detect the reflected energy and provide a measure of intensity discrimination. The radar system measures the time lag between the transmitted pulse and the received reflection. Because the propagation velocity of the RF energy is known, this time lag is a direct measure of target range. For example, it takes 12.359 microseconds for a radar signal to travel one nautical mile and return. By synchronizing the antenna sweep with the display sweep, the pilot is presented with a radar picture of the target in terms of intensity, range, and azimuth.

In order to provide a reasonably bright and useable display, the higherquality analog systems use a CRT indicator with a high-persistence phosphor. This has the characteristic of storing the target information on the face of the CRT for a time slightly less than one sweep period. The detected target information is applied to the vertical deflection circuit, and a sweep signal is applied to the horizontal deflection circuit. The horizontal sweep circuit in the indicator, which consists of a sawtooth oscillator, controls the time required for the electron beam to sweep across the face of the CRT. The sweep then “paints” the targets to be displayed.

The digital radar differs in that target information is first processed and then stored electronically in a memory circuit. The display function then becomes a process of scanning these memory circuits and writing the information on the CRT. Because the process of scanning the memory circuits is independent of the memory data input, the rate at which data is moved to the CRT is much higher than the input rate. This high rate of data repetition achieves a bright display

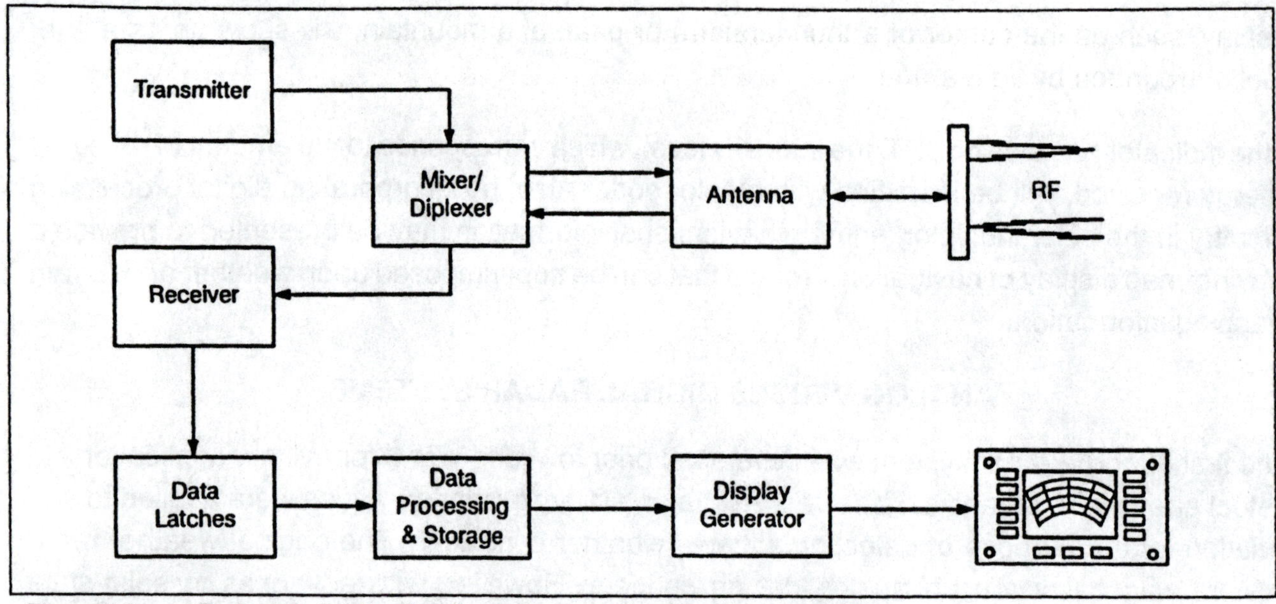

Figure 8-2. ***Typical Digital Weather Radar System Simplified Block Diagram.***

without the need for a high persistence CRT. A block diagram of a typical digital weather radar system is shown in Figure 8-2.

A more recent refinement of the digital radar concept involves the use of a color CRT. This concept uses the many advantages of the digital technique but, in addition, introduces color to differentiate between a number of target intensities. A typical color radar display, as shown in Figure 8-3, indicates the center of the storm with a red color surrounded by a yellow pattern to indicate less severe rainfall. The outer fringes of the storm front will be outlined in green. A similar arrangement is used in the ground-mapping mode to indicate the elevation of various terrain features. Color radar indicators are also sometimes employed as multifunction displays to provide additional features, such as an electronic checklist or a moving-map enroute navigation display.

As previously illustrated in Figure 8-1, the transmitted radar beam is relatively narrow and illuminates an area that depends on the aim or tilt elevation of the antenna and the distance between the aircraft and the target. Factors that affect the transmitted radar signal strength at the target include the transmitter output power, beam-width, propagation attenuation, and radome transmissivity.

WEATHER RADAR THEORY

Obviously, the more peak RF power that can be applied from the radar transmitter to a given target, such as a storm cell or mountain peak, the better will be the reflection back to the radar receiver. However, the output power of modern radar systems has been significantly decreased compared to earlier systems. For example, up until a few years ago, a typical Weather Radar System had a power output of several hundred kilowatts. Today, due to increased receiver sensitivity and digital signal processing, a typical radar system will radiate less than 100 watts.

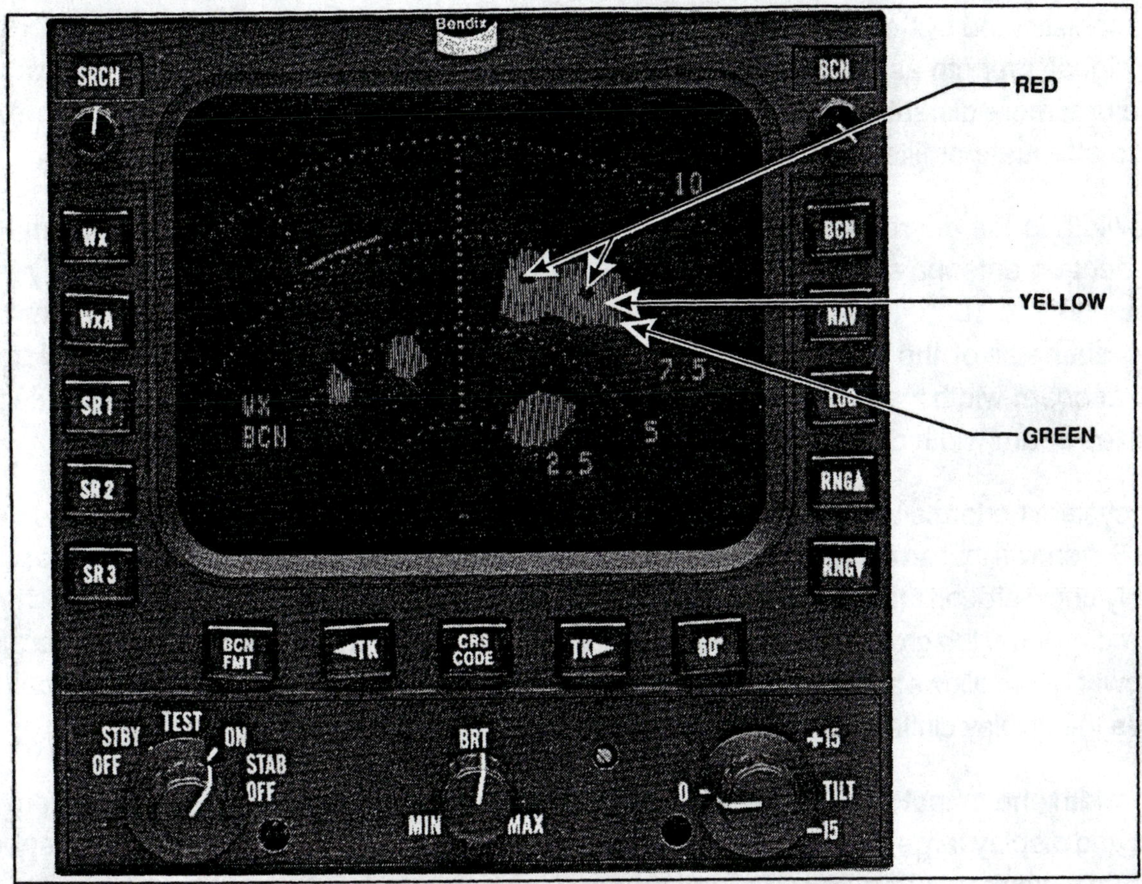

Figure 8-3. ***Bendix/King RDR-1400C Color Radar Indicator.***
(Courtesy of Allied-Signal Aerospace Company)

> ***WARNING***
>
> ***Regardless of the power output of the Radar System, care should be taken not to stand within 20 feet of the radiating antenna when transmitting. Excessive exposure to microwave energy can be lethal!***

A transmitter's average power output is a function of the pulse width and pulse repetition frequency (PRF). A narrow pulse width allows the radar to define targets in greater detail, thus resulting in better resolution on the radar indicator. PRF is adjusted according to the desired range to allow enough time for the radiated beam to strike the target and be reflected back to the antenna before another pulse is sent.

Since returns from nearby targets appear much brighter on the indicator than distant targets, radar receivers use a Sensitivity Time Control (STC). The STC eliminates the effect of range on target intensity by increasing the gain for distant returns, much like the operation of an AGC.

More sophisticated systems employ path attenuation correction as well to maintain the proper return signal strength as a function of both range and intervening rainfall attenuation. Without this feature, more distant storm cells will appear less intense only because the closer storm cell tends to attenuate or filter out the distant return.

Beam-width is the degree to which the transmitted energy is concentrated into a beam. It is dependent on antenna size and is specified by the angle which encloses the RF energy level between 50% to 100% of the peak power level of the beam. The larger the flat-plate antenna (or the diameter of the reflector on a parabolic antenna), the narrower the beam-width. A narrower beam-width improves the ability of the radar system to detect targets at longer ranges, minimizes beam-width distortion, and improves display resolution.

Radar system performance is also improved when a larger antenna (narrow beam-width) is used. Figure 8-4 shows that an aircraft flying at 40,000 feet with a 12-inch antenna and a 0° tilt angle would probably show ground targets at less than 100 nautical miles. However, with an 18-inch antenna at the same altitude, the ground return would not appear until well over 100 nautical miles. Therefore, a narrower beam allows a more direct observation at the lower and mid-sections of a storm cell and reduces the display clutter caused by ground targets.

Beam-width and output power determine the effective range that a Weather Radar System can detect and display targets. However, as the transmitted beam radiates through the atmosphere, it gradually weakens. This loss of signal strength is called propagation attenuation and is due primarily to distance and intervening atmospheric conditions.

Radar detection distance is specified in terms of "avoidance range", which is the distance required to detect returns from a storm that contains approximately 1/2 inch of rainfall per hour. A typical modern Weather Radar System with 24-watts of average power and a 12-inch flat-plate antenna will have an avoidance range of up to 340 nautical miles.

Although not part of the Weather Radar System, the radome contributes significantly to overall system performance. A poorly designed, constructed, repaired, or maintained radome can blind the radar antenna, resulting in poor transmission and reception of signals. There are many instances where the crew will report the radar as being inoperable or weak when the problem is actually the radome.

PRINCIPLES OF WEATHER RADAR SYSTEM OPERATION

The operation of the weather radar system is primarily controlled and synchronized by the clock circuits. The synchronizer provides the timing signals for the transmitter pulse and coordinates the antenna scan motion with the sweep trace on the CRT indicator. These timing signals are generated at different frequencies by dividing down the output from a master crystal oscillator.

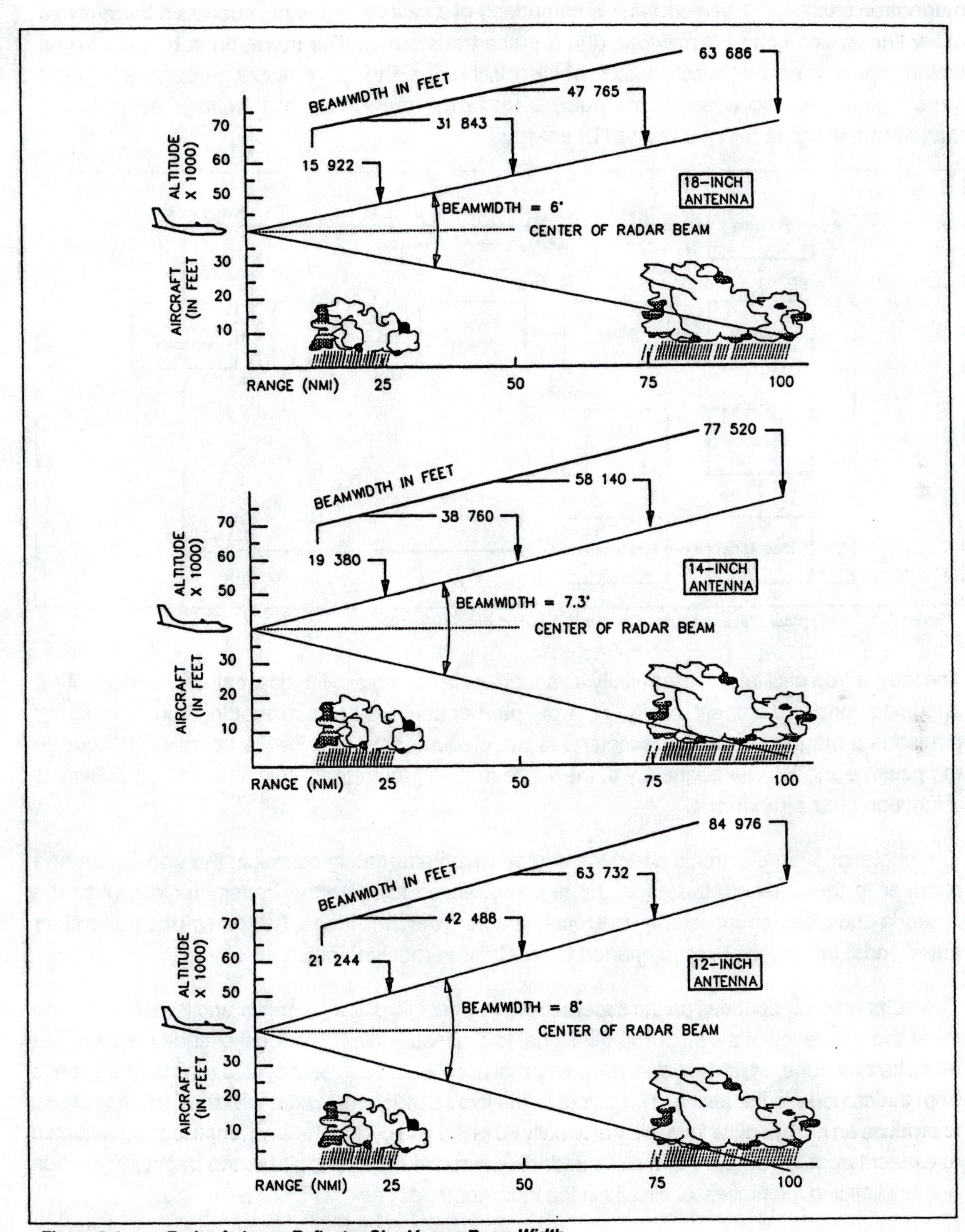

Figure 8-4. Radar Antenna Reflector Size Versus Beam Width.

As shown in Figure 8-5, the transmitter portion of the RT consists of a modulator and a magnetron oscillator. The modulator is comprised of a switching device, such as a thyratron; a delay line, which is an LC network; and a pulse transformer. The thyratron is triggered by a timing signal from the synchronizer which discharges the LC network through the pulse transformer. The secondary of the pulse transformer provides a high-voltage pulse to the magnetron, which in turn generates RF energy.

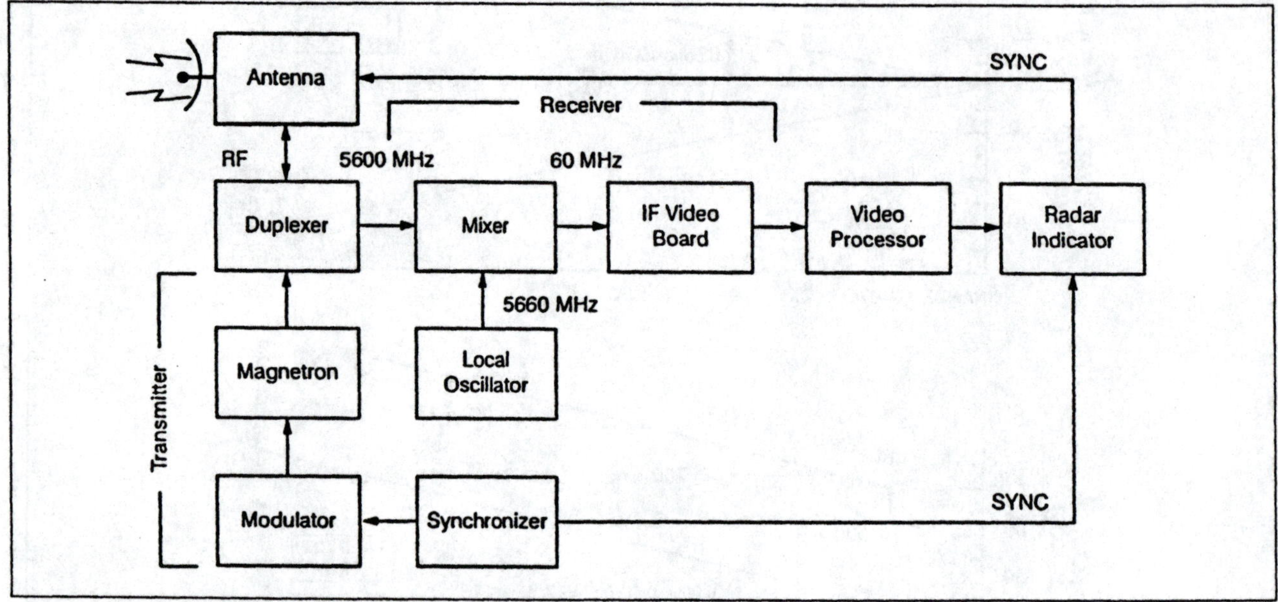

Figure 8-5. Typical Weather Radar System Expanded Block Diagram.

The magnetron oscillator is essentially a vacuum tube having a cylindrical cathode and a slotted anode to form a resonant cavity. A large permanent magnet surrounding the magnetron furnishes a magnetic field that accelerates the electron stream from the negative cathode to the positive anode. The frequency of the wave-motion generated within this resonant cavity is a function of its dimensions.

The duplexer is an electronic switching device which alternately connects the transmitter and receiver to the antenna. It detects the signal direction to route the transmitter output to the antenna and block it from entering the receiver during transmit mode. As soon as the transmitter pulse ends, the receiver line is opened to receive the reflected pulse.

The radar receiver operates on the superhetrodyne principle utilizing a mixer and local oscillator to lower the frequency of the incoming RF signal to a useable level. The local oscillator is usually a reflex klystron tube, which is a resonant cavity that accelerates a stream of electrons through space from the cathode to the anode. The output of the local oscillator is mixed with the incoming signal to produce an intermediate frequency of usually 60 MHz. The IF signal is then amplified and detected to extract target information. In a digital radar system, the receiver digitizes the target information and applies it to the processor circuits in the indicator (refer back to Figure 8-2).

In the processor, the data latches serve as short-term data storage locations until the data can be processed and shifted into the memory circuits. The scan circuits control the position of the write beam in the CRT. The data from the memory circuits is written on the CRT by essentially enabling one or more of the color guns in the color CRT, or by modulating the intensity of the write beam if using a monochrome CRT indicator.

ROCKWELL COLLINS WXR-300 WEATHER RADAR SYSTEM OPERATION

A typical first generation Weather Radar System, such as the Rockwell Collins WXR-300 consists of a WXT-250A RT, IND-300 Indicator, and an ANT-310/312/318 Antenna. The WXT-250A RT generates 5 kW RF pulses at a frequency of 9,345 MHz. The pulse length is a function of the range and selected mode. For close-in targets, the pulse length is 1.0 microsecond; for long range operation it is 5.5 microseconds. The sweep rate of the antenna causes one pulse to be radiated for every 0.234° of azimuth rotation. It takes the antenna about 4.5 seconds to complete one sweep of 120° (60° from either side of the longitudinal center-line of the aircraft).

The transmitter section of the RT, consisting of a modulator, magnetron, TR limiter, and duplexer, generates RF pulses at the rate of 120 times per second. These pulses are waveguide-coupled to the antenna, which radiates the RF pulses in a narrow pencil-like beam. When the beam strikes a target (rain or ground features), a portion of the energy in the beam is reflected back to the antenna. This returning energy is directed back through the waveguide to the receiver.

The function of the modulator and modulation transformer is to develop the high potential pulse to drive the magnetron. The modulator oscillator, operating through a pulse transformer, charges a bank of 11 capacitors. When the charge reaches about -330 volts, the reference sensor circuit disables the oscillator. The modulator trigger input causes the capacitors to discharge through the primary of the pulse transformer. The resulting induced voltage on the secondary is the 4,500-volt magnetron drive pulse. The duration of the transmitted pulse is determined by the number of capacitors that are charged during the charge cycle.

The magnetron, located on the lower left side of the mixer/duplexer shown in Figure 8-6, is responsible for generating high-energy RF pulses that are radiated by the antenna. The magnetron operates at a frequency of 9,345 MHz. When the 4,500-volt modulation transformer pulse is applied to the anode of the magnetron, a 5-kW, 9,345-MHz pulse is applied to the duplexer. The pulse width can be either 5.5 or 1.0 microsecond, depending on range and selected mode. The duplexer directs the energy to the antenna during transmit and the reflected signal from the antenna to the receiver during receive. The directed process is accomplished by circulators HY1 and HY2. The TR limiter is a gas-filled cavity located between the mixer and duplexer. The gas in the TR limiter ionizes in the presence of high RF energy. This provides a short across the TR limiter opening so that the high-level RF energy is blocked from the receiver.

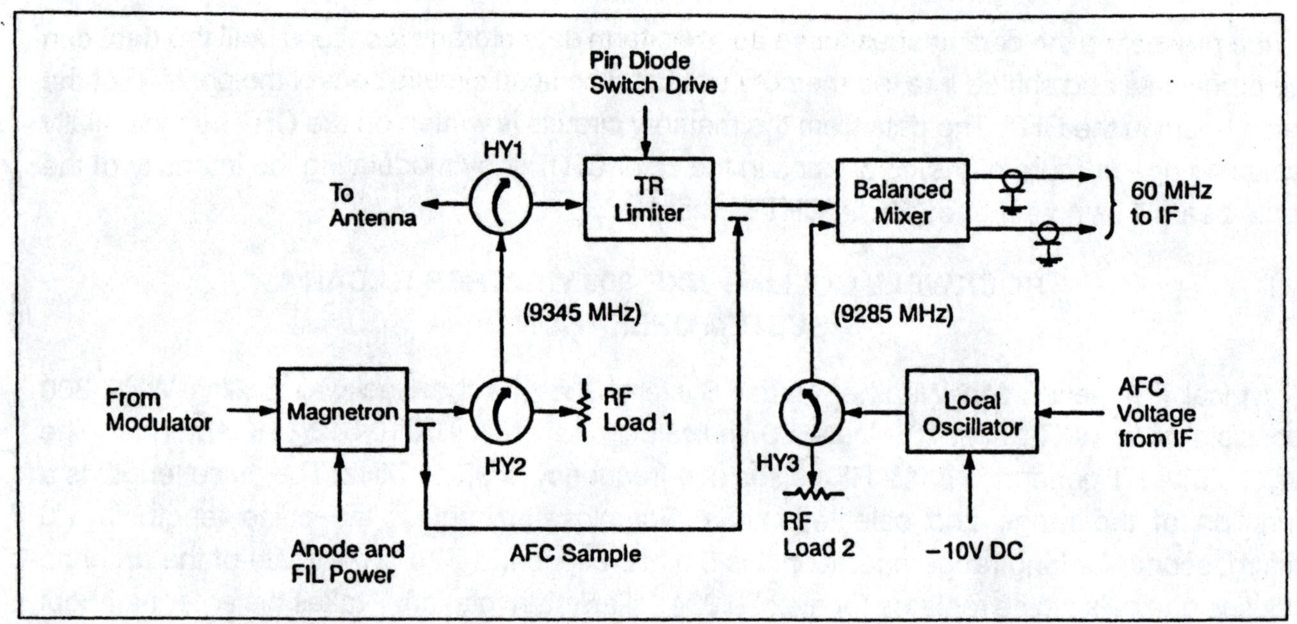

Figure 8-6. Collins WXT-250A RT Mixer/Duplexer Block Diagram.

The pin diode switch is enabled during transmit to expand the isolation period thereby ensuring good isolation of the receiver from the transmit energy.

The receiver consists of a mixer, local oscillator (LO), and an intermediate frequency (IF) amplifier. In the mixer, the received signal, which is at a frequency of 9,345 MHz, is mixed with the LO signal at 9,285 MHz to produce the 60-MHz IF signal. In the IF circuits, the 60-MHz signal, which contains the reflected target signal, is detected, amplified, and converted into three video output signals. These video signals represent three levels of return signal strength and are used in the indicator to show relative storm intensity.

When the RF pulse strikes a target, a signal is reflected to the antenna. The signal (echo) is applied to the duplexer, and circulator HY1 directs the return signal through the TR limiter to the mixer.

Because the return signal is time-delayed and space-attenuated, it sufficiently follows pin diode blocking and lacks the intensity necessary to ionize the TR limiter gas. Therefore, it is free to pass through the limiter to the mixer. In the mixer, it is mixed with the LO signal to produce the IF signal.

The primary function of the IF circuits is to detect and extract the radar target information from the IF signal and process that data into a form suitable for use in the indicator. In order to isolate the target information from the rather broad spectrum of signals that can be present in the IF, it is necessary that the IF circuits be tuned to a very narrow band of frequencies. The IF in the WXT-250A is tuned to provide peak response at 60 MHz.

The 60-MHz mixer output is applied to the 60-MHz preamplifier. The preamplifier is a double-stage amplifier with each stage tuned for maximum response at 60 MHz. The output from the preamplifier is applied to the STC amplifier and the Automatic Frequency Control (AFC) mixer. The return signal is amplifier in the STC amplifier and applied to the second mixer. Here the 60-MHz IF is mixed with 50.4 MHz from the second LO to produce the second IF of 9.6 MHz. The return signal, now at 9.6 MHz, is applied to the first and second IF amplifiers that are controlled by the AGC circuits. From the amplifiers, the signal is applied to the video detector. The video detector extracts the video information from the signal and supplies an output in the form of pulses which represent the target return signals. These pulses are applied to the three video level processors, in the Collins IND-300 Indicator, which are amplitude sensitive.

The purpose of the Collins IND-300 Indicator, shown in Figure 8-7, is to present to the pilot a radar picture of the weather conditions in the flight path. The presentation must allow him or her to readily determine as safe and comfortable a flight path that is consistent with his mission objectives.

The CRT electron beam deflection technique is quite conventional and familiar to those with a reasonable knowledge of color television operation. The data is written on the screen from top to bottom and from left to right, in an X-Y deflection pattern, and by energizing one or more of the CRT red, green, or blue color guns.

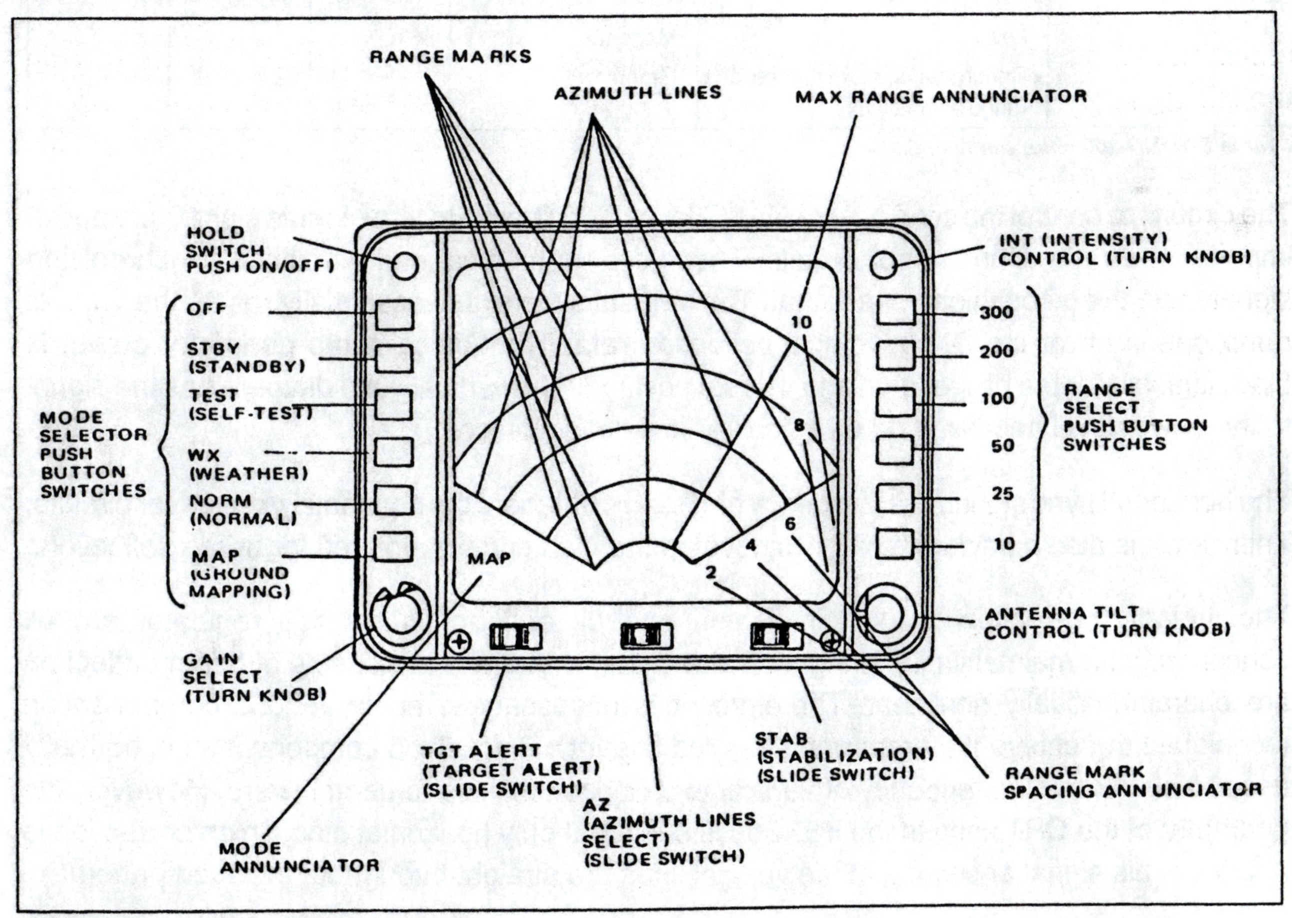

Figure 8-7. *Collins IND-300 Indicator.*

Horizontally, the data bits are written at a frequency of about 16 kHz, while the vertical scan frequency is about 60 Hz. The particular combination of colors for any given display item is determined by the mode selected. Table 8-1 illustrates the color combinations.

Level-1 video is the first level of return signal strength and represents a weather radar target of minimum intensity. Level-2 is the next higher level, and level-3 is the most intense level. A moisture density, capable of reflecting the radar signal at a level-3 intensity, is considered dangerous to flight and should be avoided. This level is displayed in red for all weather modes, and in WX mode, it is cycled between red and black to further highlight its identity. Each horizontal line of displayed data consists of 256 bits. Each vertical column of data consists of 512 bits (one from each of 512 lines).

Mode	Video	Display Color	Color Guns Blue	Green	Red
Weather Test, and Normal	Level 1	Green		X	
	Level 2	Yellow		X	X
	Level 3	Red			X
	Range marks, azimuth lines, and range annuciators	Cyan	X	X	
Ground Mapping	Level 1	Cyan	X	X	
	Level 2	Yellow		X	X
	Level 3	Magenta	X		X
	Range marks, azimuth lines, and range annunciators	Green		X	

Table 8-1. IND-300 color combinations.

The circuits to control the scan are shown in Figure 8-8. There are three inputs signals of special importance to the scan control circuits. These are the vertical and horizontal synchronizing signals and the pincushion drive signal. The vertical sync signal essentially resets the vertical ramp generator at the 60-Hz vertical deflection rate. The vertical ramp generator output is bias-adjusted for the proper gain and vertical position. The vertical yoke driver raises the signal to the power level necessary to drive the vertical deflection coil.

The horizontal sync signal, at a frequency of 16.2 kHz, triggers the horizontal yoke driver circuits. This circuit is also a power amplifier to develop the high current required for beam deflection.

The linearity correction circuit, pincushion control, and horizontal size regulator are all concerned with maintaining a straight display pattern. Virtually all aspects of beam deflection are characteristically nonlinear. Therefore, it is necessary to apply various compensation techniques to achieve the proper and desired linearity. Pincushion compensation is normally used to overcome the tendency of vertical and horizontal lines to bend inward. However, the geometry of the CRT used in the IND-300 is such that only horizontal pincushion correction is needed. This signal ensures that the vertical lines are straight throughout the scan pattern.

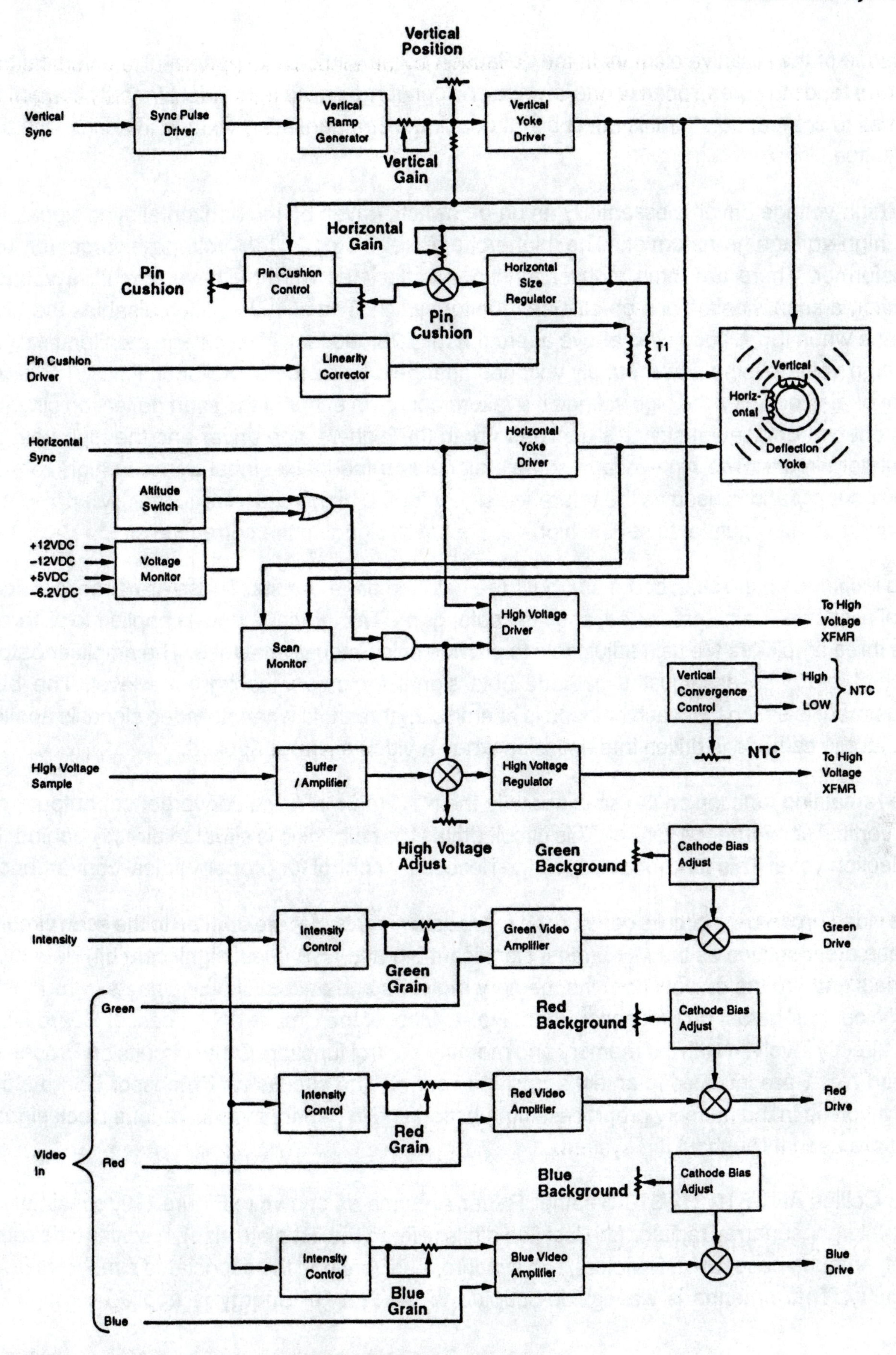

Figure 8-8. Collins IND-300 Scan Control Circuit Block Diagram.

Because of the resistive element in the deflection coil, a vertical line pattern (like a vertical bar pattern) tends to spread apart on the left side and converge on the right side. Linearity correction is used to compensate for this effect by introducing a compensating voltage in series with the deflection signal.

The high-voltage driver is essentially an on-off switch, driven by the horizontal sync signal, for the high-voltage transformer. The high-voltage regulator is the voltage source for the transformer. There are three monitor functions associated with the driver circuit: a voltage monitor, a scan monitor, and an altitude monitor switch. The altitude switch disables the high voltage when the altitude rises above approximately 25,000 feet. The voltage monitor disables the high voltage if the power supply voltages change outside the established limits. The scan monitor also disables the high voltage if a failure occurs in either of the scan deflection circuits. The enable (or disable) signal is applied to both the high-voltage driver and the high-voltage regulator circuits. The high-voltage sample signal is a feedback signal from the high-voltage power supply and is used as the reference signal for the high-voltage regulator. Adjustment is provided at the regulator to set the high-voltage drive signal at the correct level.

Also included on the scan board are the three cathode drive circuits. These are three identical, parallel circuits for the green, red, and blue color guns. The intensity input is applied to all three. The three amplifiers are gain adjustable to provide color intensity balance. The amplifier output is combined with an adjusted cathode bias signal for proper background level. The bias adjustment is set so that each cathode is at emission threshold when no video signal is applied so that the cathode is driven into emission when a video signal is applied.

The remaining function on the scan board is the NTC (normal trace convergence) output from the vertical convergence control. This circuit drives the coil which is situated directly behind the deflection yoke. This function provides the necessary control for proper vertical convergence.

The video processing circuits develop the radar data signals that are applied to the scan circuits. These are identified as the MSB (most significant bit) and LSB (least significant bit) data input signals and are the outputs from the memory registers and data multiplier circuits (through the cyclic contour gates) on Processor Board No. 1. Most of the circuits on Processor Board No. 1 are directly involved with the memory and memory control function. Other circuits on Processor Board No. 1 are involved in antenna position control. The circuits on Processor Board No. 2 are involved in the memory preprocessing function and in generating the various clock signals that are used throughout the system.

The Collins ANT-310/312/318 Weather Radar Antenna as shown in Figure 8-9, consists of a flatplate phasedarray radiator (the last two digits refer to the flat plate size), a waveguide rotary joint, antenna base, two DC motors, two synchro resolvers and the associated synchronization circuitry. That antenna is waveguide-coupled to the RT. RF energy is applied through the

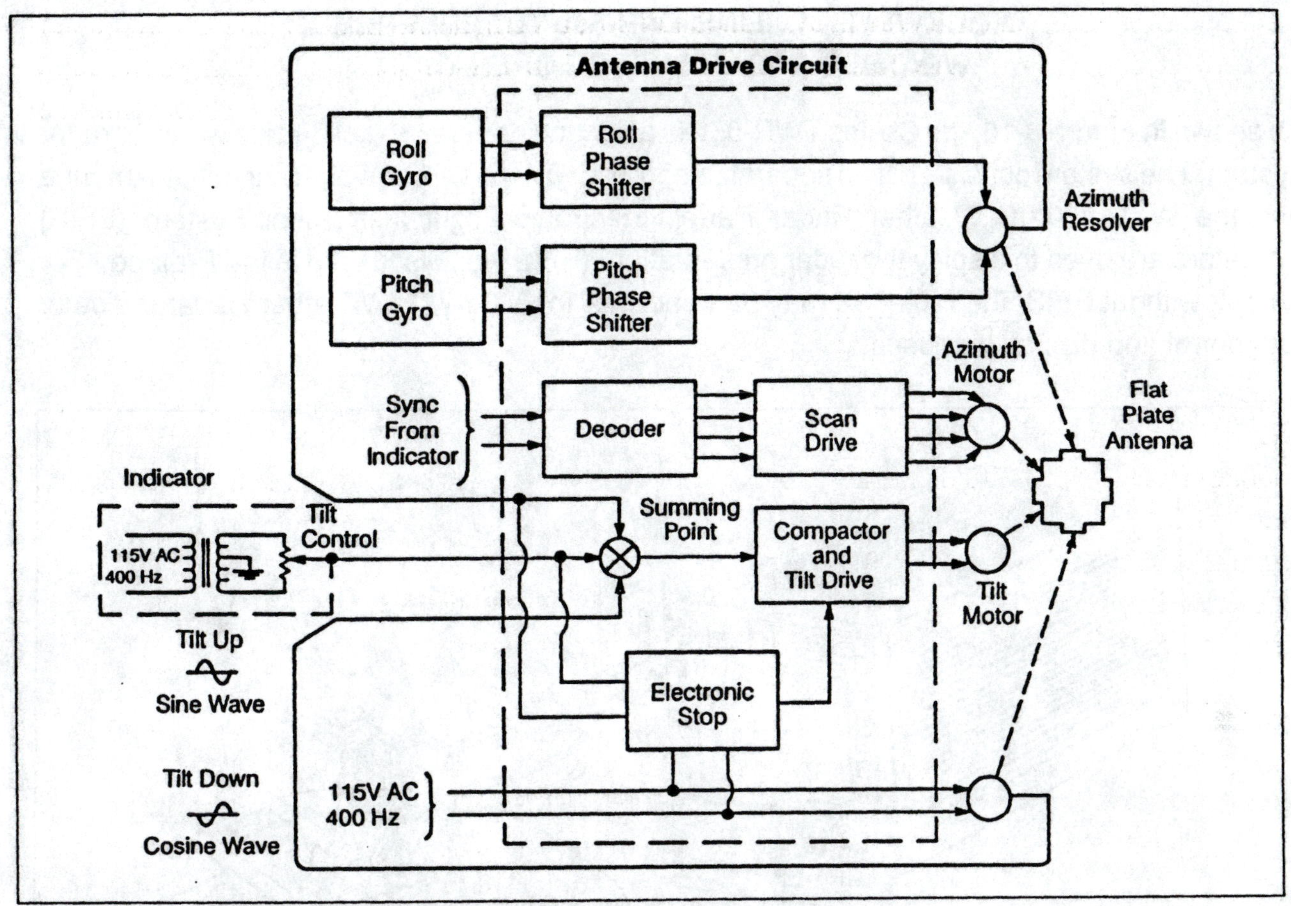

Figure 8-9. *Collins ANT-310/312/318 Weather Radar Antenna.*

antenna base to the waveguide rotary joint. From the rotary joint, the RF energy is directly coupled to the flat plate, which radiates the energy outward.

The waveguide rotary joint is driven plus or minus 60° in the azimuth plane by an azimuth motor. The tilt motor drives the rotary joint plus or minus 15° in elevation. The azimuth resolver sums the pitch and roll signals from the vertical reference gyroscope and applies the results to the tilt drive circuits to keep the antenna scan constant when the aircraft changes pitch or roll attitude. The tilt resolver nulls out tilt commands when the antenna is in the proper tilt attitude. The tilt resolver receives a 400-Hz reference signal from the aircraft AC power bus. Also, a tilt control is provided on the indicator control panel that provides a signal which is summed with the outputs of the azimuth and tilt resolvers.

In summary, the antenna azimuth scan is controlled by the timing signal from the indicator to keep it in sync with the CRT sweep; the antenna elevation position is controlled manually by the tilt control and automatically by signal inputs received from the vertical gyro. A vertical gyro stabilization inhibit switch is usually provided to disable the input signal to the radar system in the event that the VG is inoperative.

ROCKWELL COLLINS TWR-850 TURBULENCE WEATHER RADAR SYSTEM OPERATION

As shown in Figure 8-10, the Collins TWR-850 is a second generation solid-state weather radar system. The system consists of two separate units: the RTA-85X Receiver/Transmitter/Antenna and the WXP-850A/B Weather Radar Panel. Electronic Flight Instrument System (EFIS) indicators are used to display the radar presentation. (EFIS was discussed in the Preface.) For aircraft without EFIS, the RTA-85X may be used with the WXI-711A Weather Radar Indicator for control and display purposes.

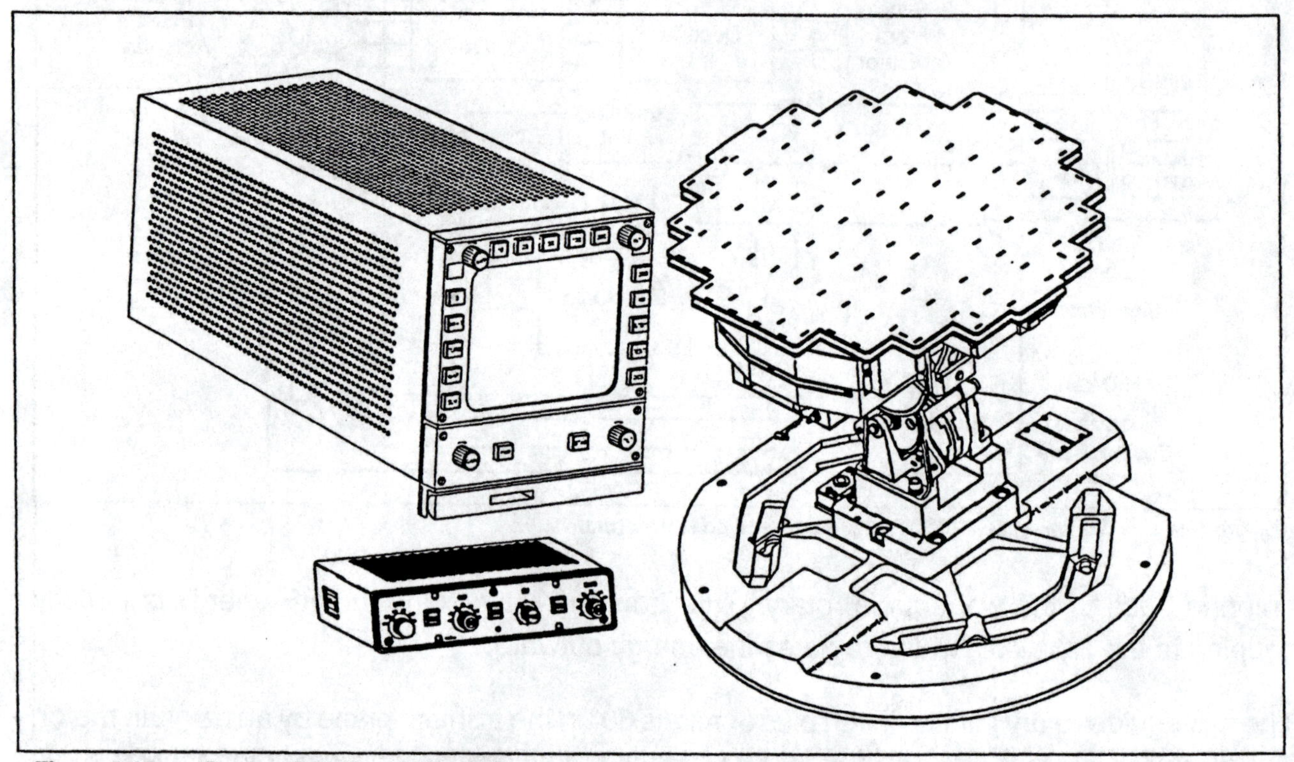

Figure 8-10. Rockwell Collins TWR-850 Turbulence Weather Radar System.

The TWR-850 Turbulence Weather Radar System and the RTA-85X with WXI-711A system provide the flight crew with a display of radar-detectable precipitation within 60° of the flight path. In addition to wet precipitation detection, the systems feature a turbulence detection capability which alerts the crew to the location of precipitation-related turbulent conditions. Turbulence detection is made possible by the use of an extremely stable transmitter which allows detection of very small shifts in frequency indicating a relative velocity of the detected precipitation. The systems cannot detect or display turbulence that is not precipitation related. Users must recognize that X-band weather radar can detect only wet precipitation. Precipitation like snow and some hail, which is typically dry, is generally not detectable by X-band radar.

The mechanical feature that distinguishes the RTA-85X unit from earlier generation weather radar designs is its one-piece construction, combining the receiver, transmitter, and antenna into a single unit as shown in Figure 8-11. The forward portion of this unit is the flat-plate antenna, which is available in either 12-, 14-, or 18-inch diameter models. Directly behind the antenna is the RF assembly, consisting of the transmitter and the receiver. Mating the antenna and the receiver-transmitter eliminates the need for a waveguide. The RTA (receiver/transmitter/antenna) assembly is mounted on the drive assembly and therefore swings from left to right as the system scans and points up or down for tilt. The drive assembly contains the motors and mechanical system for the scan and tilt functions and is attached to the base assembly.

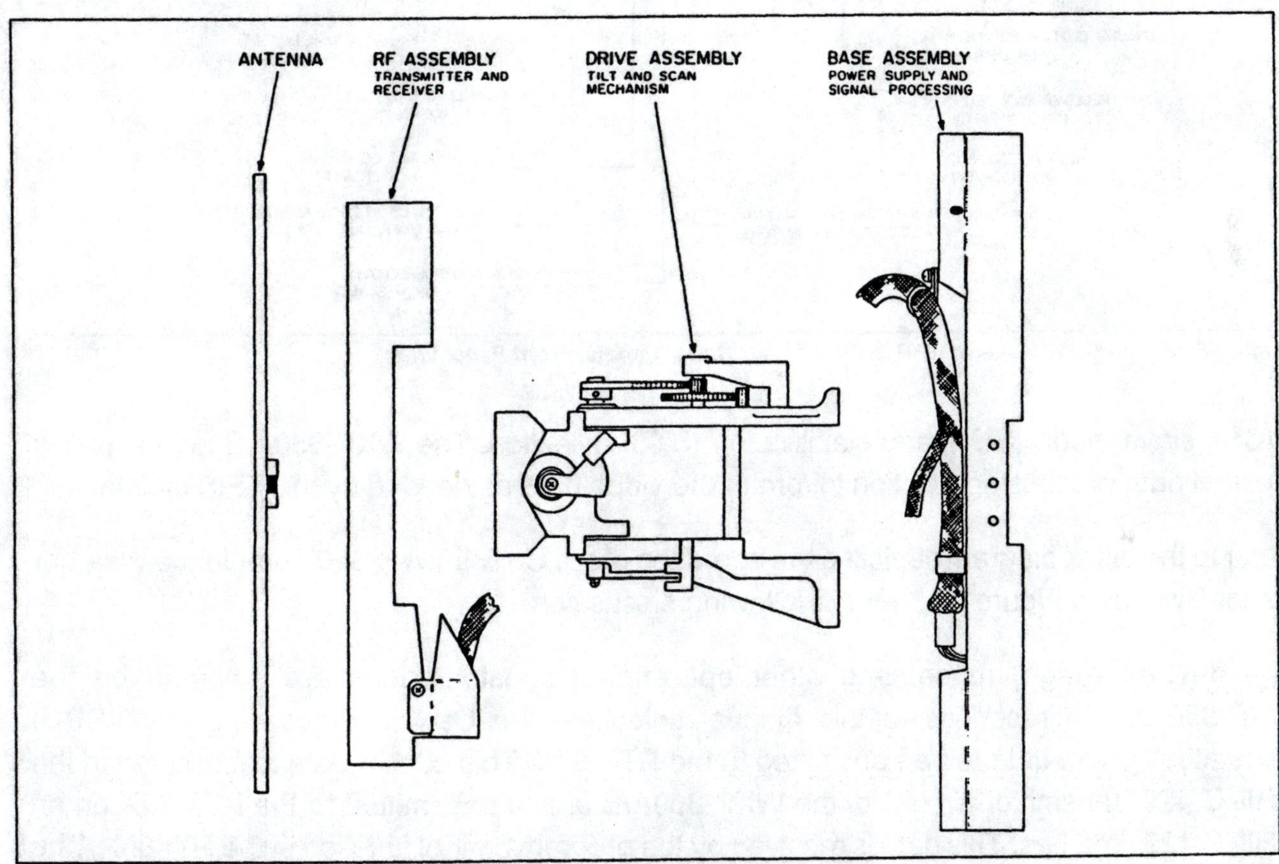

Figure 8-11. RTA-85X Receiver/Transmitter/Antenna/Major Subassemblies.

The base assembly is cylindrical in shape, about 15 inches in diameter and slightly less than two inches in depth. The base assembly contains the power supplies and signal processing functions.

As shown in Figure 8-12, the WXP-850A (panel mount) or WXP-850B (Dzus mount) Weather Radar Panel provides the operating control functions for the TWR-850 system. These units are normally mounted in the instrument panel. The WXP-850A/B control functions include MODE, GAIN, TILT, and RANGE selection knobs, plus pushbuttons for ground clutter suppression

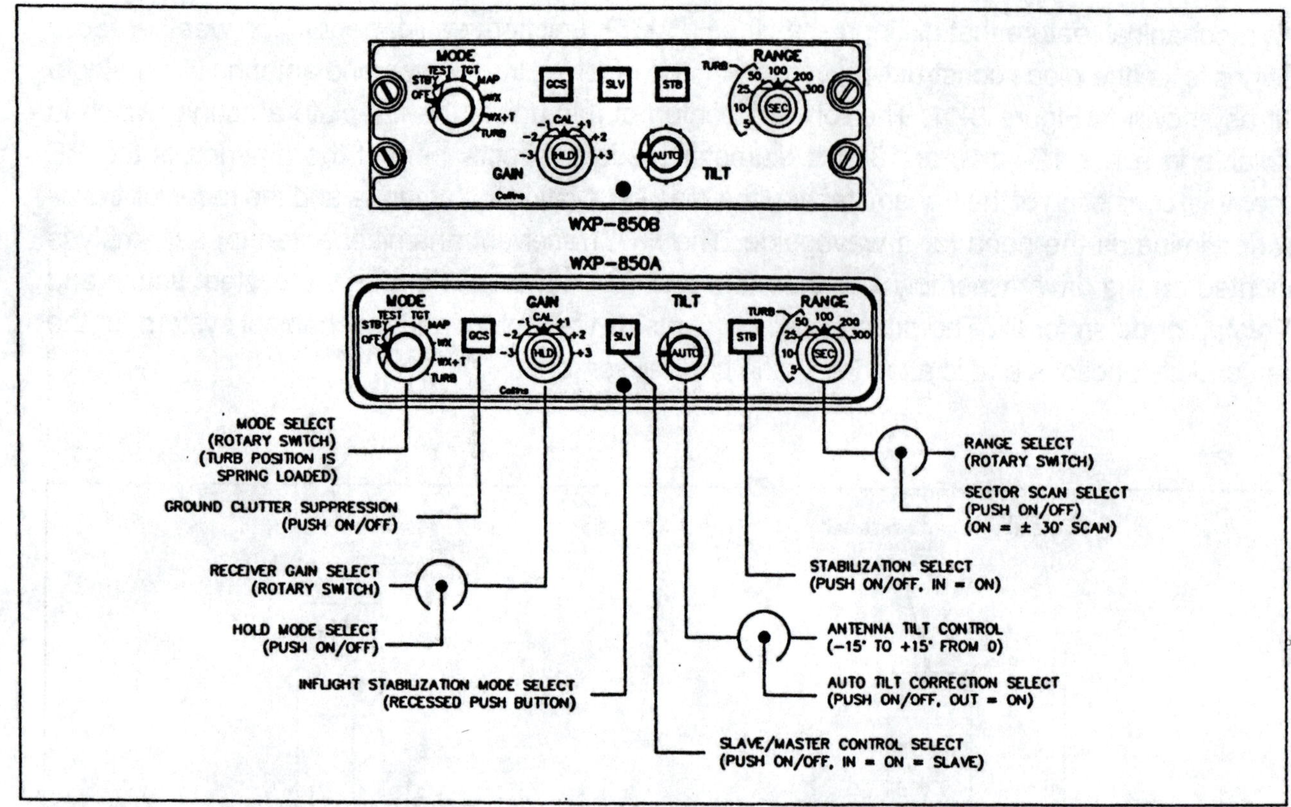

Figure 8-12. ***WXP-850A and WXP-850B Weather Radar Panels, Front Panel View.***

(GCS), slave mode (SLV), and stabilization (STB) selection. The WXP-850A/B also supplies the final data processing function to format the video data as needed by the EFIS system.

Refer to the block diagram depicting the operation of the Collins TWR-850 Turbulence Weather Radar System in Figure 8-13 for the following discussion.

The desired range, mode, and other operational considerations are selected on the WXP-850A/B. In response to this manual selection, the Central Processing Unit (CPU) generates control data to be transmitted to the RTA-85X. This control data is formatted in the ARINC 429 transmitter circuits of the WXP-850A/B and is transmitted to the RTA-85X on an ARINC 429 data bus. This data is received by the program control function in the RTA85X. The program control function in the RTA-85X also receives information from a stabilization source, such as a gyro or a digital horizontal reference system. The program control function is microprocessor controlled and, depending on the data it receives from the WXP-850A/B and the stabilization source, sends control information to the antenna scan and tilt control circuits, transmitter, receiver, and data processing circuits in the RTA-85X.

The scan and tilt control circuits have a mechanical link to the antenna which sweeps the antenna horizontally for scan and vertically for tilt. Horizontal motion is a constant sweeping movement. The vertical movement depends on manual tilt and/or automatic stabilization. The objective of automatic stabilization is to keep the antenna scan line horizontal; it automatically

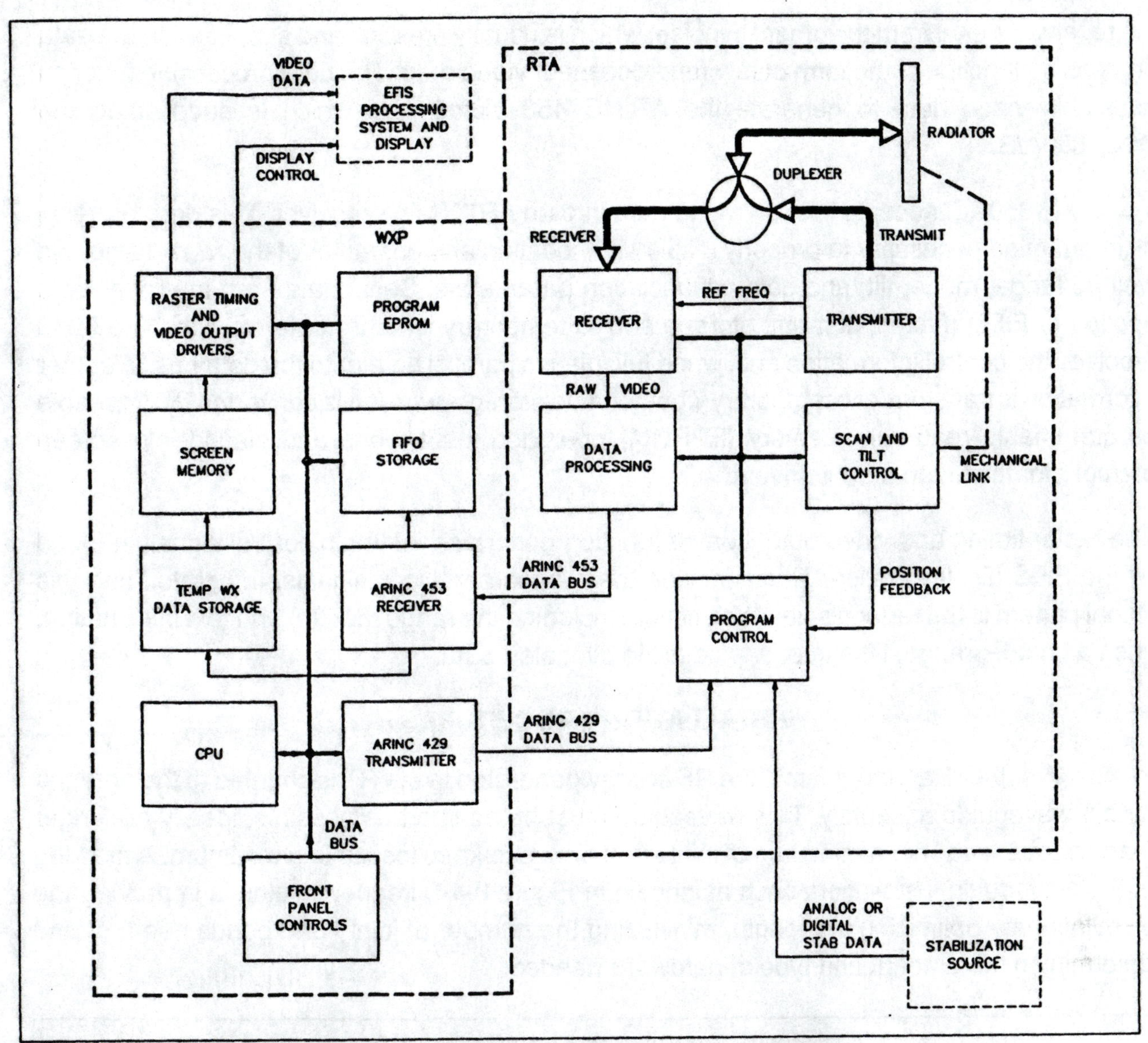

Figure 8-13. Collins TWR-850 Turbulence Weather Radar System Block Diagram.

adjusts the antenna tilt (vertical aim) as needed in response to aircraft pitch and roll attitude changes. The manual tilt is supplied by the operator and depends on the need to observe a particular precipitation pattern at an antenna aim other than straight ahead.

The transmitter function generates the energy bursts that are radiated by the antenna. These energy bursts are in the form of X-band pulses of various widths, depending on the mode and range selected, and at a repetition rate of 208 to 324 pules per second in weather detection modes and 1,456 pulses per second in turbulence detection mode. The energy level is a nominal 24 watts.

When the radiated energy encounters a sufficient amount of wet precipitant (liquid water), a portion of the energy is reflected back to the antenna. This returned energy is directed into the receiver by means of a special RF switching device known as a duplexer. The receiver extracts the received signal from the other RF noise, which is usually present, and supplies it to the data processing function in the form of raw (unprocessed) video data. The data processing function uses this video data to generate the ARINC 453 digital data which is supplied to the WXP-850A/B.

The WXP-850A/B accepts this raw video data into an ARINC 453 receiver. This data contains all information necessary to properly display the location and character of the radar target, as well as range, mode, tilt, and gain annunciation parameters. Serial data from the receiver is applied to FIFO (first-in, first-out) storage and to temporary weather data storage. The FIFO receives the control information and, when full, places parallel data onto the data bus. Weather information is transferred to temporary weather data storage, where it is converted by eraseable programmable read-only memory (EPROM) operating instructions and placed into screen memory for the video output drivers.

The raster timing and video output driver function generates the main control signals needed by the EFIS display device. These include the retrace and clock signals. Integrated into this signal pattern is the radar display information, including the range mark(s) and text information, such as mode, range, tilt angle, and possible pilot alert data.

INSTALLATION PROCEDURES

In most Weather Radar Systems, the RF energy generated in the RT is coupled to the antenna by the waveguide assembly. This waveguide must be as short as possible, ideally no longer than about 5 feet, with a minimum of joints and bends to keep losses to a minimum. A drawing of the waveguide installation, such as shown in Figure 8-14, frequently aids in improving the installation by optimizing the length, minimizing the number of joints and bends needed, and determining the amount and type of hardware needed.

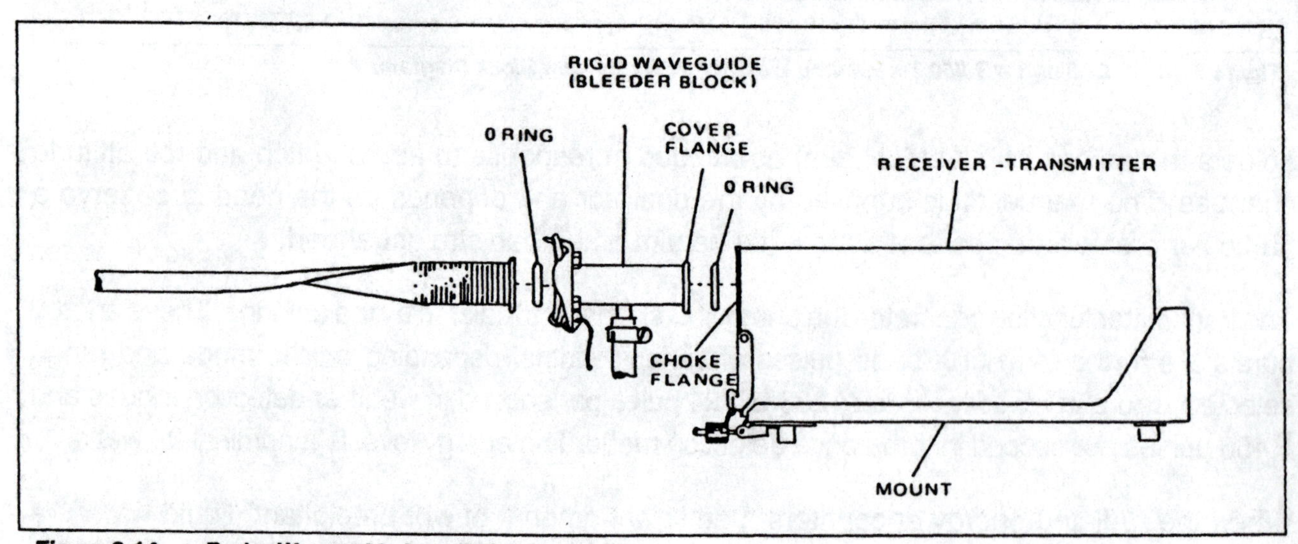

Figure 8-14. Radar Waveguide Installation Drawing.

If any particular installation is likely to be operated at altitudes above 40,000 feet, the waveguide assembly should be pressurized. The relatively high-energy concentrations in radar transmission lines can result in arcing when atmospheric pressure is reduced during high-altitude operation. Usually pressurization is provided by venting cabin air into the waveguide assembly. This practice is satisfactory, however a filter/dehumidifier should be installed in series with the source since humidity present in the cabin atmosphere will condense in the relatively cold waveguide, which usually resides in unpressurized sections of the aircraft. Water can not be tolerated in the waveguide because it introduces losses and increases the standing wave ratio which is the ratio of energy reflected back to the RT to the amount of energy received by the antenna.

The following recommended guidelines should be adhered to when performing a radar waveguide installation.

1. Minimize the number of joints and bends. Where short bends are required, use an exact radius of 2.214 inches to the waveguide center in either the "E" or "H" planes. For larger bends, use a 7-inch minimum radius to the waveguide centerline.

2. Make axial twists in the waveguide only where necessary. One such twist is usually required to accommodate the 90° difference in waveguide orientation between the RT and the antenna.

3. Use adel clamps to secure the waveguide to the aircraft structure, but be certain that the clamps do not distort the waveguide wall. Flexible waveguides should be clamped at intervals not to exceed 18 inches.

4. Each waveguide joint should have one choke flange, one cover flange, and an Oring for a pressure seal. Usually, only one RF window is necessary at the antenna end since the RT waveguide components are normally sealed for pressurization.

It is advisable to achieve as good a waveguide pressure seal as possible. The system leakage rate should be less than 18 cubic inches per minute, when pressurized to 10 pounds per square inch. Obviously, the greater the leakage rate, the shorter the service life expectancy of the dehumidifier desiccator crystals. The desiccator crystals are normally a deep blue color when dry. The need for replacement, due to moisture saturization, is indicated by a pale pink color.

The aircraft radome, which houses the radar antenna, must allow a minimum RF transmissivity (transmission efficiency) of 90%. Any factor less than 90% will cause reduced range and contour performance. Average radome paint causes a loss of about 3.5% to 5% in transmissivity. Rubber boots, when properly installed, present about 5% loss. If a boot is used on the nose of the radome for protection, it must be adequately sealed to prevent moisture accumulation between the boot and the radome surface. Plastic boots usually present a loss of 20% to 50%, which makes them totally unacceptable. In selecting a new radome, or in evaluating an existing one, the technician should be guided by FAA Advisory Circulars 90-20 and 43-202.

The antenna must be installed on a mounting plane that is perpendicular with respect to the aircraft centerline, both horizontally and vertically. Once the aircraft is leveled, the antenna can be aligned by installing shims or spacers under the mounting surface. The antenna may then be calibrated for stabilization by applying vertical gyro signals to the antenna provided an RF dummy load is first connected to the RT waveguide mounting flange. With respect to the VG input signal, the greatest accuracy is achieved by removing the VG from the aircraft and installing it on a tilt table. The tilt table is then adjusted so the VG provides a 10° pitch-up and pitch-down signal and a 10° roll-right and roll-left input signal to the antenna. With the stabilization switch on, the antenna should move the same angular distance from center, but in the opposite direction from the movement of the tilt table. This angular movement is measured by means of a protractor and is adjusted internally by trimmer resistors in the antenna stabilization circuitry.

The installation is completed once the RT mounting tray is rigidly secured in the nose radio rack, the indicator/control is installed in the instrument panel, and the wiring cables are terminated with connectors that mate with the system components. The Weather Radar System may then be tested on the ground provided the precautions concerning radiation safety, described in FAA Advisory Circular 2068A, are followed. The area within the antenna scan arc and within at least 20 feet of an operating radar constitutes a hazardous area to ground personnel. It is advisable to test the radar with the antenna tilted at least 5° upward with the aircraft pointed away from the hanger or any other obstructions.

PASSIVE WEATHER DETECTION SYSTEMS

As previously discussed, the weather radar system detects severe weather by correlating rainfall intensity to storm severity. However, the weather radar detects only precipitation and cannot detect other phenomena related to thunderstorm activity, such as lightning. However, the Ryan Stormscope, relies on the detection of lightning strikes rather than the presence of raindrops to warn of severe weather. Unlike radar, which transmits a beam of microwave energy to illuminate the rainfall for its own receiver, the Stormscope requires only a receiver and processor to convert electrical disturbances to a useful display. The Stormscope, therefore, is described as a passive weather-mapping system since it does not transmit energy.

Lightning discharges are of enormous strength (typically 100 volts per meter at 100 miles) which distinguishes it from other electromagnetic emissions. Lightning discharges, centered around 50 kHz, are picked up by a 360° direction finder in the Stormscope. The DF antenna senses the electrical and magnetic components of the discharge. The magnetic field is used for determination of both azimuth and range in mapping the storm. The electrical field is used as a correlation function that is analogous to the sense antenna on an ADF receiver.

The received signals are then sent to analog and digital processors. The analog processor handles the analysis of sensor data, while the digital processor translates the information into digital format for display. The display unit consists of a small CRT, similar to a PPI display format, which indicates a lightning discharge by a bright green dot on a 360° screen in which the aircraft is at the center. Azimuth and range marks also appear on the CRT. The intensity of the storm can be determined by observing the size of the cluster of green dots appearing on the Stormscope indicator.

CONCLUSION

We began this chapter with an introduction into the theory of radar system operation, compared analog versus digital systems, and discussed the operation of two typical systems, the Collins WXR-300 and the newer Collins TWR-850. The TWR-850 not only provides weather avoidance and ground-mapping information, but also detects turbulent air movement to give pilots an advance warning of windshear conditions ahead.

Windshear detection systems are becoming more prevelant. Allied-Signal's Windshear Detector, a modified version of their RDR-4A Weather Radar System, measures the speed of water particles in the air and displays the magnitude of the windshear by color codes on the indicator. Hazard factors are color coded red meaning "danger". In addition, any windshear detected up to three nautical miles from the aircraft triggers an aural "Windshear Alert". Below 2,500 feet, the system operates in weather radar mode during the clockwise scan of the antenna and in turbulence mode during the counterclockwise scan. Tilt is adjusted automatically for windshear detection through an azimuth coverage of plus or minus 40°.

Another popular form of an airborne radar system is the radar altimeter. This device functions in the same manner as the weather radar except that the transmitted beam of energy is directed downwards instead of ahead of the aircraft. By measuring the time required from when the transmitted beam is sent to when the reflected signal is received, the radar altimeter system calculates the aircraft's height above the terrain. This system will be discussed in more detail in the next chapter.

REVIEW QUESTIONS

1. Define the primary and secondary functions of a airborne Weather Radar System.
2. Describe the principle upon which the weather radar derives information concerning the distance and relative position of the target.
3. What is the frequency range in which most Weather Radar Systems operate?
4. List the principle components of the weather radar system and describe their functions, including the control functions available on the indicator.
5. What is the purpose of radar antenna stabilization and what is the source of the signals used to derive this function?
6. How is the intensity level of the radar signal return displayed on both monochromatic and color indicators? What is the purpose of the contour control?
7. Compare the differences in operation between analog and digital Weather Radar Systems.
8. What four factors determine the effective range of a Weather Radar System?
9. What are the advantages of having a narrower beam-width? How is a narrower beam-width obtained?
10. Describe the function of the STC and attenuation correction circuits.
11. What is the purpose of generating timing signals during weather radar system operation?
12. Describe the operation of the modulator and magnetron contained within the radar transmitter.
13. What is used to inject a signal frequency into the mixer to produce an IF output?
14. Describe the operation of a Weather Radar antenna drive circuit.
15. Describe the proper procedure for the installation of a waveguide assembly.
16. When does a waveguide need to be pressurized and for what reason?
17. Why is it important that the radome be electrically and structurally intact?
18. Describe the proper procedure for leveling and calibrating a radar antenna installation.
19. What precautions should be taken prior to applying power to the weather radar system while the aircraft is on the ground?
20. Compare the principles of operation and the differences in display format between the active and passive weather detection systems.

Chapter 9

Radar Altimeter Systems

INTRODUCTION

The purpose of the Radar Altimeter System is to measure the vertical distance from the aircraft to the ground to provide the pilot with an indication of the absolute altitude above the terrain. The Radar Altimeter receiver-transmitter functions in the same manner as the Weather Radar RT, in that it directs a narrow beam of high-energy RF pulses towards a reflecting target. The target in this case is the terrain directly beneath the aircraft.

The altitude of the aircraft above the terrain is directly proportional to the time required for the transmit signal to make a round trip to the terrain and return to the airborne Radar Altimeter System. Since the Radar Altimeter is sensitive to variations in the terrain, it will provide the pilot with an accurate indication of the absolute altitude above the ground, rather than altitude relative to sea-level pressure which is derived from a standard barometric altimeter.

The Radar Altimeter is used in-flight to monitor absolute altitude within the maximum range of the indicator (usually 2,500 feet). The Radar Altimeter is also used for displaying ground separation and climb conditions during night or instrument takeoffs, as well as indicating ground clearance during approaches. Most Radar Altimeter indicators have a Decision Height (DH) control which is used by the pilot to select a pre-determined altitude. When the aircraft reaches the altitude set by the DH control, the Radar Altimeter automatically energizes an indicator light and audio tone generator to alert the pilot.

RADIO ALTIMETER (RAD ALT) SYSTEM DESCRIPTION

A typical Radar Altimeter System, as shown in Figure 9-1, consists of a receiver-transmitter, two antennas (one for receive and one for transmit) and an indicator. Radio altitude information may be displayed from either a motordriven analog indicator or from a digital readout indicator, and may also be displayed on the flight director indicator.

The analog Radar Altimeter indicator provides an absolute altitude display from zero to 2,500 feet and includes a red warning flag to indicate when the system is inoperative. At altitudes within the usable range of the indicator, proper system operation is indicated by the absolute altitude pointer being in-view . Once the aircraft has flown above the usable range of the indicator, the pointer hides behind the mask in the top left corner and the warning flag remains out-of-view. Momentary signal loss within the usable range of the indicator will cause the pointer to disappear from view temporarily.

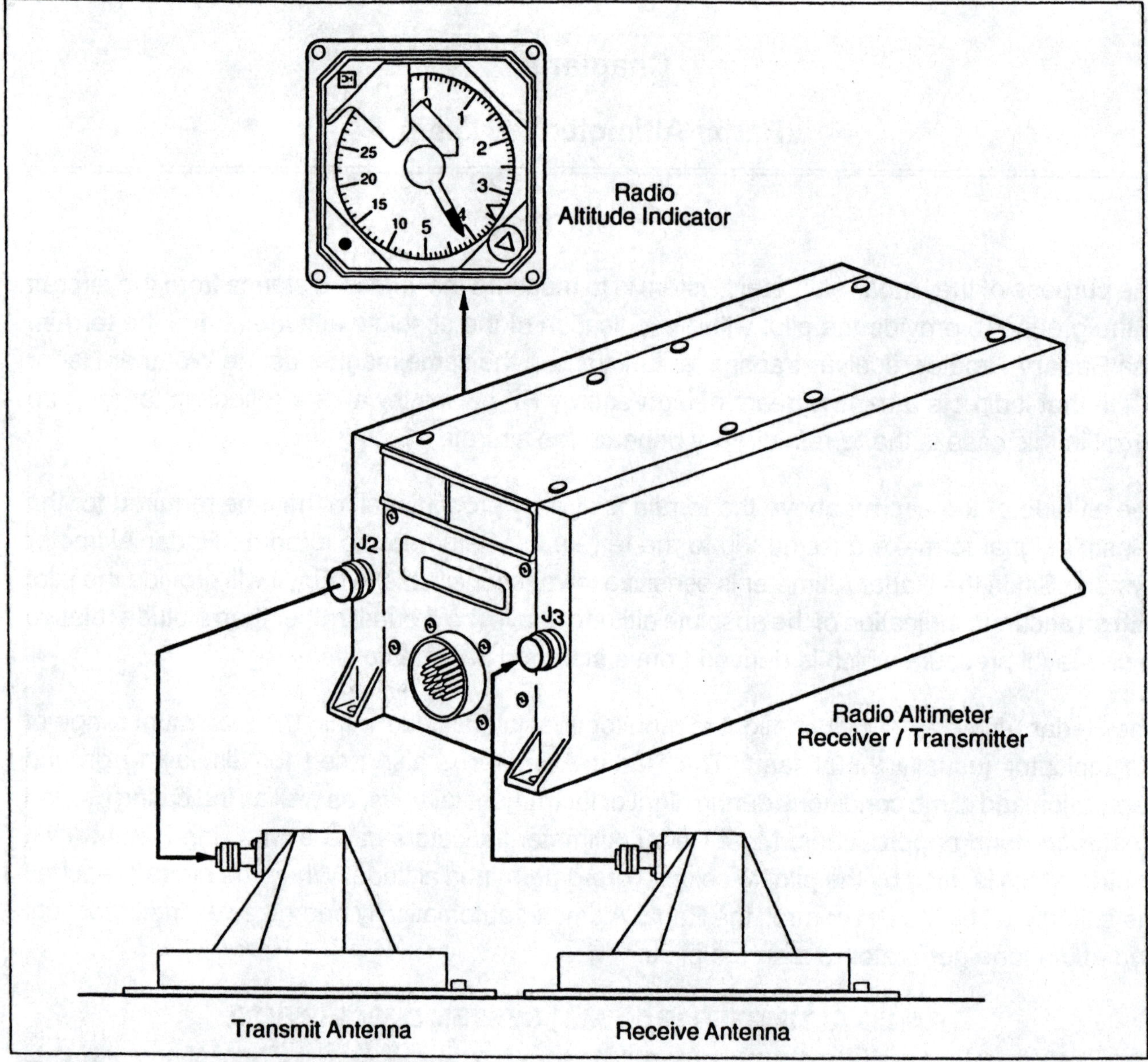

Figure 9-1. Typical Radio Altimeter System.

The digital Radar Altimeter indicator does not incorporate the use of a warning flag, but instead, the LEDs display dashes in place of numerals when a system fault occurs or the when the aircraft altitude is above 2,500 feet. Both the analog and digital indicators, as shown in Figure 9-2, have a system self-test switch, a knob for setting the desired decision height, and a DH indicator light. Pressing the "test button" will cause the indicator to read a pre-determined altitude and the warning flag to appear, and if the aircraft altitude is less than the pre-selected DH setting, the DH annunciator will light.

In addition to radar altitude information, preset radar altitude "trip points" may be provided from the RT to be used as signal inputs to the flight control system and Ground Proximity Warning System (GPWS). (The GPWS monitors the Radar Altimeter output and provides the pilot with an aural (e.g., synthesized voice says "PULL UP") and visual warning when the aircraft is closing

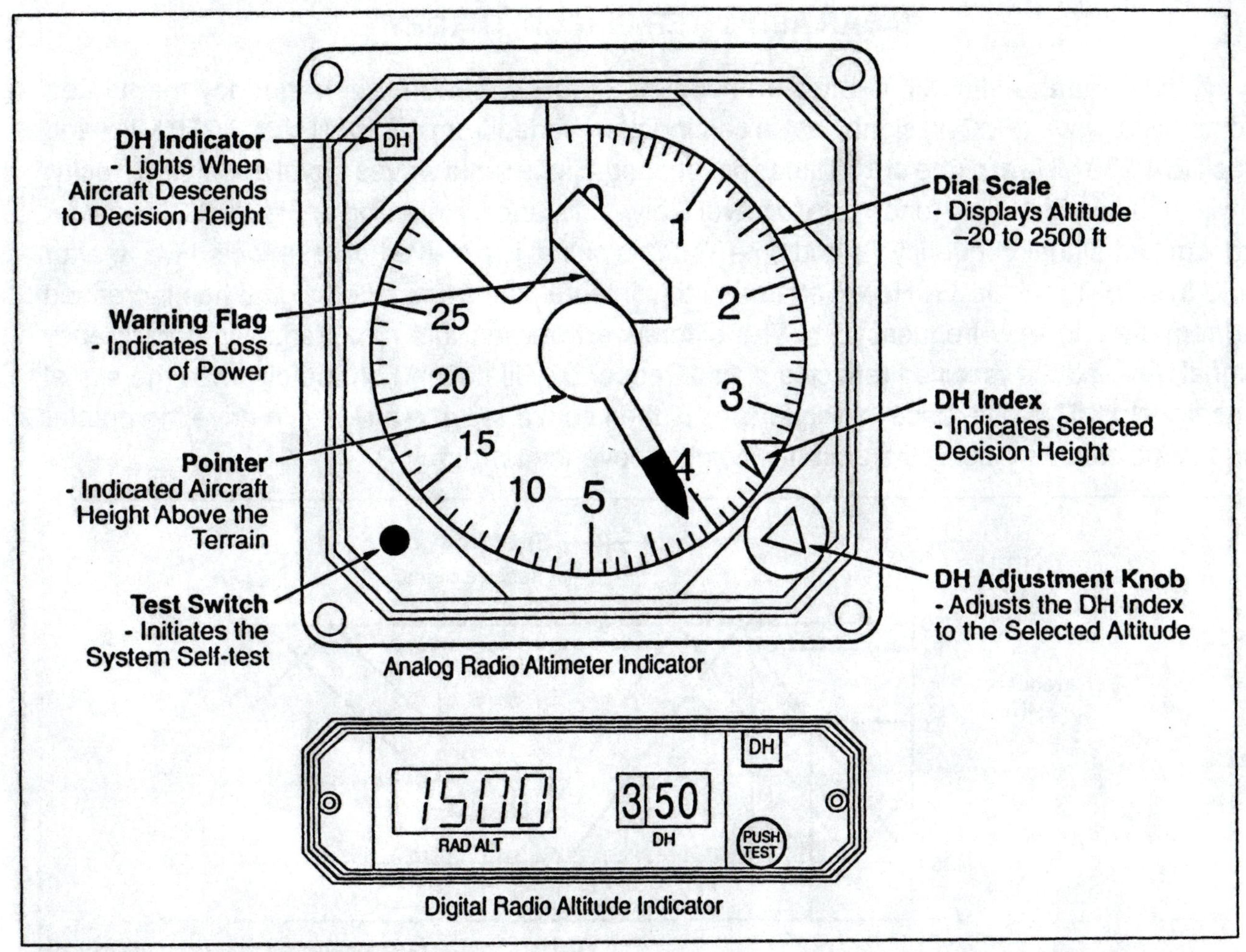

Figure 9-2. Radar Altimeter Indicators.

in on the terrain, such as approaching a mountain peak.) If a failure occurs at anytime or the system is turned off, the red warning flag will appear and the warning signal will be applied externally to other avionic systems that use radio altitude information to indicate that usable information is not available.

The Radar Altimeter System is usually turned on prior to take-off and may be left on for the duration of the flight. In normal in-flight operation, the Radar Altimeter indicator pointer comes into view when the aircraft flies within the useable range of the indicator in respect to the aircraft's absolute altitude above the terrain. Before initiating an approach, the pilot should test the system and then adjust the DH knob to set the DH index on the published decision altitude for the airport facility being used. When the aircraft descends to the selected decision height, the DH annunciator lights and an external audio tone generator is energized to alert the pilot that a continue-approach or go-around decision must be made.

RADAR ALTIMETER CIRCUIT THEORY

A typical Radar Altimeter C-band transmitter outputs a 4.3-GHz frequency-modulated continuous-wave (FMCW) signal that is continuously varied from 4,250 MHz to 4,350 MHz and back to 4,250 MHz at a rate of 100 times per second. Since radio waves travel at a finite velocity, it will require a finite time for a wave to travel a given distance. Referring to Figure 9-3, consider transmitted signal frequency F_1 and time T_1. This signal is received as a reflected wave after time interval T_2 minus T_1. However, during this interval, the transmitted signal has increased in frequency to new frequency F_2. The difference between this new transmitted frequency signal, F_2, and the received reflected signal frequency will depend on the distance the signal has traveled. This difference in frequencies is then converted into a signal to drive the pointer on the indicator to display the aircraft's height above the terrain.

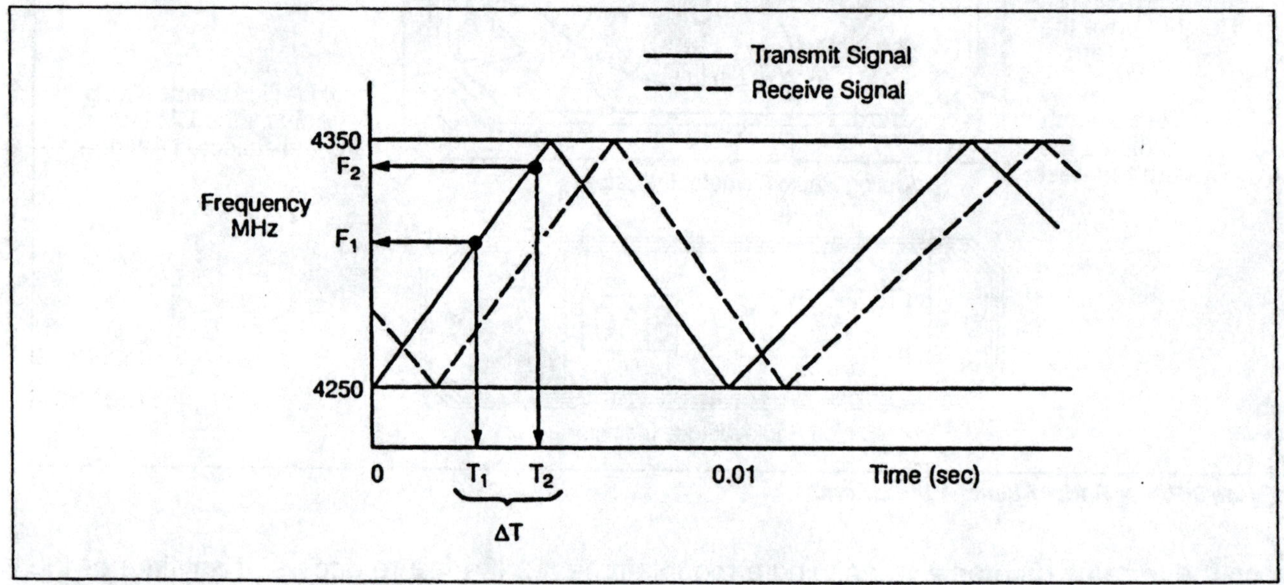

Figure 9-3. Frequency Shifting Technique for Determining Altitude.

A simplified block diagram of a typical Radar Altimeter System is shown in Figure 9-4. Here the transmitter output is sent to the transmitting antenna, with a small sample coupled to the mixer. The transmitted signal is reflected from the ground, picked up by the receiving antenna, filtered, and applied to the mixer as an injection frequency. The mixer output is the difference between the transmitting frequency being coupled directly to the mixer and the frequency of the reflected transmitted signal picked up by the receiving antenna. Therefore, the mixer output is directly proportional to the time required for the signal to travel to the ground and back, and is proportional to the distance from the ground. The difference frequency output from the mixer is measured and converted into a DC analog voltage which is used to drive the altitude pointer motor in the Radar Altimeter indicator.

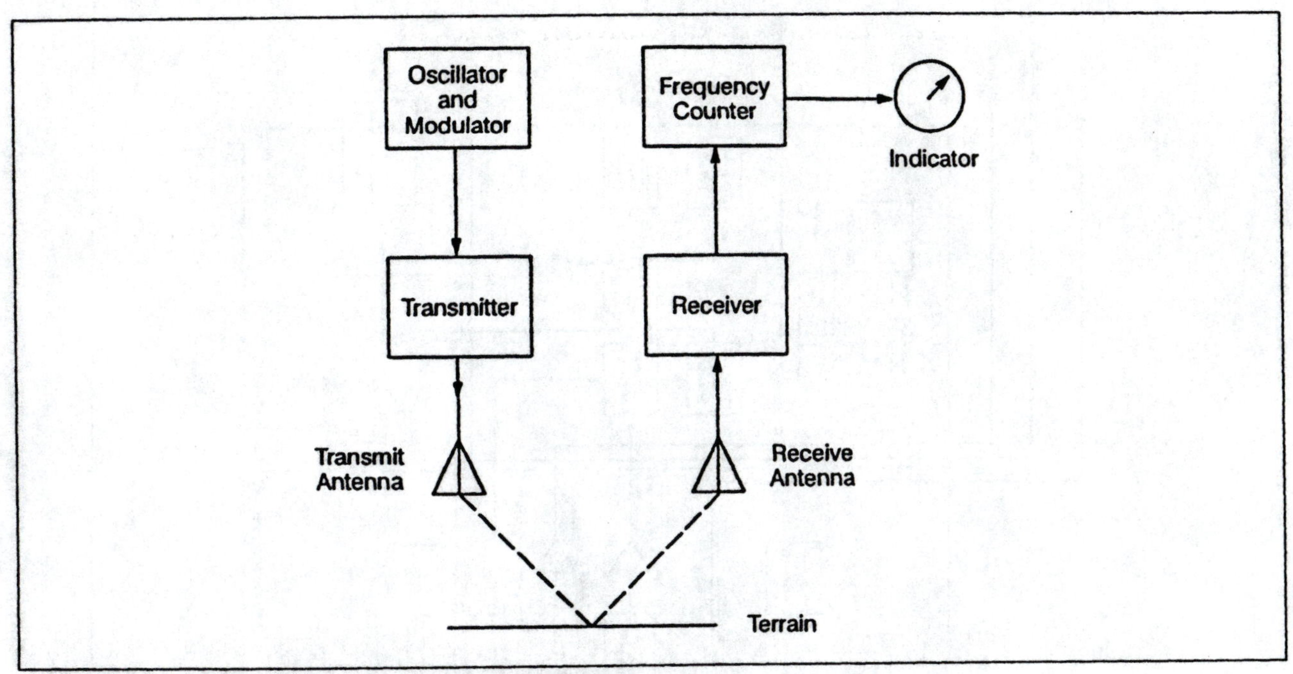

Figure 9-4. Radar Altimeter System Block Diagram.

ROCKWELL COLLINS 860-F4 RADAR ALTIMETER RT OPERATION

The Rockwell Collins 860F-4 Radar Altimeter RT is part of the Rockwell Collins AL-101 Radio Altimeter System. The 860F-4 was designed to ARINC Specification No. 552 which makes it directly interchangeable with other Radar Altimeter RTs designed to that specification. The following discussion describes the internal operation of the 860F-4. Refer to the 860F-4 block diagram, shown in Figure 9-5, throughout this discussion.

Modulator and Transmitter

The modulator is a 100-Hz triangle-wave generator that consists of an eight-bit binary up/down counter and Digital-to-Analog (D/A) converter. The count continually cycles from zeros to ones and back, with 256 counts making a half cycle. The eight bits are summed in the D/A converter to make a staircase waveform, which is filtered to make a smooth waveform. This waveform, which is the complement to the capacitor voltage curve of the transmitter tuning varactor, allows frequency modulation to be linear.

The transmitter consists of an oscillator and power amplifier, both using 1 watt, 4-GHz transistors. Frequency modulation is achieved by applying the triangular voltage waveform to the varactor in the oscillator circuit. The mean voltage of this waveform, adjusted by a control on the modulator, determines the center frequency of the oscillator. The 4,300-MHz FMCW oscillator output passes through a directional coupler and is amplified by the power amplifier. The amplifier output is routed through a second directional coupler and an isolator, or terminated three-port circulator, to the transmit antenna.

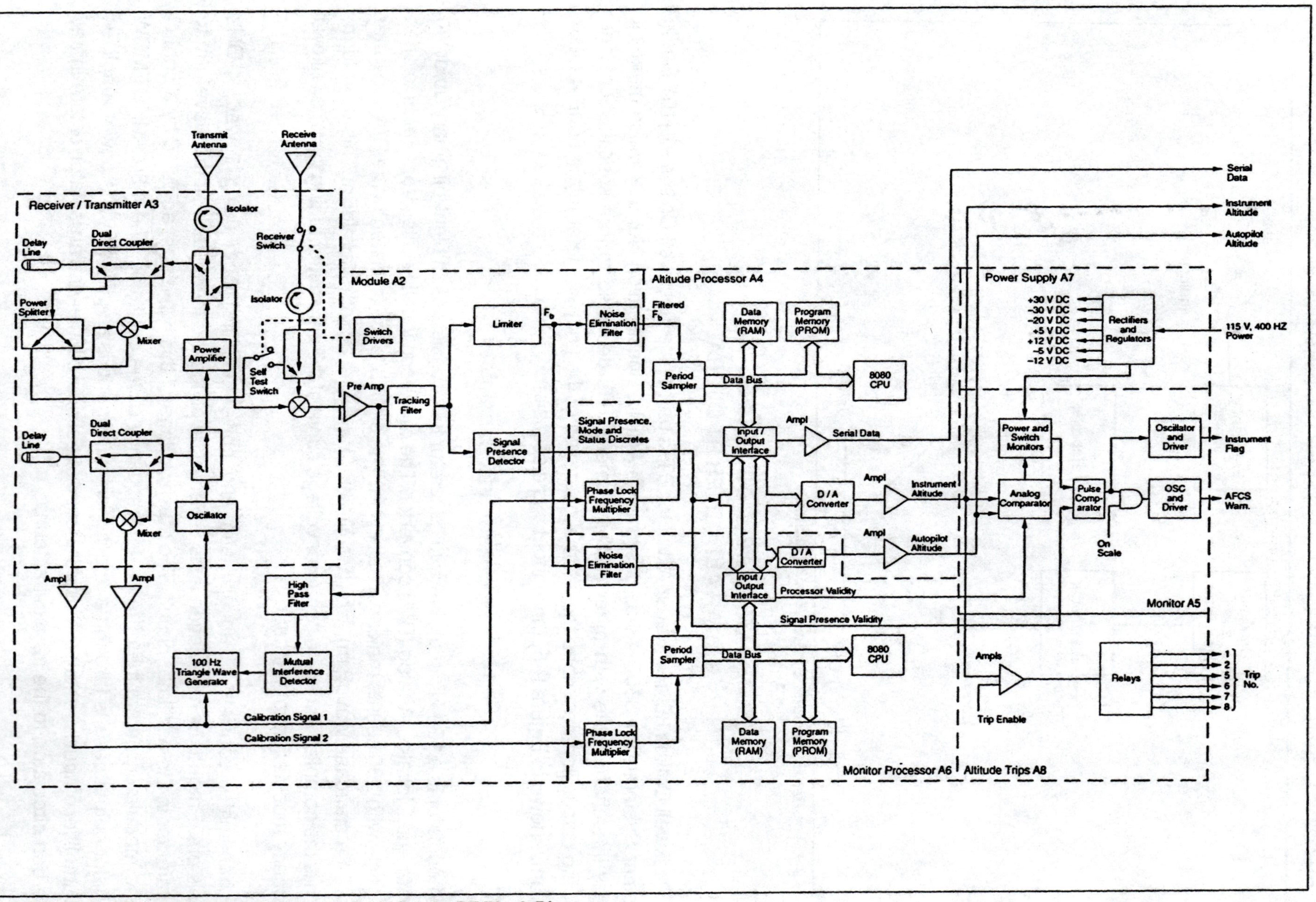

Figure 9-5. Rockwell Collins 860F-4 Radio Altimeter RT Block Diagram.

The isolator provides non-reciprocal power transfer from one port to another. When one of the three ports is terminated with a load, the device will transfer power with minimum loss from one port to another (as indicated by the arrow) while in the reverse direction, the power is transferred to the load with a high resultant loss, thus providing isolation.

Each directional coupler is connected to a quartz bulk acoustic delay line that is the equivalent of 300 feet of altitude. The delay signal is heterodyned in a mixer with the non-delayed signal applied from the transmitter. The mixer output is the difference frequency of approximately 12 kHz, which is amplified and limited into a square wave. One delay line circuit produces CALIBRATION SIGNAL 1, which is used for comparison in the altitude processor. The other delay circuit produces CALIBRATION SIGNAL 2, which is used in the monitor processor.

Receiver

The delayed reflected signal picked up by the receive antenna is routed through an isolator to a mixer. The other input to the mixer is a sampling of the transmitter signal from a 25 dB directional coupler. The mixer heterodynes the two signals to produce a difference or beat frequency. Because of the modulation, this frequency is proportional to the time required for the round trip to the terrain by the transmitted signal. The difference of altitude signal frequency changes at the rate of approximately 40 Hz per foot of altitude.

The mixer output is applied to a preamplifier on the modulator whose output is routed through a tracking filter to a limiter. The tracking filter and associated amplifiers are controlled by the altitude processor that selects the correct receiver passband shape for the altitude being tracked. The shaped amplification increases the gain with increasing altitude. That is, the high-frequency response is increased to make the amplifiers more responsive to the high altitude frequencies. The output of the limiter is a relatively constant amplitude altitude beat frequency (Fb).

The altitude beat frequency is then applied to two different noise-elimination filters, one for the altitude processor and one for the monitor processor. These filters convert the beat signal into a 60/40% duty cycle waveform to remove extraneous noise and provide additional signal stabilization. The tracking filter output is monitored by a signal presence detector. This detector provides a discrete signal to both the altitude and monitor processors when a signal is present.

In order to prevent interference between redundant radio altimeters on the same aircraft, the phase of modulation must be adjusted when mutual interference is detected. In the 860F-4, the mixer signal from the receiver is routed through a high-pass filter to a mutual interference detector that contains a threat comparator. When two altimeters approach modulation phase alignment, the threat detector is activated and causes a binary addition to the up/down counter in the modulator. The number added to the count is dependent upon the jumper connection on the RT rack-mounted connector.

Altitude Processor

Altitude processing is accomplished by a general purpose microcomputer. The microcomputer consists of an 8,080 CPU, a Programmable Read-Only Memory (PROM) for program memory, a Random-Access Memory (RAM) for data memory, and Input/Output (I/O) interface circuits.

The processed receive signal or filtered beat frequency signal, along with the 12-kHz CALIBRATION SIGNAL 1, are applied to the altitude processor. As previously mentioned, the processed receive signal is proportional to the altitude and the calibration signal is proportional to the reference delay line. These two signals are applied to a ratio circuit or period sampler whose output is in the form of parallel digital words.

The digital data words are continuously loaded into the working memory of the microcomputer. The microcomputer operates on the data to derive a running average, applies the correction factors for the aircraft installation delay, controls the bandwidth of the receiver, computes the logarithm of the altitude, and couples the final altitude answer to the digital-to-analog converters. The D/A converters and associated amplifiers convert the digital answer to the DC altitude analog signals required for compatibility with the indicator and the other avionic systems that require radio altitude information.

The basic equation solved by the microcomputer is:

$$h = 300 \text{ feet } (F_b/F_{cal})$$

Where:

h = indicated altitude plus AID (in feet)

F_b = beat frequency

F_{cal} = calibration frequency (approximately 12 kHz)

The DC analog signals are linear up to 480 feet and logarithmic from 480 feet to 2,500 feet to coincide with the indicator scale divisions. This computation is derived from the following equations:

$$-20 \text{ feet} < h < 480 \text{ feet: Volts} = 0.02\,(h+20)$$

$$480 \text{ feet} < h < 2{,}500 \text{ feet: Volts} = 10 \ln (2.7183\,(h+20)/500)$$

Monitor Processor

As previously mentioned, the 860F-4 has two reference delay lines that produce independent calibration frequencies: CALIBRATION SIGNALS 1 AND 2. The CALIBRATION SIGNAL 2 is applied to the monitor processor, which is essentially identical to the altitude processor except for the stored program used. The monitor processor computes the altitude in the same way as the altitude processor and compares the two computed altitude answers. If the two answers are within a predetermined comparison threshold, the PROCESSOR VALIDITY signal is interrupted, which will cause a warning to be given to the Radar Altimeter indicator and any other associated avionic systems.

The monitor processor also initiates periodic test routines to ensure that the RF, IF, and signal presence functions are operating properly. Failure of the test cycles will result in a warning output being produced.

RADAR ALTIMETER INSTALLATION AND TEST PROCEDURES

As previously mentioned, Weather Radar antennas are typically mounted in the nose of the aircraft so that the radar can "paint" a picture of the targets that lie ahead of the aircraft's projected flight path. Radio altimeter antennas are mounted on the underside of the fuselage of the aircraft to send and receive radar signals from the ground directly beneath the aircraft. Ideally, a typical radio altimeter installation would have the antennas mounted in an area on the belly of the aircraft that was entirely free of extraneous protrusions so that no object is visible to either antenna within a plus or minus 45° conic area below the aircraft. Also, the antennas should be spaced 36 inches apart (center-to-center) with 0° squint angle.

Squint angle is defined as the angle between the vertical axes of the two antennas and is a measure of the ability of the antennas to illuminate the same area of ground. The angle of tilt is defined as the angle between the plane containing the front face of both antennas (assuming 0° squint angle) and the ground. To minimize variations in altitude readout due to changes in the angle of tilt, the antenna mounting should be as close to the aircraft center-of-gravity as possible.

The antenna must be mounted "H-plane" coupled in order to maintain adequate isolation. This consideration requires that the connectors be perpendicular to a line connecting the center of both antennas. The antennas must be mounted on a conductive surface for proper operation. A conductive gasket is usually provided to give a good electrical bond between the antenna and the fuselage, and in addition, provide an airtight seal which is essential when the antennas are installed in pressurized compartments.

Most Radar Altimeter Systems require an adjustment after initial installation to read zero foot altitude at the point of touchdown. Compensation for antenna cable lengths and the antenna height above the ground at the point of touchdown may be made by applying an aircraft installation delay (AID) bias. The AID bias is selected from a chart, such as the one shown in Figure 9-6, and may be applied by jumpering connector pins on the rear of the RT rack connector on some typical installations.

In determining the antenna height, it is important to jack-up the aircraft so that the landing gear is unloaded in a simulated touchdown configuration. Using the chart shown, the coaxial cable (RG-214/U) for both the transmit and receive antennas may be cut to any one of three different

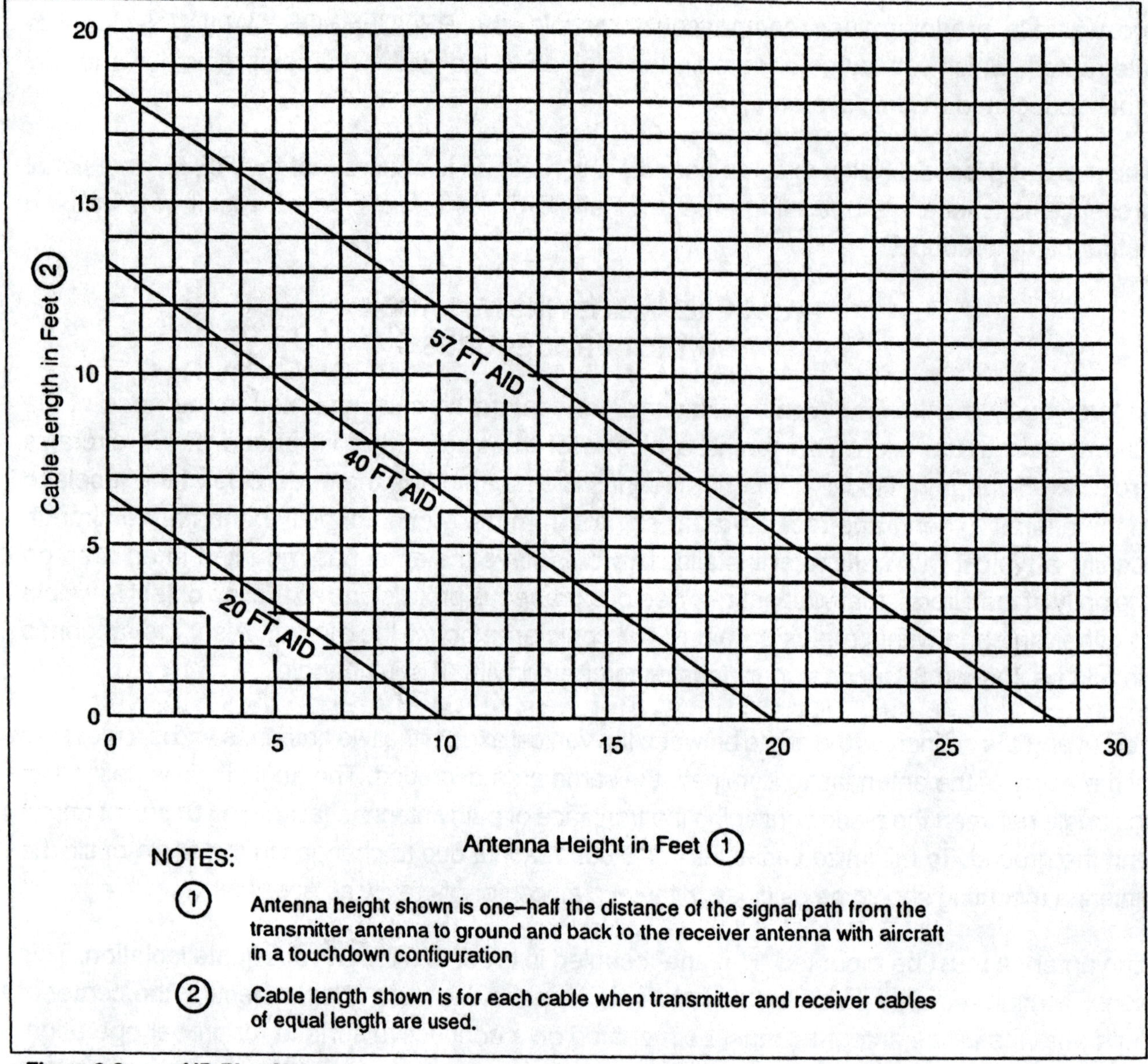

Figure 9-6. AID Bias Chart.

lengths for a particular antenna height depending on the AID selected. Cable lengths are critical and should be measured accurately. Any excessive cable should be coiled up and secured to the aircraft. The cables must also be continuous without any intervening connectors or breaks.

It is important to note that in some installations, the altimeter indicator may read less than zero altitude while the aircraft is on the ground. This is a normal phenomenon in aircraft whose altimeter antennas are nearer the ground on the ramp than at touchdown.

Prior to take-off, the Radar Altimeter System should be tested using the following procedure:

1) Apply power to the Radar Altimeter System.

2) Press and hold the test switch on the indicator and observe that the pointer indicates the proper test altitude (usually 50 or 100 feet) and that the warning flag is in view.

3) With the test switch depressed, slowly adjust the DH setting outside the test altitude range. The DH light should remain off until the DH setting is adjusted to within the test altitude range.

4) Release the test switch and observe the indicator. The warning flag should be outofview and the altitude indication should be zero foot nominal.

CONCLUSION

This chapter on the operation and maintenance of Radar Altimeter Systems was placed after the chapter that discussed Weather Radar Systems since both systems operate on the principle of transmitted RF energy being reflected off a target to provide information to the receiver onboard the aircraft. The primary difference between the Weather Radar and the Radar Altimeter is that the former paints a picture of its target, while the Radar Altimeter calculates the time required for the reflected wave to return to provide an indication of altitude above the terrain. As such, Radar Altimeters are very useful instruments for low-altitude flight in that they are not dependent on barometric pressure as are conventional altimeters.

This chapter provides a transition to the next chapter which discusses other types of flight instruments commonly used, such as the barometric altimeter, airspeed indicator and vertical speed indicator. In addition, Chapter Ten will present both conventional pitot-static systems used to drive pneumatic flight instruments and central air data computers which drive electric flight instruments.

REVIEW QUESTIONS

1. Compare the similarities and differences in the Weather Radar and Radar Altimeter Systems.
2. What is the distinct advantage of using the Radar Altimeter over a standard barometric altimeter and when is the Radar Altimeter of no use whatsoever?
3. List some of the uses of the Radar Altimeter in-flight.
4. Compare the differences in the operation of the analog and digital Radar Altimeter indicators when the aircraft is flying at an altitude above the useable range of the system.
5. What action is initiated upon depressing the "test" switch on the indicator?
6. What is the purpose of the Radar Altimeter trip outputs?
7. Describe the operation of the Decision Height control.
8. How should the Radar Altimeter system be tested prior to take-off?
9. What is the desired spacing between the Radar Altimeter antennas during installation?
10. What is the difference between squint angle and tilt angle and how can these two variables be reduced during a Radar Altimeter antenna installation?
11. How is H-plane coupling and grounding of the antennas achieved during installation?
12. What is the purpose of the applying the AID bias following the initial installation of the Radar Altimeter system?
13. Referring to the AID Bias Chart in Figure 9-6, if a Radar Altimeter RT is installed in an area that is located seven feet from the antennas, and the distance from the antennas to the ground is five feet in touchdown configuration, what should be the length of each coaxial cable and what AID bias should be selected?
14. Describe the characteristics of the Radar Altimeter transmitted signal.
15. What is the relationship between the transmitted frequency and response time in determining absolute altitude?
16. Why does the mixer in the receiver sample the transmitter output?
17. Describe the modulation drive signal applied to the transmitter in the Rockwell Collins 860F-4 Receiver-Transmitter.

18. How is mutual interference prevented in the operation of a dual radio altimeter system installation?

19. What is the purpose of the delay line in the 860F-4?

20. If an aircraft is flying at an absolute altitude of 1,500 feet, what should be the value of F_b from the receiver mixer output in the 860F-4?

Chapter 10

Flight Instrumentation

INTRODUCTION

As we have seen in the preceding chapters, RF signals received by the airborne navigation and radar systems are transformed into usable information which is displayed on the instrumentation located within the cockpit. Already we have studied the operation of the ADF bearing indicator, VOR/ILS course deviation indicator, DME distance and groundspeed indicator, Weather Radar display, and Radar Altimeter indicator. These indicators, with the exception of the Radar Altimeter, are classified as navigational instruments since they provide information which enables the pilot to guide the aircraft accurately along a given course.

Instruments that aid the pilot in controlling the altitude, attitude, airspeed, and direction of the aircraft are known as flight instruments. These instruments are classified as either pneumatic, electropneumatic, or electrically driven from a central air data computer. The primary flight instruments required on all aircraft include the Barometric Altimeter, Airspeed Indicator, Turn-and-Bank Indicator, and Magnetic Compass. Small private aircraft will usually have, in addition to the primary flight instruments, an Artificial Horizon Indicator to display the attitude of the aircraft, and possibly a Vertical Speed Indicator to indicate the aircraft's rate of climb or dive.

Larger aircraft come equipped with flight director Attitude Direction Indicators (ADI) and Horizontal Situation Indicators (HSI). The ADI is essentially an artificial horizon with lateral bars superimposed to display computer-generated pitch and bank steering commands from the flight director system. The HSI is similiar to a CDI, except that it combines VOR/ILS information or long-range navigation (e.g., Global Positioning System, Very Low Frequency Navigation System, and/or Inertial Navigation System) information with a gyro-slaved compass card to provide heading, actual track, desired track, track angle error, drift angle, crosstrack deviation, and distance information on a single instrument. These instruments will be discussed in Chapter Twelve.

Other instruments that are commonly found installed in the instrument panels of large transport and corporate jet aircraft include the Radar Altimeter, Altitude Alerter, Angle-of-Attack, True Airspeed (TAS), Static Air Temperature (SAT), and Radio Magnetic Indicators. The location of some of these instruments are illustrated in the McDonnell Douglas DC-10 First Officier's instrument panel illustration shown in Figure 10-1. These instruments and more will be discussed in detail in the following paragraphs. But first we begin this discussion with the operation of two instruments not shown in Figure 10-1, the Turn and Bank and Angle-of-Attack indicators.

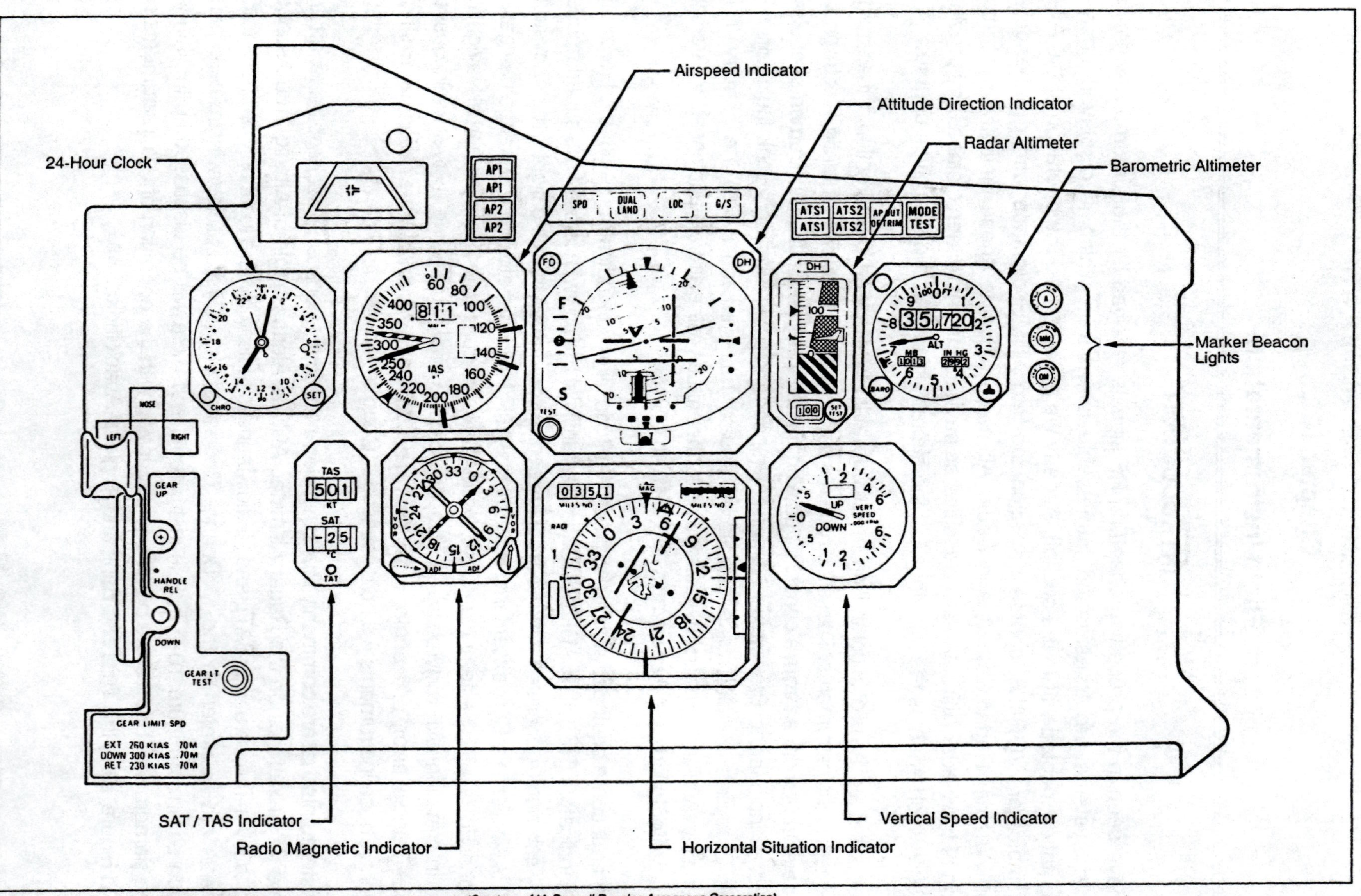

Figure 10-1. DC-10 First Officer's Instrument Panel. *(Courtesy of McDonnell Douglas Aerospace Corporation)*

TURN-AND-BANK INDICATOR OPERATION

The Turn-and-Bank indicator shown in Figure 10-2 shows the correct execution of a bank and turn and indicates the lateral attitude of the aircraft in level flight. The turn needle is operated by an electric gyroscope. A gyroscope consists of a rotor mounted on moveable frames. When the rotor spins at high speed, the axle (spin axis) on which it turns continues to point to the same direction, no matter how the frames are moved. The spin axis is automatically positioned parallel to the earth's axis. A gyro turns over slowly as the earth rotates. This turning, called precession, can cause the spin axis to change direction. The turn needle indicates the rate, in number of degrees per second, at which an aircraft is turning about its vertical axis. It also provides information on the amount of bank.

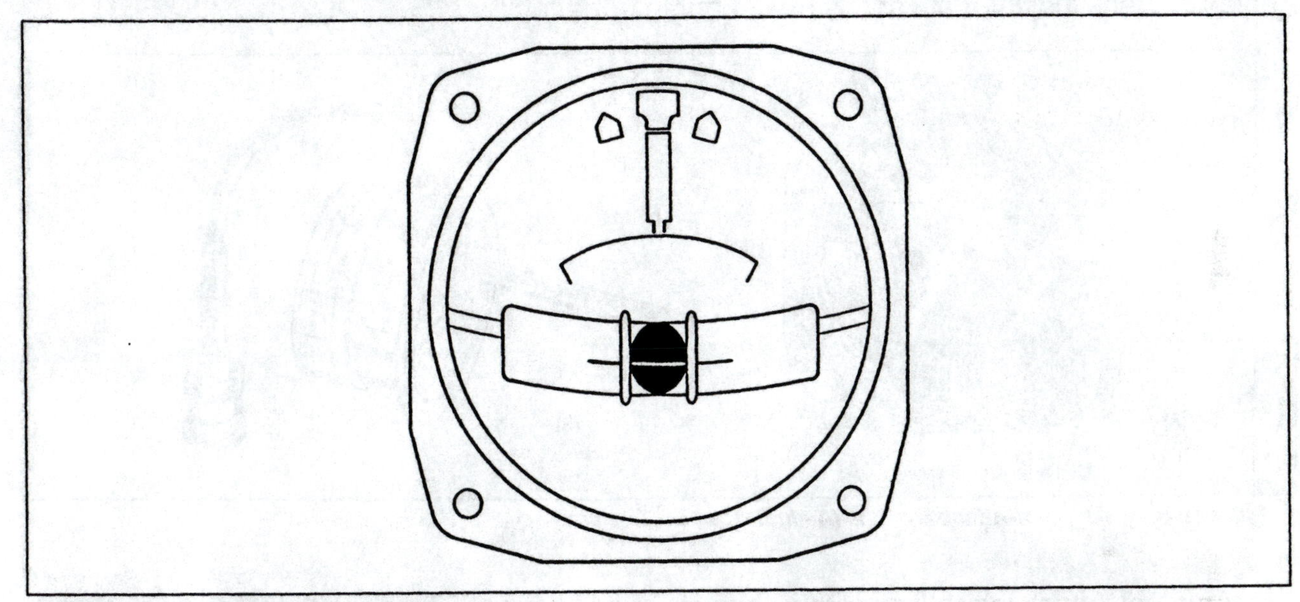

Figure 10-2. Turn-and-Bank Indicator.

The gyro axis is horizontally mounted so that the gyro rotates up and away from the pilot. The gimbal around the gyro is pivoted fore and aft. Gyroscopic precession causes the rotor to tilt when the aircraft is turned. Due to the direction of rotation, the gyro assembly tilts in the opposite direction from which the aircraft is turning. This prevents the rotor axis from becoming vertical to the earth's surface. The linkage between the gyro assembly and the turn needle, called the reversing mechanism, causes the needle to indicate the proper direction of turn. A single needle-width deflection shows when the aircraft is turning at one and one half degrees per second, or half standard rate (four minutes for a 360° turn).

The slip indicator (ball) part of the instrument is a simple inclinometer consisting of a sealed, curved glass tube containing kerosene and a black agate or a common steel ball bearing, which is free to move inside the tube. The fluid provides a damping action, ensuring smooth and easy movement of the ball. The tube is curved so that in a horizontal position the ball tends to seek the lowest point. During coordinated turns and straight-and-level flight, the force of gravity causes the ball to rest in the lowest part of the tube, centered between the reference wires.

ANGLE-OF-ATTACK SYSTEM OPERATION

The Angle-of-Attack indicating system detects the local angle of attack of the aircraft from a point on the side of the fuselage and furnishes reference information for the control and actuation of other units and systems in the aircraft. Signals are provided to operate an Angle-of-Attack indicator located on the instrument panel, where a continuous visual indication of the local angle-of-attack is displayed.

A typical Angle-of-Attack system, shown in Figure 10-3, provides electrical signals for the operation of a rudder pedal shaker, which warns the pilot of an impending stall when the aircraft is approaching the critical stall angle-of-attack. Electrical switches are actuated at the Angle-of-Attack indicator at various preset angles-of-attack.

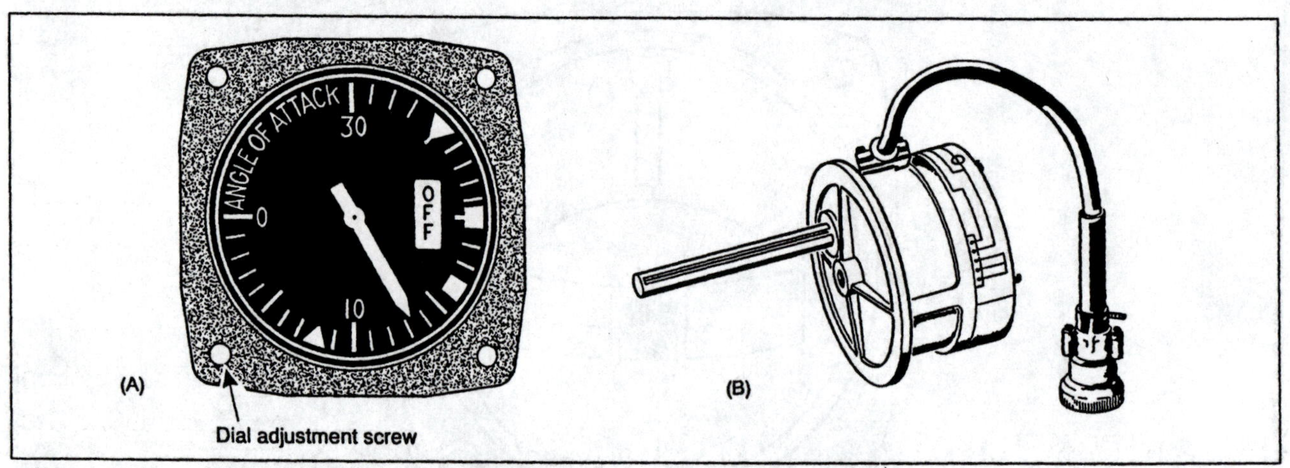

Figure 10-3. Angle-of-Attack System. (A - Indicator, B - Transmitter)

The Angle-of-Attack indicating system consists of an airstream direction detector (transmitter), and an indicator located on the instrument panel. The airstream direction detector contains the sensing element which measures local airflow direction relative to the true angle-of-attack by determining the angular difference between local airflow and the fuselage reference plane. The sensing element operates in conjunction with a servo-driven balanced bridge circuit which converts probe positions into electrical signals.

The operation of the Angle-of-Attack sensing element is based on detection of differential pressure at a point where the airstream is flowing in a direction that is not parallel to the true angle-of-attack of the aircraft. This differential pressure is caused by changes in airflow around the probe unit. The probe extends through the skin of the aircraft into the airstream.

The exposed end of the probe contains two parallel slots which detect the differential airflow pressure, as shown in Figure 10-4. Air from the slots is transmitted through two separate air passages to separate compartments in a paddle chamber. Any differential pressure, caused by misalignment of the probe with respect to the direction of airflow, will cause the paddle to

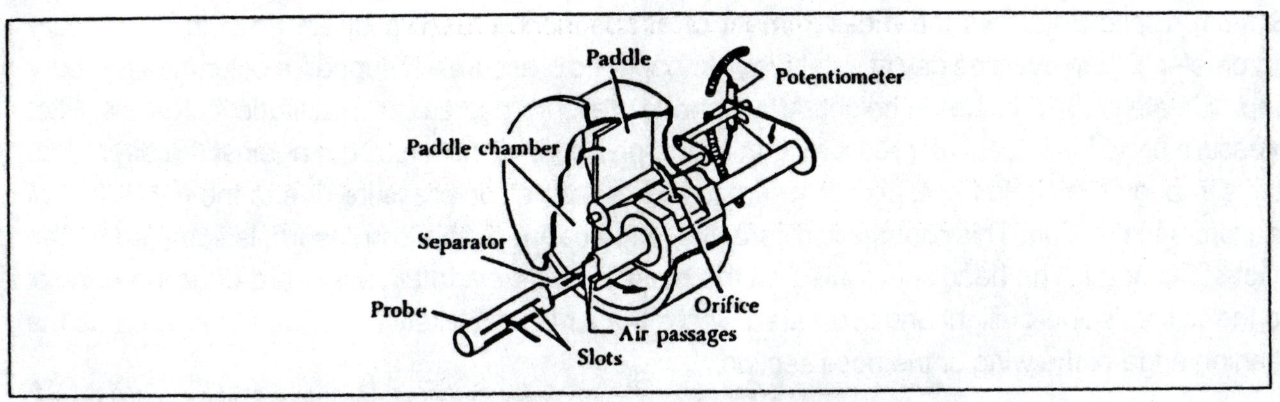

Figure 10-4. ***Airstream Direction Detector.*** *(Courtesy of the Federal Aviation Administration)*

rotate. The moving paddles will rotate the probe, through mechanical linkage, until the pressure differential is zero. This occurs when the slots are symmetrical with the airstream direction.

Two electrically separate potentiometer wipers, rotating with the probe, provide signals for remote indications. Probe position, or rotation, is converted into an electrical signal by one of the potentiometers which is the transmitter component of a self-balancing bridge circuit. When the angle-of-attack of the aircraft is changed and, subsequently, the position of the transmitter potentiometer is altered, an error voltage exists between the transmitter potentiometer and the receiver potentiometer indicator. Current flows through a sensitive polarized relay to rotate a servomotor in the indicator. The servomotor drives a receiver/ potentiometer in the direction required to reduce the voltage and restore the circuit to an electrically balanced condition. The indicating pointer is attached to, and moves with, the receiver/potentiometer wiper arm to indicate on the dial the relative angle-of-attack.

When the aircraft is in the approach configuration, VREF speed will be achieved when the indicator pointer is in the center of the approach band index, located at the three o'clock position on the indicator dial. With the pointer center in this position, VREF speed will be obtained regardless of gross weight or "G" forces.

The continuous presentation of the margin from stall is always valuable information, but particularly when minimum energy landings are required (wet, icy, or short runways). Indices are placed on the Angle-of-Attack Indicator dial for optimum approach, maximum endurance (maximum lift/drag), approximate maximum range, and low-speed buffet onset. The Angle-of-Attack Indicating System is flap compensated, therefore capable of presenting a valid lift indication in all flap configurations. It interfaces with the FAST-SLOW indicator on all flight directors and a glareshield-mounted indexer providing the deviation from the optimum approach reference.

INTRODUCTION TO PITOT-STATIC SYSTEMS

The FAA requires that the most essential flight instruments, such as the Altimeter, Airspeed Indicator, and Rate-of-Climb Indicator, operate from outside air pressure inputs provided by the Pitot-Static

System. Static air allows the measurement of atmospheric pressure by an aneroid or mercury barometer. On an average day at sea-level, atmospheric pressure will support a column of mercury approximately 29.92 inches in height. Atmospheric pressure decreases as altitude increases. Pitot pressure provides impact air pressure; that is, the pressure of the airstream against the aircraft as it flys through the air. In flight, pitot pressure is higher than static pressure due to the ram effect of the aircraft in motion. This pitot-static differential, often called dynamic pressure, is sampled by the pitot-static head. The head is mounted on the outside of the aircraft in a forward direction parallel to the aircraft's line-of-flight and in an area where the air is least likely to be turbulent, such as the leading edge of the wing or the nose section.

As shown in Figure 10-5, the front end of the head is open to receive the full force of the impact (pitot) air pressure. A baffle plate is installed just inside this opening to prevent moisture and debris from entering the pitot tube. Moisture can escape through a small drain hole located at the bottom of the unit. The pitot tube leads back to a chamber near the rear of the assembly. A riser, or upright tube, routes the air from this chamber through tubing in the aircraft to the airspeed indicator and auxiliary equipment, such as the Central Air Data Computer.

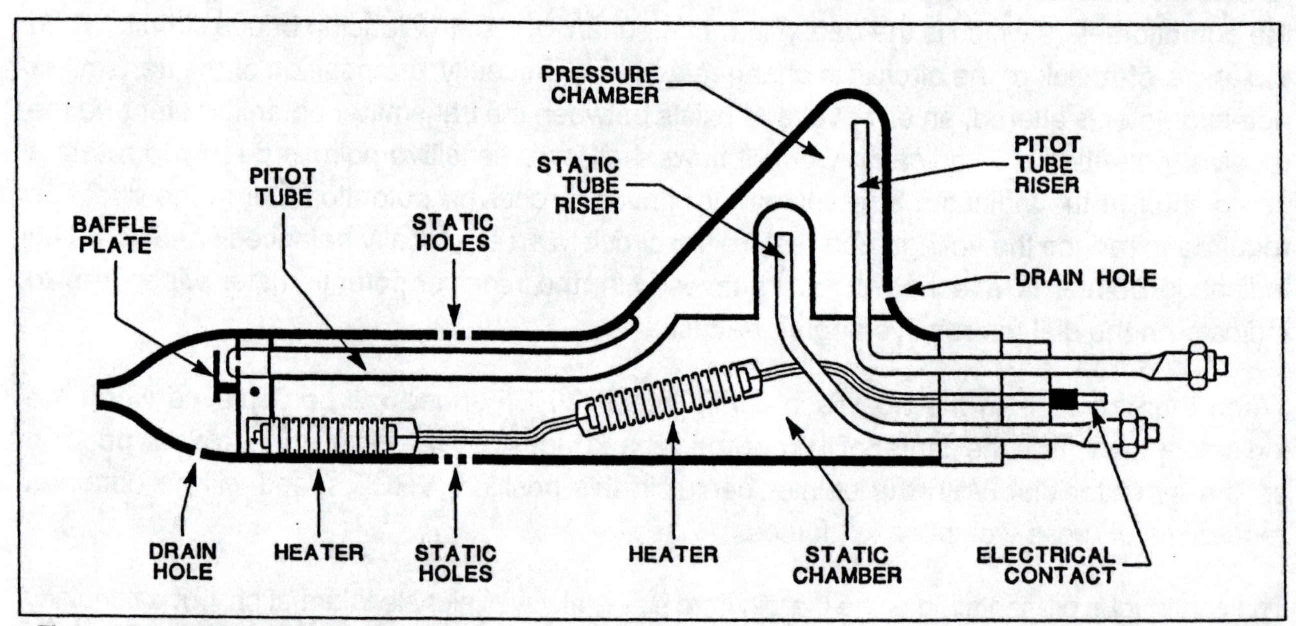

Figure 10-5. Pitot/Static System Head.

The outside of the pitot-static head is pierced with small openings on the top and bottom surfaces that allows still (static) air to enter into the static air chamber. A riser tube in the static chamber provides a sample of this atmospheric pressure to the Altimeter, Airspeed Indicator, Rate-of-Climb Indicator, and other units that require this information, such as the Cabin Pressure Controller and the Cabin Differential Pressure Gage used in pressurized aircraft. A heating element is located within the head to prevent the unit from becoming clogged due to icing conditions experienced during flight. The heater is controlled by means of a switch in the cockpit.

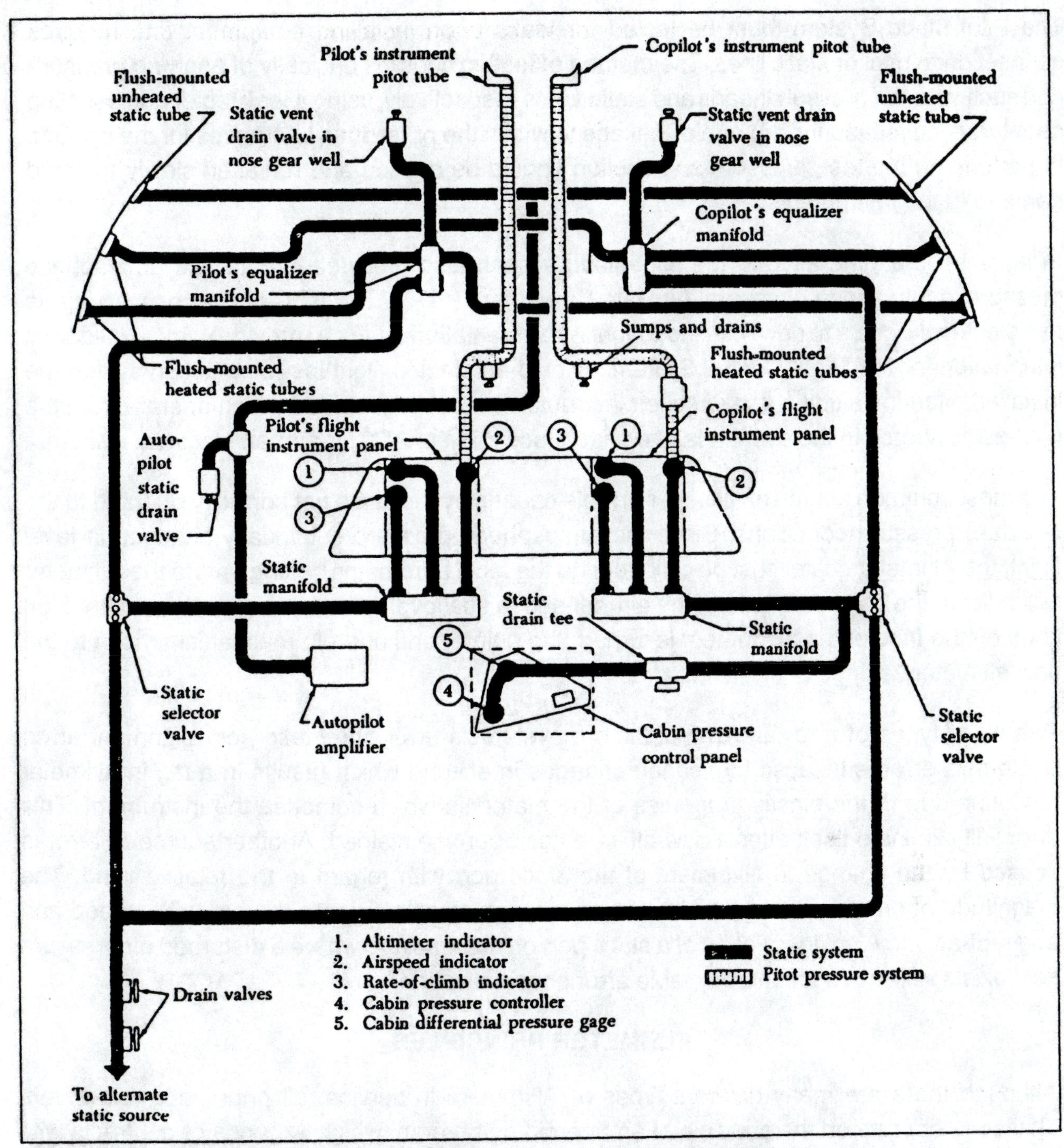

Figure 10-6. ***Typical Pitot-Static System Diagram.*** *(Courtesy of the Federal Aviation Administration)*

An alternate method of deriving impact and static pressure may be obtained from separate pitot tubes and static vents installed in different locations on the aircraft. A schematic diagram of a Pitot-Static System of this type is illustrated in Figure 10-6. Note that the left and right static ports are paralleled together to compensate for any variation in static pressure due to erratic changes in the aircraft's attitude.

The Pitot-Static System must be tested for leaks upon installing equipment that requires connection to pitot or static lines. The method of testing consists basically of applying pressure and suction to the pressure heads and static vents respectively, using a leak tester and coupling adapters, and measuring the rate of leakage to within the prescribed tolerances for the system. In performing the test, pressure and suction should be applied and released slowly to avoid damage to the instruments.

In large multi-engine aircraft, the pitot-static system also provides impact and atmospheric pressure inputs to one or several Air Data Computers (ADC). The ADC converts these inputs into electrical signals to drive electrical flight instrumentation and/or to provide vertical guidance information for the Flight Control System. Should an electric flight instrumentation system be installed, standby altitude and airspeed instruments must be provided which operate from the pitot-static source in the event a failure should occur in the ADC or aircraft electrical system.

The most common source of altimeter error is caused by the scale not correctly oriented to the standard pressure conditions. Since the atmospheric pressure continually changes in level flight, the Altimeter scale must be calibrated to the local barometric setting before the Altimeter will indicate the correct altitude of the aircraft above sea level. When the barometric correction knob on the face of the instrument is turned, the pointer and aneroid mechanism move to the new altimeter setting.

Two other types of error are also common: hysteresis error and static port alignment error. Hysteresis error is induced by sudden changes in altitude which results in a lag in altimeter response due to the elastic properties of the materials which comprise the instrument. This error will eliminate itself after a new altitude has been maintained. Another source of error is caused by the change in alignment of the static port with regard to the relative wind. The magnitude of error caused by static port misalignment will vary with the aircraft's speed and angle-of-attack. Also, installation of a static port on the fuselage where a disturbed air pressure field exists will cause an unpredictable erroneous indication.

ALTIMETER PRINCIPLES

Although there are many different types of Altimeters in service, all pnuematic (air-driven) altimeters operate on the principle of an aneroid diaphragm, which expands or contracts with changes in atmospheric pressure. As the aircraft altitude increases, the static source senses the decrease in atmospheric pressure and causes the aneroid to contract. This movement of the aneroid actuates mechanical linkages which drives the pointer to the proper altitude scale reading on the indicator. A bi-metallic yoke surrounding the aneroid compensates for inaccuracies caused by variations in temperature. An illustration of a pneumatic altimeter mechanism is shown in Figure 10-7.

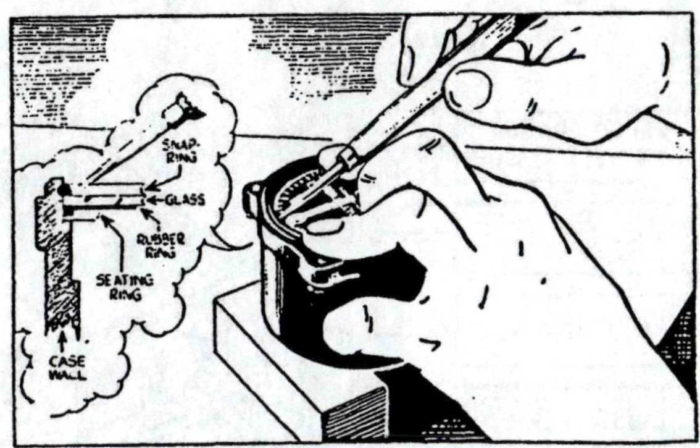

To take out the outer snap-ring, put a screw-driver or the blade of a knife under one end of the ring where its ends come together, lifting it out.

To take out the glass, a suction-cup is pushed hard against the glass, and lifted up.

Figure 10-7. Typical Pneumatic Altimeter Mechanism. *(Courtesy of Kollsman Division of Sequa Corporation)*

RADBAR ENCODING ALTIMETER SYSTEM OPERATION

The Kollsman RADBAR (RADio-BARometric) Electropneumatic Encoding Altimeter System, as depicted in Figure 10-8, is designed to provide both barometric and radar altitude information in a single instrument. The barometric section of the Altimeter measures and displays static-defect-corrected pressure altitude, and encodes that altitude into a binary code output that is sent to the onboard radio beacon Transponder for automatic altitude reporting (when operating in Mode C). Synchro outputs of the baro-corrected altitude are provided for the Altitude Alerter, Flight Management System and/or Long-Range Navigation System, and an altitude rate signal is provided to drive an electric Vertical Speed Indicator.

The RADBAR Indicator is shown in Figure 10-9. The basic mechanism of the barometric section of the indicator is a servo operated, diaphragm displacement, follow-up transducer. By sensing the diaphragm motion through an inductive pickoff, very accurate repeatable altitude

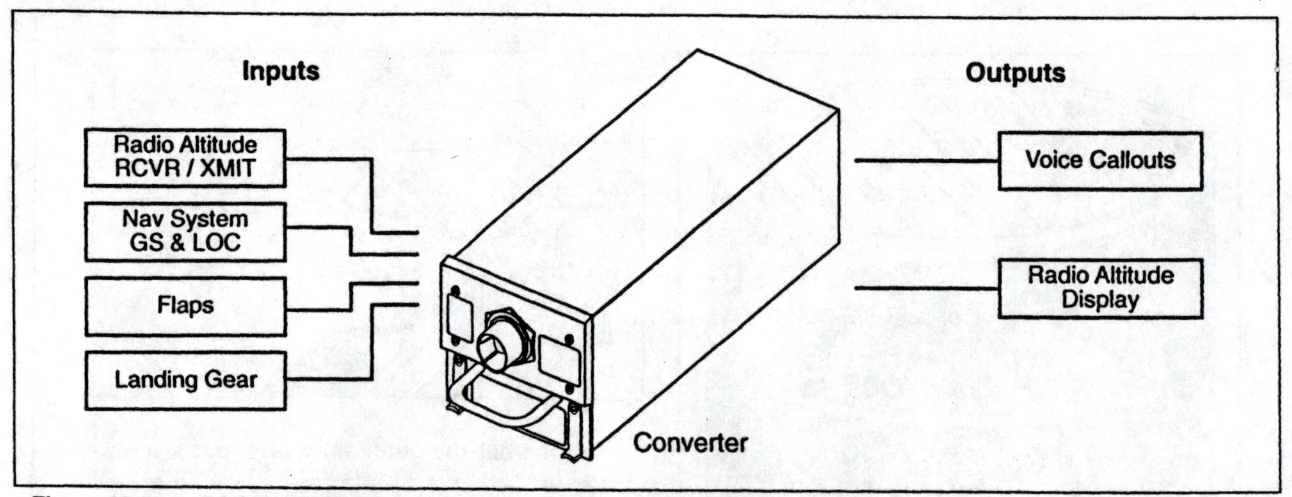

Figure 10-8. ***RADBAR Altimeter System.*** *(Courtesy of Kollsman Division of Sequa Corporation)*

measurements are made and presented on the altimeter counter-drum pointer display. The pointer makes one revolution per 1,000 feet and graduations are at 20-foot increments. The counter reads altitude in thousands and hundreds of feet.

The knob located in the lower right-hand corner of the RADBAR Altimeter permits setting of the local barometric pressure in both inches of mercury and millibars. Setting of the barometric scale knob does not affect the reported altitude information sent to the Transponder, but only the indicated altitude reading on the Altimeter. Reported altitude is always the altitude relative to standard barometic pressure at sea level (29.92 inches of mercury).

Figure 10-9. ***RADBAR Altimeter.*** *(Courtesy of Kollsman Division of Sequa Corporation)*

DH and Self-Test Operation

Depressing the decision height (DH) knob on the top right-hand corner of the RADBAR indicator will cause the radio altitude display to readout the preset decision height. Depressing and rotating the knob allows the new DH to be set Into the system. Whenever the DH knob is depressed to display the DH altitude, the visual and aural DH warning circuit is self-tested. A self-test button is also included to test both the barometric and radio altitude sections of the altimeter. Depressing the self-test button displaces the pointer approximately 300 feet, and causes the digital readout to display all "8s". Releasing the button initiates a test of the complete radio altimeter system.

Altitude Digitizer

The optical encoder in the barometric section of the altimeter uses light-emitting diodes as light sources and photo-darlington transistors as light sensors. An encoding disk with nine concentric tracks of clear and opaque sections is positioned between the LEDs and photo sensors. As the disk is rotated by the altimeter aneroid mechanism, the light passing through the clear sections of the disks tracks the altitude changes. For every 100 feet of altitude, a new combination of light patterns reaches the sensors to produce the coded altitude information which is processed into a binary code output and sent to the Transponder.

Static Defect Correction (SDC) Module

The airflow past the static port will cause the pressure in the static system to be different from the undisturbed air. At low speeds, this effect is insignificant, but at higher speeds accustomed to jet aircraft, static source error correction is required. This correction is a function of Mach number and is performed by a Static Defect Correction module. The SDC module measures pitot and static pressures and, through the use of distinct calibration, provides the necessary information to the altimeter to correct for static source error. This is accomplished though an airspeed servo network that provides outputs to automatically adjustable calibration potentiometers to eliminate or reduce the static error of the particular aircraft at designated altitude-airspeed points in the flight envelope.

The basic mechanism of the SDC module is a servo operated, diaphragm displacement, follow-up airspeed transducer; servomotor generator; servo amplifier; and failure monitor circuit. The servo shaft of the airspeed transducer in the SDC module operates with tapped potentiometers that work in conjunction with a bank of adjustable potentiometers corresponding to the altitude-airspeed points distributed over the complete flight envelope.

A change in airspeed produces an output signal in the SDC airspeed transducer. This signal is fed into a servo amplifier which drives the servo motor to position the shaft of the transducer so that its output is nulled. The shaft also positions a multi-deck, multi-tapped potentiometer, which is connected to a bank of stationary trimming potentiometers forming the airspeed static error correction circuit. During flight, the signal from the trimming potentiometers permits independent correction of the static source error at altitudes between zero and 50,000 feet, and airspeeds between 175 and 400 knots.

The knob in the lower left-hand corner of the encoding altimeter positions the STANDBY-NORMAL switch (see Figure 10-9). When the instrument is used with a SDC module, the NORMAL position allows the altimeter to accept static defect corrections. In the event of a failure in the SDC module, the failure flag on the indicator will appear and the common lead of the altitude encoder will open to prevent the Transponder from reporting an inaccurate altitude. When the failure flag appears, the pilot should select the STANDBY position to stow the flag and allow the altimeter to operate without static error correction. Once in the STANDBY mode, the flag remains armed to signal a failure in the altimeter.

Radio Altitude Readout with Voice Terrain Advisory (VTA)

The radio altimeter section of the RADBAR altimeter is connected to a radio altitude digital-to-analog converter to provide a three-digit, seven-segment, incandescent digital readout of the absolute altitude or height of the aircraft above the terrain. The radio altitude converter receives analog signals from the radio altimeter RT, and glideslope signals from the ILS navigation receiver. The radio altitude signals are converted to digital form to drive the radio altimeter display in the RADBAR altimeter, and provide voice terrain advisory outputs of radio altitude. The localizer and glideslope signals are also converted to provide voice output warnings of excessive localizer and/or glideslope deviation from center on ILS approaches.

In actual operation, when the aircraft is on the ground, the radio altitude section will display zero feet and a placard will show the legend "RADIO ALT". After take-off, the radio altitude display will indicate the height of the aircraft above the terrain up to 990 feet. The voice function remains silent during take-off unless a negative radio altitude rate is sensed. The illuminated placard, "RADIO ALT", will remain on up to an altitude of 2,500 feet above ground level (AGL) and then turn off. As the aircraft descends to the approach altitude, the placard will illuminate at 2,500 feet AGL, and the radio altitude converter synthesized voice output will announce "TERRAIN" at 2,000 feet AGL. Radio Altitude will be displayed in 100-foot increments down to 1,000 feet AGL. The "TERRAIN" announcement will be repeated at the 1,000-foot level, and shortly thereafter, the radio altitude numerical display will illuminate at 990 feet above the terrain and begin to count down in 10-foot increments.

At 900 feet AGL, the VTA will announce "NINE HUNDRED"; at 800 feet AGL, the announcement "EIGHT HUNDRED" will be heard. The announcements will continue at each 100-foot interval (700, 600, 500, etc.) until touchdown. When the aircraft reaches the minimum decision height (DH), or the altitude at which the pilot must declare a missed approach ("go-around") and abort the landing if the runway is not yet visable, the external DH warning light will illuminate, the aural warning from the external tone generator will sound, and the VTA announcement

"MINIMUM" will be heard over the cockpit speaker. The system also provides voice advisories for reminding the pilot to extend the flaps and lower the landing gear.

If the aircraft deviates below the glideslope less than one dot on the CDI or HSI, the announcement "GLIDESLOPE" will be heard at a nominal voice level at two second intervals. Should the aircraft deviate below the glideslope more than one dot, the announcement "GLIDESLOPE" will be heard at above nominal voice levels at one second intervals. Similar operation is provided for localizer deviations.

ALTITUDE ALERTER OPERATION

The altitude alerter, shown in Figure 10-10, is an optional instrument used to alert the pilot of an approach to, or departure from, a preset flight level by means of automatic audio and visual signals. The input signals required to operate the altitude alerter consist of a two-speed, baro-corrected, synchro altitude signal from a servoed altimeter (such as the one previously discussed), and a 26-volt AC 400-Hz reference voltage from the aircraft electrical bus instrumentation transformer.

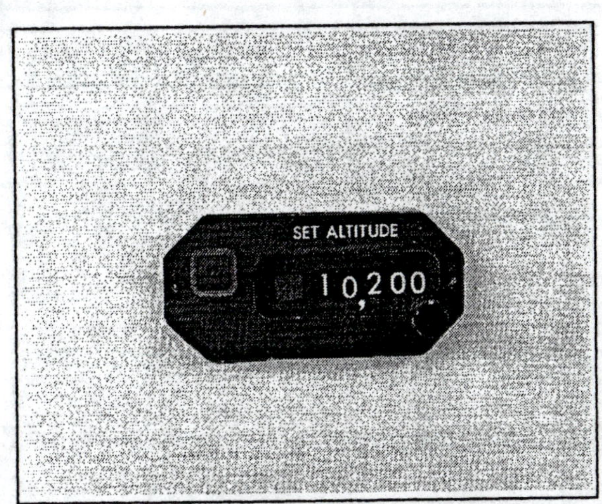

Figure 10-10. Altitude Alerter/Preselector.
(Courtesy of Kollsman Division of Sequa Corporation)

The altitude alerter incorporates a five-digit display to indicate the preset flight level, which can be adjusted by means of a knob located in the lower right-hand corner of the instrument. A three-digit counter displays thousands and hundreds of feet. Fixed zeros are placed in the tens and units digits. An altitude alert light is located on the left side of the instrument to indicate approach to or departure from the preset flight level.

In addition to altitude performance data, the alerter displays an "OFF" flag to indicate a loss of either the altitude valid signal from the altimeter or 26-volt AC power. The alerter also contains dual coarse and fine differential resolver synchros to provide an altitude preselect output for

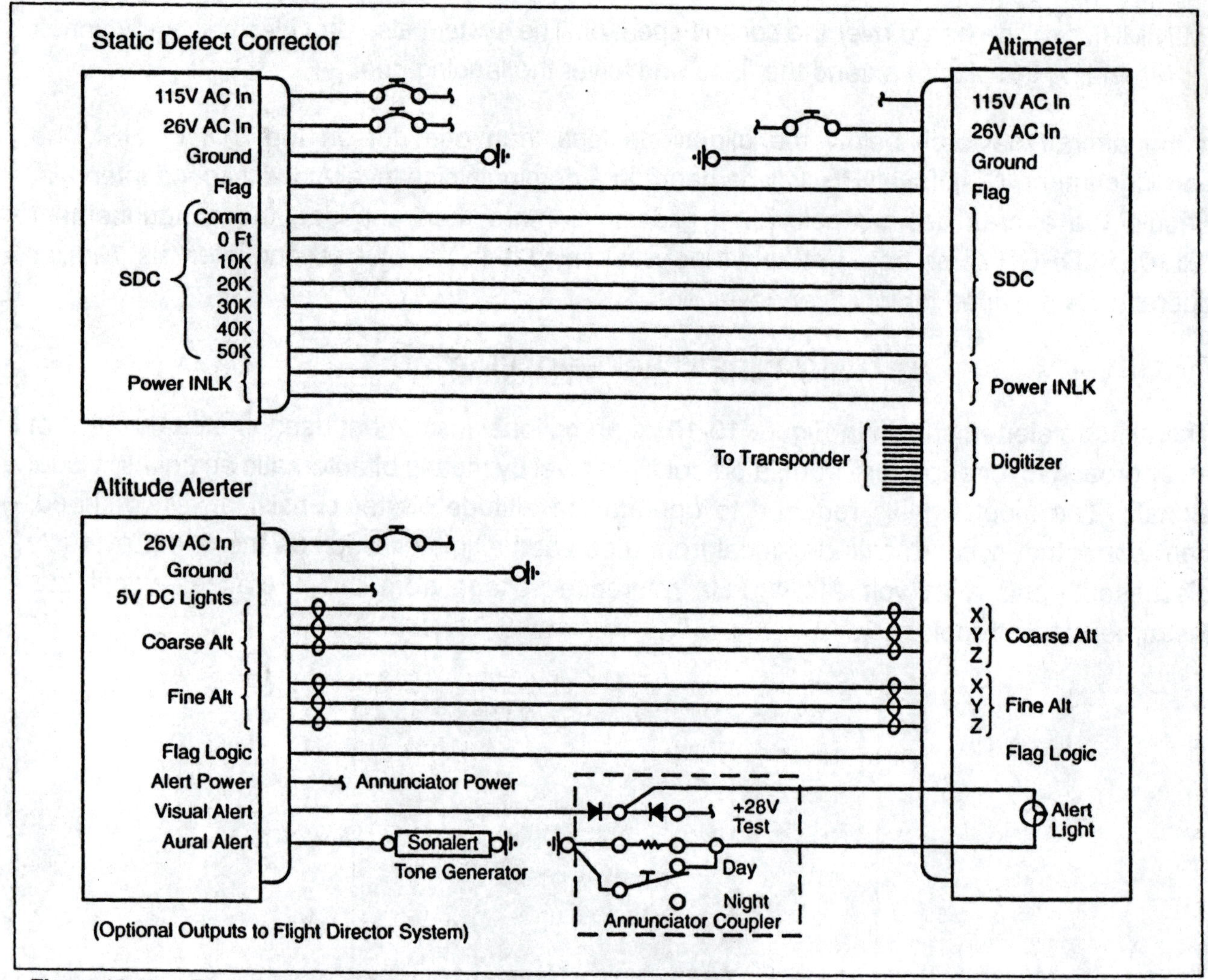

Figure 10-11. Typical Altimeter/Alerter Wiring Diagram.

the flight control system. An interconnect wiring diagram of a typical altimeter/alerter system installation is shown in Figure 10-11.

Flight Test Procedures

In actual operation, as the aircraft approaches within 1,000 feet of the preset flight level, the altitude alert light comes on and an aural tone is momentarily heard (for approximately two seconds) through an external tone generator, such as the Mallory Sonalert. The light, however, remains on until the aircraft crosses 300 feet of the preset altitude, at which time it extinguishes. Should the aircraft deviate from the preset altitude by more than plus or minus 300 feet, the aural tone and alert light will again be activated. The light will then remain on until the aircraft returns to within 300 feet of the preset altitude, or until the pilot selects a new altitude setting on the alerter indicator.

AIRSPEED INDICATOR PRINCIPLES

The airspeed indicator is basically a differential pressure gage which measures and indicates the difference between impact and static air pressures. As shown in Figure 10-12, the airspeed indicator consists primarily of a sensitive metallic diaphragm that responds to pitot-static pressure differential and multiplies this movement through mechanical linkages to impart the aircraft's velocity on the dial face in terms of knots or miles per hour.

The majority of airspeed indicators are marked to indicate speed limitations at a glance. The never-exceed velocity is designated by a red radial line; a yellow arc designates the cautionary range; and a white arc is used to indicate the range of permissible limits for flap extension.

Maximum Allowable Airspeed Indicators

Another type of airspeed indicator commonly used is the maximum-allowable airspeed indicator. This indicator includes a maximum-allowable needle, which shows a decrease in maximum-allowable airspeed with an increase in altitude. The needle operates from an extra diaphragm, which senses changes in altitude, to provide an indication of the maximum-allowable airspeed at any altitude.

Machmeters

Machmeters, installed onboard high-performance aircraft, indicate the ratio of aircraft speed to the speed of sound at the particular altitude and temperature existing at any time during flight. A machmeter usually consists of a differential pressure diaphragm which senses pitot-static pressure, and an aneroid diaphragm which senses static pressure. The actions of these diaphragms are mechanically resolved to display mach numbers on the face of the instrument.

Combined Airspeed/Mach Indicators

Combined airspeed/mach indicators are usually provided where instrument space is at a premium and it is desirable to present this information on a combined indicator. These instruments display indicated airspeed, mach, and limiting mach by means of differential pressure and aneroid diaphragms. The combined airspeed/mach indicator uses a single needle which indicates airspeed on a fixed scale and mach on a rotating scale. A knob, located on instrument, is provided to set a movable index marker to reference a desired airspeed/mach setting.

MAXIMUM-ALLOWABLE AIRSPEED/MACH INDICATOR OPERATION

The Kollsman maximum-allowable airspeed/mach indicator, illustrated in Figure 10-13, is intended for use in subsonic aircraft to provide airspeed, mach, and maximum allowable airspeed indications in a single instrument. The indicator comprises two separate capsule mechanisms, a stationary airspeed dial, a moving mach sub-dial, a single airspeed pointer, a

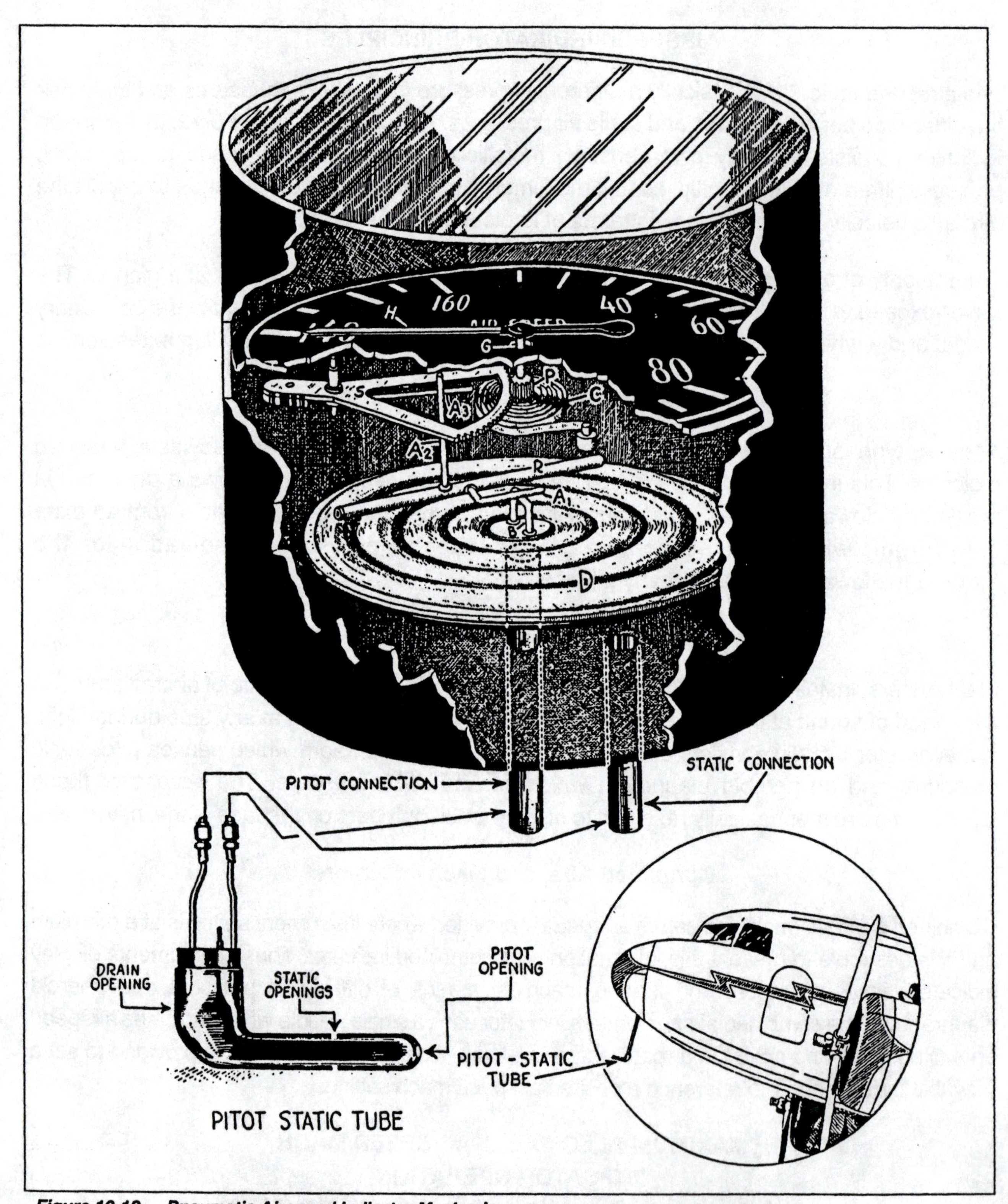

Figure 10-12. Pneumatic Airspeed Indicator Mechanism. *(Courtesy of Kollsman Division of Sequa Corporation)*

Figure 10-13. Maximum-Allowable Airspeed/Mach Indicator.
(Courtesy of Kollsman Division of Sequa Corporation)

maximum-allowable airspeed indication, and an index marker which can be rotated by means of a knob in the lower left-hand corner of the instrument.

A white pointer indicates airspeed against a fixed subdial calibrated from 60 knots to 400 knots, and mach number against a moving mach subdial calibrated from 0.3 mach to 1.0 mach. Maximum-allowable airspeed is indicated on the dial by a yellow and red marker.

The index marker is basically a manually-set command bug that permits the pilot to select a desired approach speed and provides signal outputs to drive the fast/slow speed command needle on the attitude direction indicator. The speed command output signals are determined by the difference between the airspeed pointer and index marker. The indicator also provides airspeed hold, mach hold, and AC and DC airspeed and mach analog outputs for use in the flight control system.

Functional Description

The Kollsman maximum allowable airspeed/mach indicator comprises primarily of two mechanically independent diaphragm-operated mechanisms; one responding to the differential pressure between pitot and static pressures, and the other to static pressure only. The airspeed mechanism of the indicator is driven from a diaphragm capsule with pitot pressure introduced into its interior, while the exterior of the diaphragm is subjected to the static pressure introduced into the instrument case. The capsule expands approximately linearly with the indicated airspeed. The motion of the centerpiece of the capsule is transmitted to a jewelled rocking shaft by a dual link and lever arrangement.

One link is connected to a short lever on the rocking shaft, which is at nearly right angles to the link. This lever is lengthened or shortened to provide the correct calibration of the pointer on the spread-out linear section of the dial up to 200 knots. At 190 knots, the second link comes into contact with its lever. The second lever is longer than the first, and at 190 knots, is more nearly in line with the link. This angular relation between link and lever converts the diaphragm's linear motion into rotational movement, linear with the log of differential pressure. This nonlinear portion of the dial above 190 knots is required in order to correctly indicate mach. Both links are slotted so each can overtravel while the other is active. A hairspring on the rocking shaft provides tension on the active link at all times.

The rocking shaft motion is amplified by a step-by-step gear pair to turn the jewelled pointer shaft. Damping of the airspeed pointer is accomplished by a capillary tube through which the pitot pressure is fed into the capsule. The indication is compensated for ambient temperature changes by selection of materials for the frame, linkage, and capsule. Static balance is achieved through an adjustable counterweight on the rocking shaft and a balanced pointer shaft.

The mach subdial is driven from an evacuated diaphragm capsule through a link, lever, rocking shaft, and gearup stage. The capsule expansion is linear with pressure altitude, and when modified by link and lever, the subdial is rotated in proportion to the log of static pressure. In the manner of a circular slide rule, this causes the correct mach number for any altitude to appear opposite the corresponding airspeed value. The airspeed pointer indicates both mach and airspeed over the range where the sub-dial is visible.

Automatic Maximum Allowable Airspeed

This indicator is also available with a moveable, red-striped, radial marker to show the maximum allowable airspeed. If this option is elected, the marker will be positioned over the mach subdial by the altitude capsule so that its lower edge is opposite 360 knots at all altitudes up to 19,300 feet. Above this altitude, the radial marker moves downward as if attached to the mach scale with its lower edge at 0.765 mach. The marker indicates the maximum speed for both Vmo (maximum allowable airspeed) and Mmo (maximum allowable mach).

TAS/SAT INDICATOR OPERATION

Modern commercial aircraft instrument panels often have installed a true airspeed (TAS) indicator to supplement the information provided by the airspeed/mach indicator. True airspeed is indicated airspeed corrected for variations in pressure altitude and temperature. TAS is always greater than indicated airspeed (IAS), except in extremely low temperatures at low altitudes.

Pneumatic TAS Indicator

A typical pnuematic TAS indicator consists of an aneroid, differential pressure diaphragm, and bulb temperature diaphragm; which responds to changes in barometric pressure, impact

Figure 10-14. ***TAS/SAT Indicator.*** *(Courtesy of Kollsman Division of Sequa Corporation)*

pressure differential, and free air temperature, respectively. The actions of the diaphragms are mechanically resolved to show true airspeed in knots on the dial face of the indicator.

Electric TAS/SAT System

The Kollsman TAS/SAT indicator, shown in Figure 10-14, is an electric flight instrument that provides a digital incandescent display of both true airspeed and static air temperature (SAT) from analog signals received from a TAS/SAT computer, or from an independent airspeed network contained within the static defect correction module. In poor weather, SAT indicates the potential for icing to occur. During takeoff, it affects the amount of thrust available from the engines and available lift due to air density.

The outside air temperature input for TAS computation is derived from a fuselage-mounted total air temperature probe that contains a thermistor. The thermistor has a negative temperature coefficient, or in other words, its resistance increases upon sensing a decrease in outside air temperature. At 0°C, the resistance of the temperature sensing element in the probe is typically 500 ohms.

Within the TAS/SAT computer, a matrix of trimming potentiometers, connected to pots on the altitude and airspeed servos, compute the function of mach. The combination of temperature and mach produces the required true airspeed output. Another circuit within the TAS/SAT computer uses the same temperature probe input signal to provide a static air temperature output. The TAS and SAT output signals from the computer are provided to the TAS/SAT indicator. The TAS output is also provided to the long-range navigation system or flight management system for computation of ground speed.

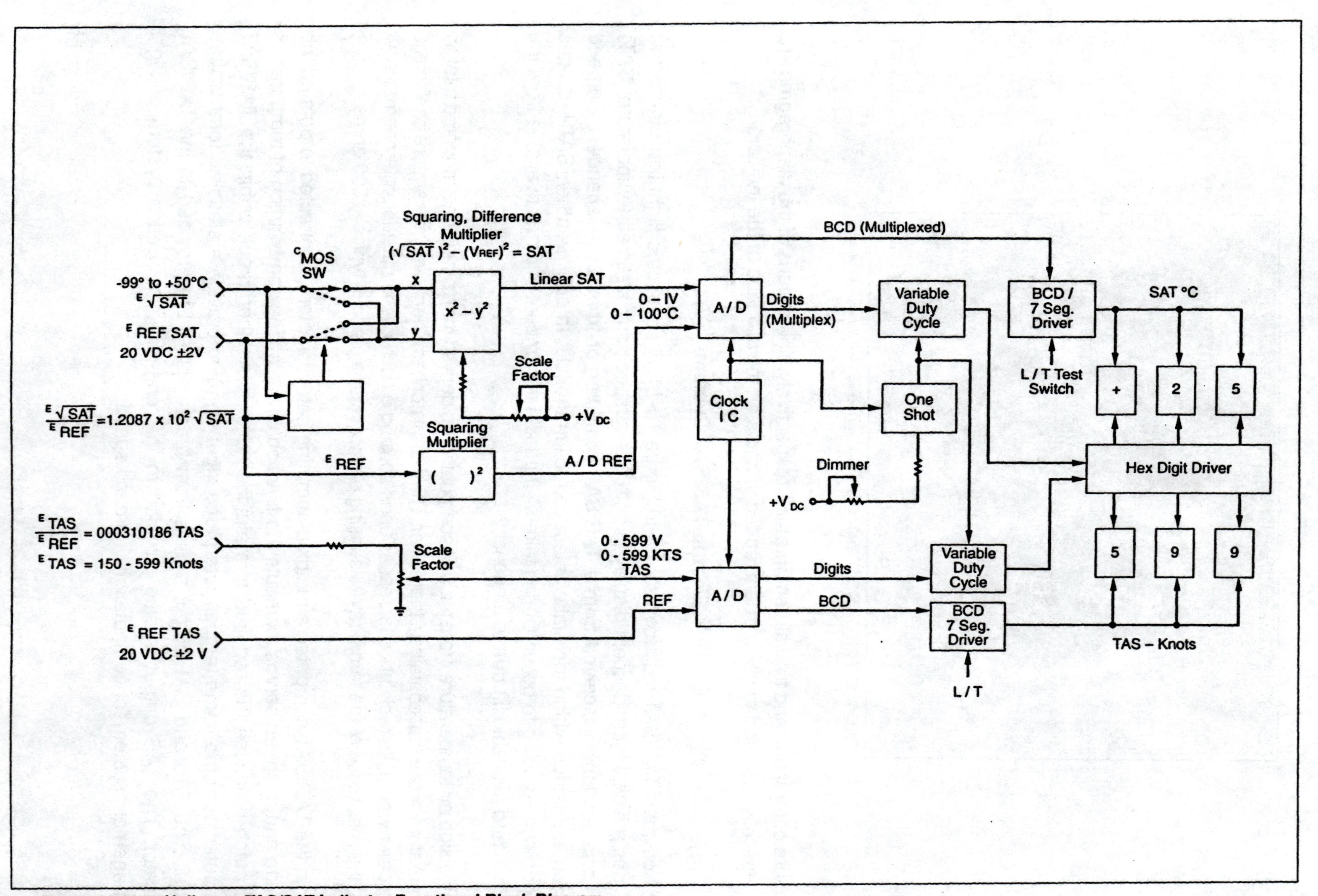

Figure 10-15. Kollsman TAS/SAT Indicator Functional Block Diagram.

A block diagram of the Kollsman TAS/SAT indicator is shown in Figure 10-15. The DC SAT voltage from the computer is applied to the indicator where a squaring operation is performed in order to linearize tthe SAT voltage with temperature. (The computer outputs the square root of SAT.) The 20-volt DC SAT reference signal is also squared, and then subtracted, to provide 0 volts at 0°C. The SAT voltage is then applied to a scaling amplifier and measured by an analog-to-digital (A/D) converter. The output results in a seven-segment digital display of plus or minus SAT in degrees centigrade. The DC TAS voltage from the computer is applied to the indicator where a scaling function is performed, and then applied to an A/D converter in conjunction with a 20-volt DC TAS reference signal. This output results in a seven-segment digital display of TAS in knots.

VERTICAL SPEED (VS) INDICATOR PRINCIPLES

The pneumatic vertical speed indicator is a sensitive differential pressure instrument that measures and displays the rate at which an aircraft is ascending or descending in altitude. The VS indicator is connected to the static system and senses the rate of change in static alr pressure. The rate of altitude change, as shown on the indicator dial, is positive (clockwize rotation of the pointer) in a climb, and negative (counterclockwize pointer rotation) in a dive. In level flight, the pointer remains at zero.

As illustrated in Figure 10-16, the case of the instrument is airtight, except for a small connection through a restricted passage to the static line. Inside the sealed case is a diaphragm with connecting linkage and gearing to the indicator pointer. Both the diaphragm and the case receive atmospheric pressure from the static line.

When the aircraft is on the ground or in level flight, the pressures inside the diaphragm and the instrument case remain the same and the pointer indicates zero vertical speed. When the aircraft climbs, the pressure inside the diaphragm decreases but, due to the metering action of the restricted passage, the case pressure will remain higher which causes the diaphragm to contract. The diaphragm movement actuates the mechanism, causing the pointer to indicate a rate of climb.

In a descent, the pressure conditions are reversed. The diaphragm pressure becomes greater than the pressure in the instrument case. This causes the diaphragm to expand which drives the pointer mechanism to indicate a rate of descent. When the aircraft levels off, the pressure in the instrument case becomes equalized with the pressure in the diaphragm causing the diaphragm to return to its neutral position and the pointer to return to zero.

When the aircraft is ascending or descending at a constant rate, a definite ratio is maintained between the diaphragm pressure and the case pressure through the calibrated restricted passage. Since it requires approximately six to nine seconds to equalize the two pressures, a definite lag persists before the proper vertical speed is indicated. Also, any abrupt changes in

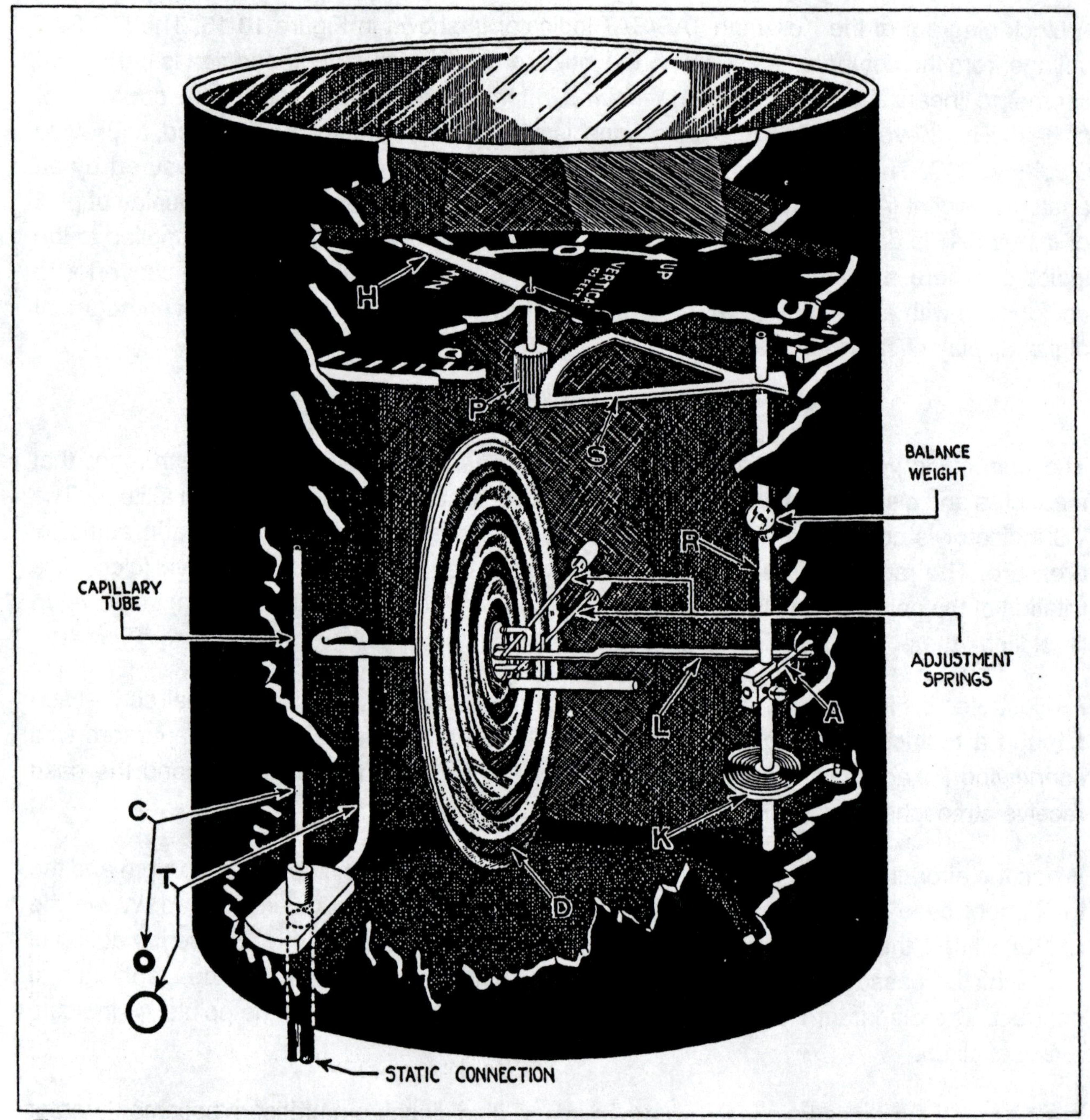

Figure 10-16. ***Pneumatic Vertical Speed Indicator Mechanism.*** *(Courtesy of Kollsman Division of Sequa Corporation)*

the aircraft's attitude may cause erroneous indications due to the sudden change of airflow over the static ports.

Instantaneous Vertical Speed Indicator (IVSI)

The instantaneous vertical speed indicator is a more recent refinement of the conventional pnuematic VS indicator. The IVSI incorporates acceleration pumps to eliminate the limitations associated with the calibrated leak. For example, during an abrupt climb, vertical acceleration

Figure 10-17. ***Electric Vertical Speed Indicator.***
(Courtesy of Kollsman Division of Sequa Corporation)

causes the pumps to supply extra air pressure into the diaphragm to stabilize the pressure differential without the usual time lag. During level flight and steady-state climbs and descents, the instrument operates on the same principle as a conventional vertical speed indicator.

ELECTRIC VERTICAL SPEED INDICATOR OPERATION

The Kollsman electric vertical speed indicator, shown in Figure 10-17, derives its inputs from an altitude rate signal supplied by a servoed altimeter or central air data computer, and a 26-volt AC, 400-Hz reference signal from the aircraft electrical bus instrumentation transformer. The altitude rate input is a phase-reversing 400-Hz signal with a scale factor of 250 millivolts per 1000 feet per minute. During ascent, the altitude rate signal is in-phase with the reference voltage; during descent, it is out-of-phase.

The rate signal input is demodulated and amplified to produce a DC voltage which operates a torque motor in the indicator. The motor shaft is used to position the indicator's pointer and to drive a follow-up potentiometer. The nonlinear scale (compressed at high rates) is achieved by diode shunt limiter circuits controlled by the follow-up potentiometer. The potentiometer has a nonlinear function so that its resistance various accordingly with the shaft angle for a specified indicator dial reading. The proper time constant for pointer damping is achieved by an RC tank circuit.

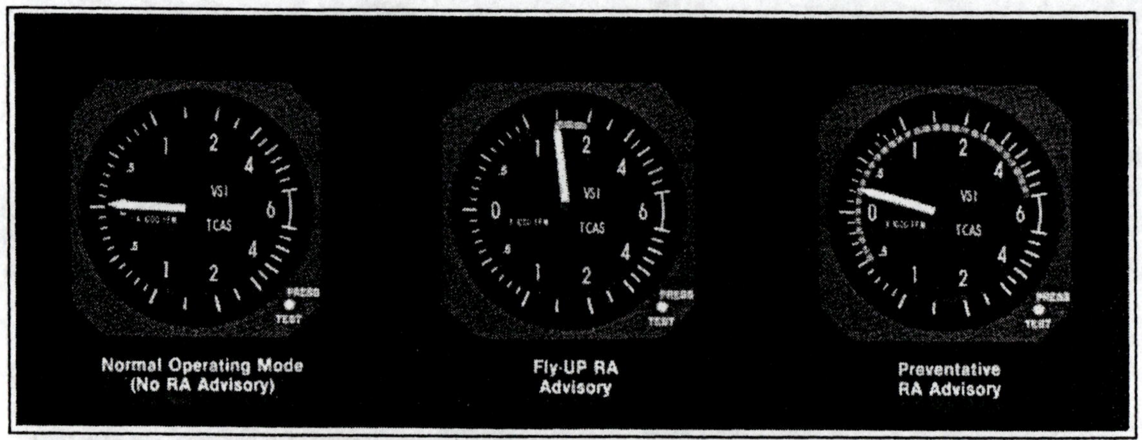

Figure 10-18. TCAS Resolution Advisory/Vertical Speed Indicator.
(Courtesy of Kollsman Division of Sequa Corporation)

TCAS RESOLUTION ADVISORY/VERTICAL SPEED INDICATOR OPERATION

As discussed in Chapter Seven, TCAS uses a Mode-S Transponder that interrogates other Mode-S Transponders of aircraft in the vicinity which might pose a collision threat. The TCAS will then compute and display the optimum flight path strategy, or resolution advisory.

The Kollsman 47174 Resolution Advisory/Vertical Speed Indicator (RA/VSI) is fully compliant with the TCAS guidelines set forth in ARINC 735, and will interface with all TCAS manufacturers' equipment. The resolution advisory display is shown in Figure 10-18. A single RA arc is composed of 52 bi-color, surface-mounted LEDs. These LEDs are individually addressed by an embedded microcontroller and provide a high resolution RA display.

The display on the left is the normal operating mode with no collision threats detected by the Mode-S Transponder. The center display is advising the pilot to climb at 1,500 feet per minute minimum to avoid an impending collision with another aircraft. The right display is telling the pilot that he/she is in danger of a possible collision if the aircraft should descend.

CENTRAL AIR DATA COMPUTER (CADC) OPERATION

The central air data computer is a sophisticated analog-digital, electro-mechanical device designed to operate in aircraft equipped with electrical flight instrumentation, and those requiring additional air data information for the operation of other onboard avionic systems. The altitude outputs supplied by the CADC are corrected for instrument scale error, and static defect correction is provided as a function of mach. A typical CADC receives both impact and atmospheric pressure from the pitot-static system and supplies the following outputs in accordance with ARINC specification 565:

1) An encoded digit pressure altitude output, in increments of 100 feet, for use by an ATC transponder for automatic altitude reporting.

2) A dual-synchro coarse and fine altitude output for driving an electric servoed altimeter.

3) Altitude rate output for driving an electric vertical speed indicator.

4) Outputs for operating electric mach, true airspeed, and static air temperature indicators.

5) Altitude, airspeed, and mach hold functions for use by the flight director system in computing vertical guidance steering commands for driving the autopilot elevator servo.

6) A failure monitor signal to actuate a malfunction indicator flag in the systems receiving information from the CADC.

The Kollsman model 621 is a typical example of a standard ARINC CADC unit presently being used on large corporate and commercial transport aircraft. The principle functional components of the Kollsman CADC, shown in Figure 10-19, consist of two servo-operated diaphragm displacement transducers, a mach follow-up servo, and the associated servo amplifier/failure monitor cards.

The altitude transducer servo drives fine and coarse output synchros, an altitude digitizer encoder, and an altitude scale error and function potentiometer used in computing mach. The altitude hold functions are also geared to this servo shaft. The airspeed transducer servo positions an airspeed scale error potentiometer and an airspeed function potentiometer, also used in computing mach. The airspeed function potentiometer provides the signal for the mach servo. The mach servo shaft positions the potentiometer which provides the static source error correction as a function of mach. It also provides the drive for the mach hold and the mach output synchros.

A change in static pressure produces an error signal in the altitude transducer. In repositioning the transducer to null the error signal, the altitude potentiometer, coarse and fine synchros, two clutched synchros, and the digitizer are driven to provide the desired outputs. The potentiometer provides an output to the altitude scale error, linear altitude, and altitude function potentiometers for the computation of mach.

Summed into the input network of the altitude servo is a static source error correction signal which is supplied by the mach servo. The airspeed transducer has a follow-up servo similar to the altitude servo. This servo produces a shaft output linear with airspeed. The shaft positions the potentiometer, which accepts the airspeed scale error correction and also supplies an output to the linear airspeed potentiometer and the airspeed function potentiometer used in computing mach. Also geared to this shaft is the airspeed hold clutch servo.

Figure 10-19. Central Air Data Computer. *(Courtesy of Kollsman Division of Sequa Corporation)*

Mach is computed by altitude and airspeed functions to provide the shaft position for the static source error correction signal, which is used in the correction of altitude. Also geared to the mach shaft is the mach output synchro and the clutched synchro for mach hold.

CONCLUSION

Many of the books written about Avionics concentrate only on aircraft radios and fail to discuss flight instruments. This is a mistake since flight instruments are essential for controlling the altitude, attitude, airspeed, and direction of the aircraft, and except for the standby instruments, are being more tightly integrated with other Avionic Systems. This chapter not only presented Turn and Bank and Angle of Attack Indicators, Pitot-Static Systems and Pneumatic Altimeters, Airspeed Indicators, and Vertical Speed Indicators, but also discussed Electric Encoding RADBAR Altitude Indicators, Altitude Alerters, TAS/SAT Indicators, Electrical Vertical Speed Indicators (including one with a TCAS resolution advisory), and Central Air Data Computers which drive the flight instrumentation system. The last chapter will present a higher level of avionic system integration beyond the CADC through the use of Flight Management Computers and Electronic Flight Instrumentation Displays, but first we turn our discussion to Long-Range Navigation Systems developed for transcontinental flight.

REVIEW QUESTIONS

1. Define the principles of operation of a gyroscope.
2. What two methods does an Angle-of-Attack system use to warn the pilot of an impending stall?
3. Describe the operation of the airstream direction detector.
4. What is the difference between Pitot and Static pressure and how is each derived?
5. Which flight instrument requires both pitot and static pressure and why?
6. Define the three common sources of pneumatic altimeter error.
7. Describe the principles of operation of a pneumatic altimeter.
8. What are the four functions of the RADBAR encoding altimeter system?
9. Explain the operation of the altitude digitizer.
10. What is the purpose of the SDC module?
11. Define the rationale for each VTA output.
12. What is the purpose of the altitude alerter?
13. Describe the interface between the electric altimeter and the transponder, alerter, and SDC.
14. Define the operation of a pneumatic airspeed indicator.
15. What is the purpose of a machmeter and a maximum allowable airspeed indicator?
16. Explain the operation of a maximum allowable airspeed indicator.
17. Compare and contrast the principles of operation of a pneumatic and electrical TAS indicator.
18. Compare and contrast the principles of operation of a pneumatic and electrical vertical speed indicator.
19. When does the TCAS provide a RA display?
20. Define the outputs provided by a typical CADC.

Chapter 11

Long-Range Navigation Systems

INTRODUCTION

The science of navigation may be defined as the application of calculations to determine the position of a vehicle and to direct it to a predetermined destination. With the advancement of radio techniques, positioning systems, such as VOR/DME, were developed to aid the navigator. As air transportation reached transcontinental proportions, radio positioning and celestial techniques were further developed to satisfy the faster and more accurate navigational requirements. However, radio systems, such as LORAN and Omega/VLF, are somewhat limited, as they require extensive networks of ground stations and are subject to both man-made and natural interference. As a result, more advanced navigation systems (found in larger aircraft) were developed, including the self-contained Inertial Navigation System (both mechanical and the newer laser gyro platforms) and the Global Navigation Satellite Positioning System. These modern long-range navigation systems evolved from the Gyrosyn Compass which is discussed first in this chapter, even though it is not typically termed a long-range navigation system.

THE GYROSYN COMPASS SYSTEM

The Gyrosyn Compass System, developed in 1946, is probably the most important advance ever made in aircraft direction keeping. The various motions of an airplane in flight cause severe fluctuations and serious errors in ordinary magnetic compasses. The early Directional Gyro (DG) panel instrument helped by giving the pilot a stable indication of heading, but it had to be reset every 15 or 20 minutes, because of its tendency to drift.

The Gyrosyn Compass System, however, combines the functions of a directional gyro and a magnetic compass. It is a gyro, such as that shown in Figure 11-1, that is synchronized continuously with the earth's lines of force which indicate magnetic North. Moreover, its magnetic sensing element, known as a flux valve, is located remotely in a wing tip or other location, where it is free from local magnetic influence. The gyro element is mounted in a low vibration area, and its signals are transmitted electrically to a repeating compass on the instrument panel. A control panel, known as a Remote Magnetic Compensator, is located in the cockpit to adjust the flux valve control signal to the gyro based on variations in the earth's magnetic lines of force at various latitudes. The control panel also has a "SLAVE/DG" switch to connect or disconnect the slaving signal from the flux valve. The Gyrosyn needs no resetting, and can be relied upon constantly for extremely accurate and stable magnetic heading information.

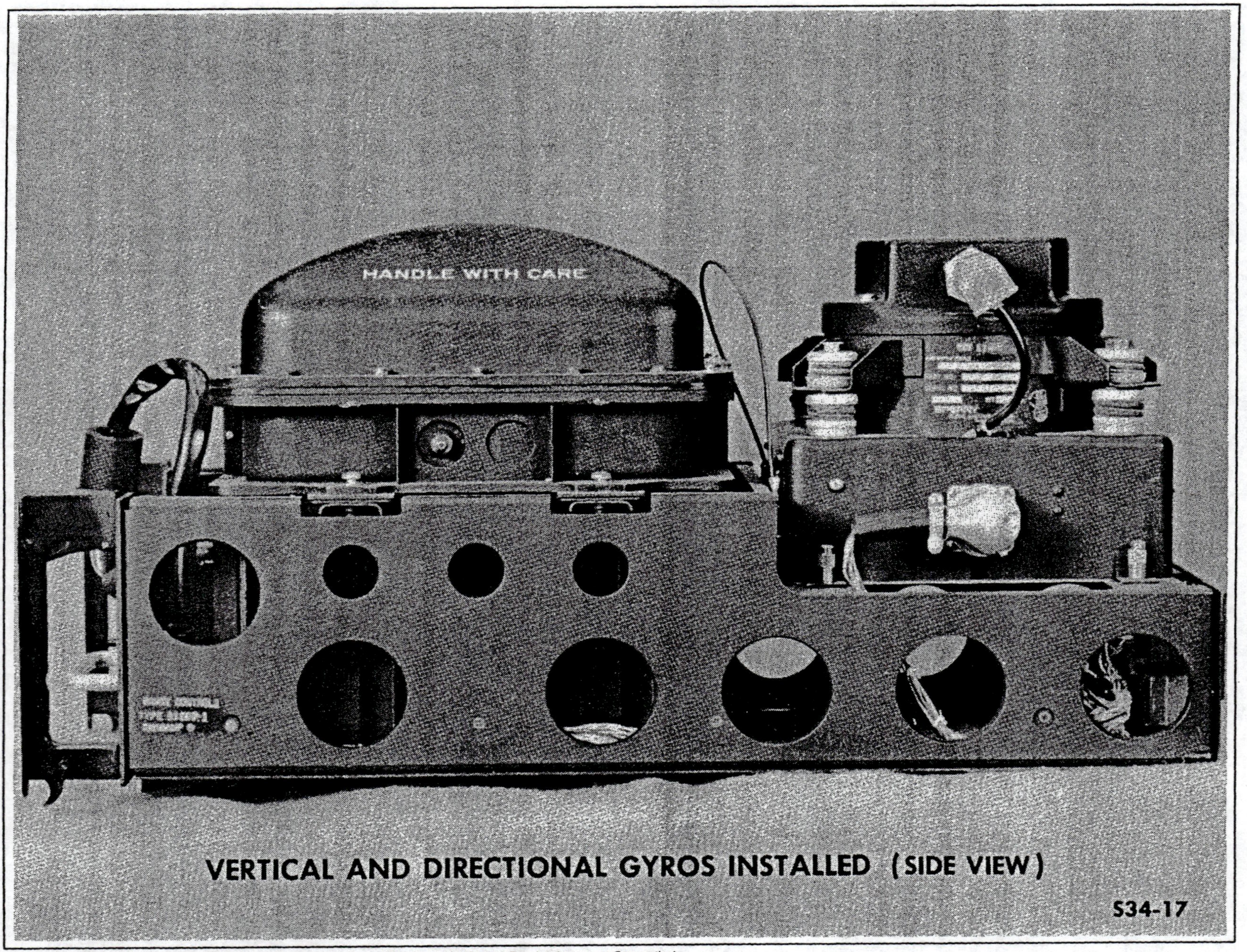

Figure 11-1. Vertical and Directional Gyros. (Courtesy of McDonnell Douglas Aerospace Corporation)

INERTIAL NAVIGATION SYSTEMS (INS)

Inertial navigation is the process of determining an aircraft's location using internal inertial sensors rather than external references. The inertial navigation system is the only self-contained single source for all navigation data. After being supplied with initial position information, it is capable of continuously updating extremely accurate displays of: position, ground-speed, attitude, and heading. In addition, it provides guidance or steering information for the autopilot and flight instruments.

The basic measuring instrument of the inertial navigation system is the accelerometer. Two accelerometers are mounted in the system. One will measure the aircraft's accelerations in the North-South directions, and the other will measure the aircraft's accelerations in the East-West directions. The accelerometer, shown in Figure 11-2, is basically a pendulous device. When the aircraft accelerates, the pendulum, due to inertia, swings off its null position. A signal pickoff device tells how far the pendulum is off the null position. The signal from the pickoff device is sent to an amplifier, and current from the amplifier is sent back into the accelerometer to the torquer motor. The torquer motor will restore the pendulum back to its null position.

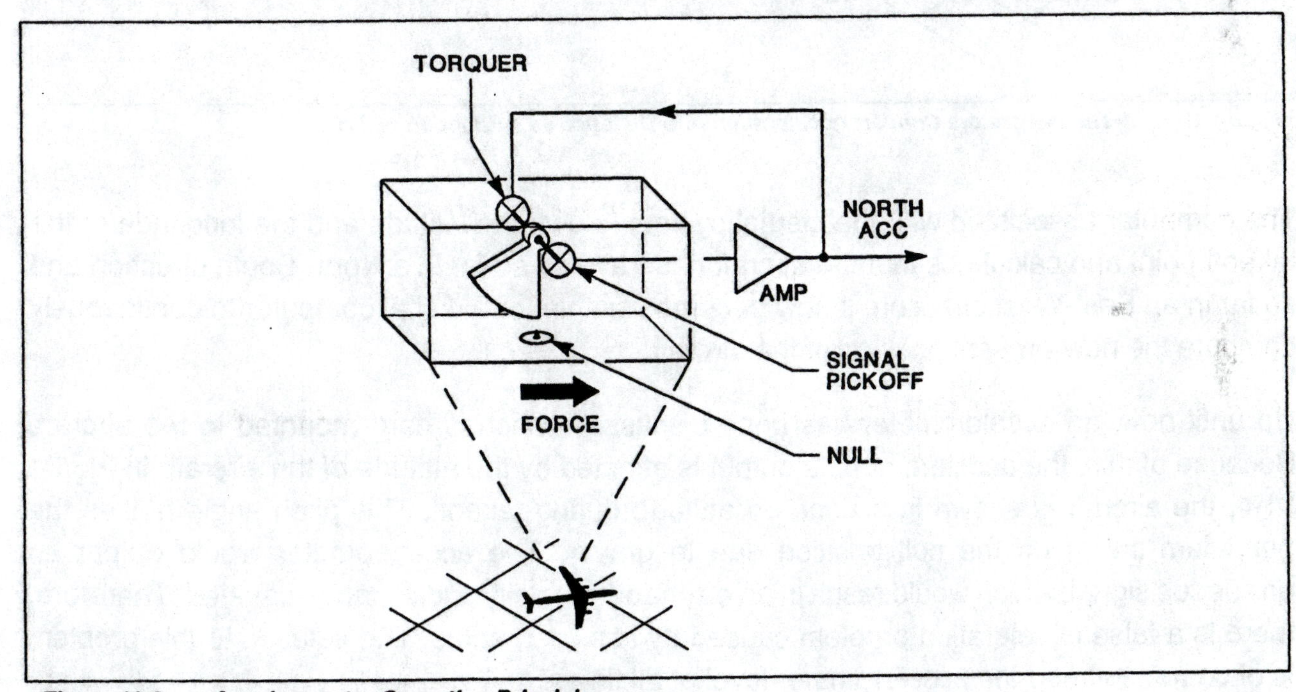

Figure 11-2. Accelerometer Operating Principles.

The acceleration signal from the amplifier is also sent to an integrator, which is a time multiplication device. It starts out with acceleration, which is in feet per second squared. In the integrator, it is literally multiplied by time and the result is a velocity in feet per second. It is then sent through a second integrator and again it is time multiplied. With an input of feet per second, which is multiplied by time, the result is a distance in feet or in miles. It can be computed that the aircraft has traveled 221 miles in a northerly direction from time of takeoff. This is shown in Figure 11-3.

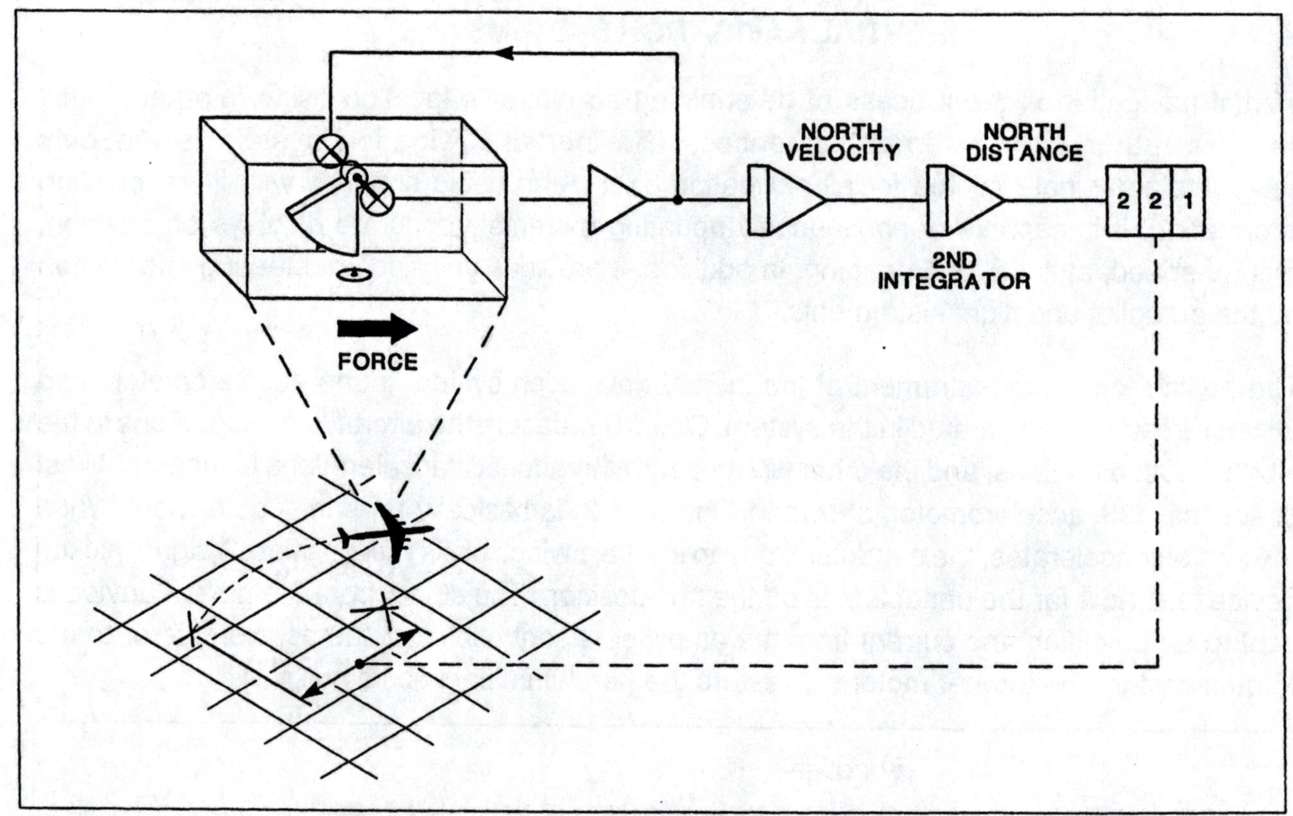

Figure 11-3. ***The Integrators Convert Acceleration into Distance as a Function of Time.***

The computer associated with the inertial system knows the latitude and the longitude of the takeoff point and calculates that the aircraft has traveled so far in a North-South direction and so far in an East-West direction. It now becomes simple for a digital computer to continuously compute the new present position of the aircraft.

Up until now, an accelerometer has been discussed which is hard mounted to the aircraft. Because of this, the accelerometer's output is affected by the attitude of the aircraft. In Figure 11-4, the aircraft is shown in a nose up attitude during takeoff. This pitch angle makes the pendulum swing off the null position due to gravity. The accelerometer would output an erroneous signal, which would result in an erroneous velocity and distance traveled. Therefore, there is a false acceleration problem caused by this pitch angle. The solution to this problem is of course to keep the accelerometer level at all times.

To keep the accelerometer level, it is mounted on a gimbal assembly commonly referred to as a stabilized platform. The platform, shown in Figure 11-5, is nothing more than a mechanical device which allows the aircraft to go through any attitude change and at the same time maintain the accelerometers level. The inner element of the platform where the accelerometers are mounted, will also mount the gyroscopes used to stabilize the platform. The gyros provide signals to motors, which control the gimbals of the platform.

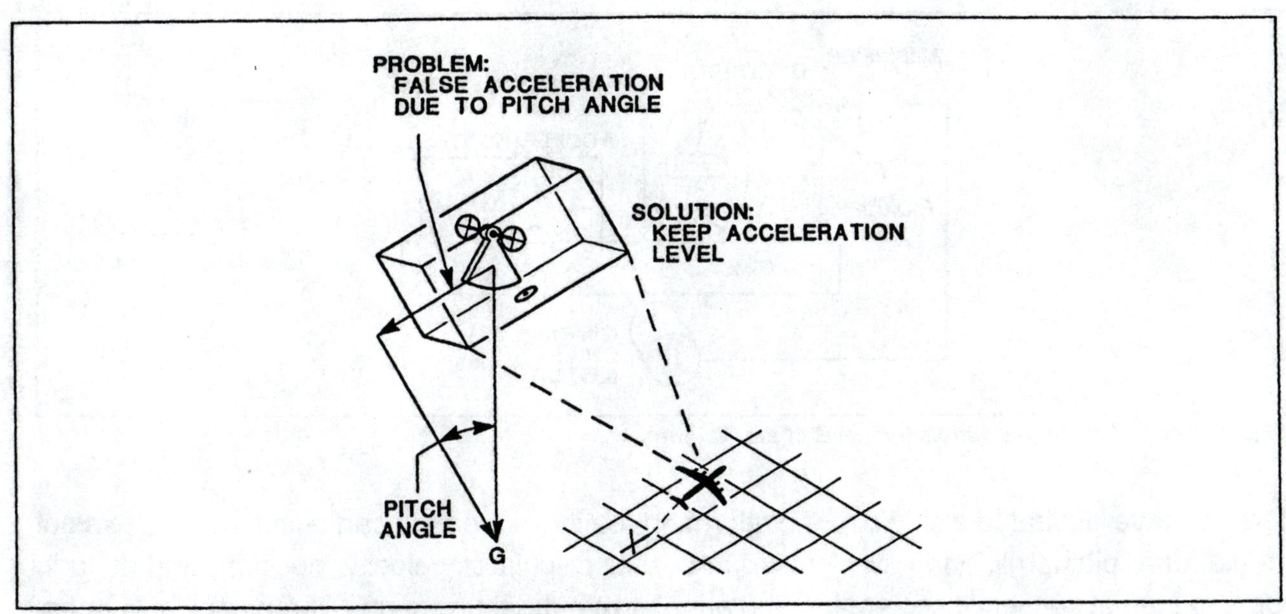

Figure 11-4. ***Accelerometer Error Due to Pitch Angle.***

Figure 11-6 shows how the gyro is used to control the level of the platform. The gyro and accelerometer are mounted on a common gimbal. When this gimbal tips off the level position, the spin axis of the gyro remains fixed. The case of the gyro moves with the gimbal, and the amount of movement is detected by the signal pickoff in the gyro. That signal is then amplified and sent to a gimbal drive motor, which restores the gimbal back to a level position. In this example, the accelerometer is going along for the ride. Since the accelerometer is just being kept level, it does not sense a component of gravity and is able to sense only true horizontal accelerations of the aircraft.

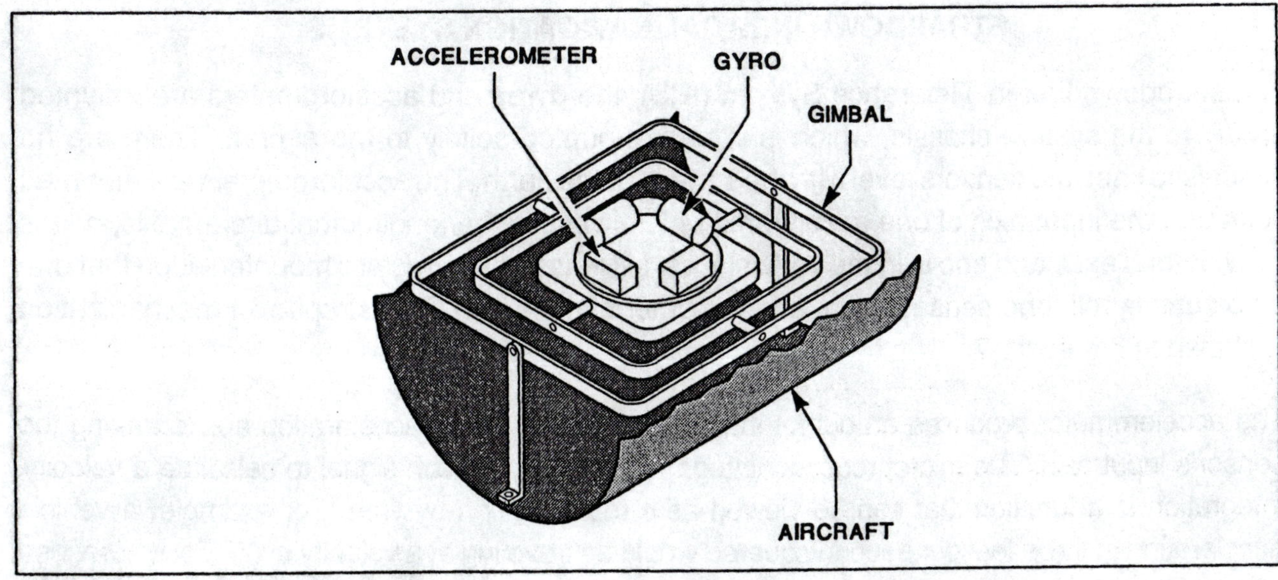

Figure 11-5. ***Stabilized Platform.***

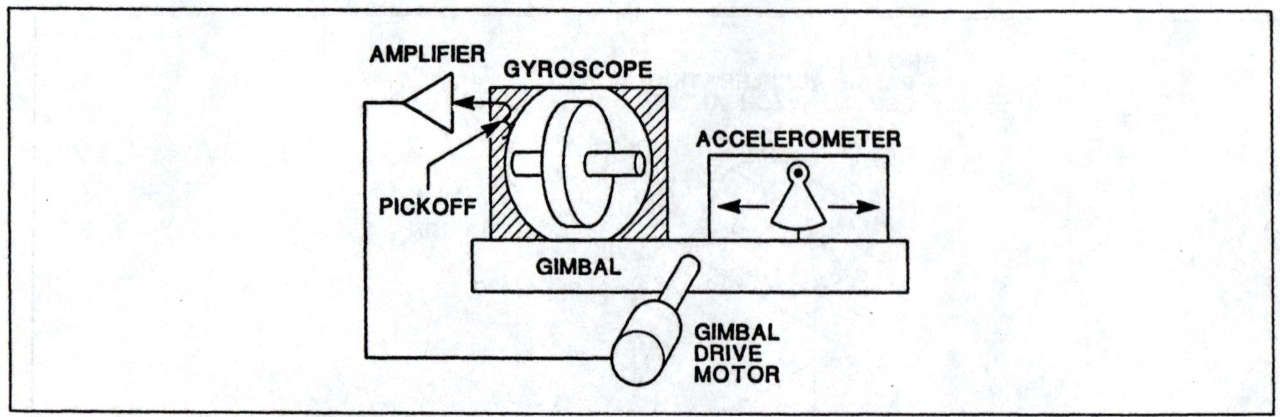

Figure 11-6. The Gyro Controls the Level of the Platform.

Here we have illustrated a single axis platform. In reality, movement can occur in three axes of the platform: pitch, roll, and yaw. A microprocessor calculates velocity, position, and attitude from the inertial sensors' acceleration and angular rate measurements. Three accelerometers and three gyros are needed because, in a three-dimensional world, an aircraft can simultaneously accelerate and rotate in three orthogonal axes - pitch, roll and heading.

Inertial navigation depends on the integration of acceleration to obtain velocity and distance. In any integration process, one must know the initial conditions, which in this case are velocity and position. The accuracy to which the navigation problem is solved depends greatly upon the accuracy of the initial conditions. Therefore, system initialization is of paramount importance. Initialization (also called alignment) of the Inertial Navigation System is accomplished while the aircraft is in a known fixed position on the ground by manually entering the present latitude and longitude on the INS Control/Display Unit. For example, at Los Angeles Airport, the aircraft's position would be entered as "N33560" and "W118234." Some systems have the last known position stored in computer memory for rapid initialization once power is restored.

STRAPDOWN INERTIAL NAVIGATION SYSTEMS

In a strapdown Inertial Reference System (IRS), the gyros and accelerometers are mounted solidly to the system chassis, which is in turn, mounted solidly to the aircraft. There are no gimbals to keep the sensors level with the surface of the earth. The accelerometers are mounted such that the input axis of one accelerometer is always in the longitudinal aircraft axis, one is in the lateral axis, and one is in the vertical axis. Likewise, the gyros are mounted such that one gyro senses roll, one senses pitch, and the other senses yaw. The strapdown mechanization is shown in Figure 11-7.

The accelerometer produces an output that is proportional to the acceleration applied along the sensor's input axis. The microprocessor integrates the acceleration signal to calculate a velocity. Integration is a function that can be viewed as a multiplication by time. For example, a vehicle accelerating at three feet per second squared would be traveling at a velocity of 30 feet per second after 10 seconds have passed. Note that acceleration was simply multiplied by time to get a velocity.

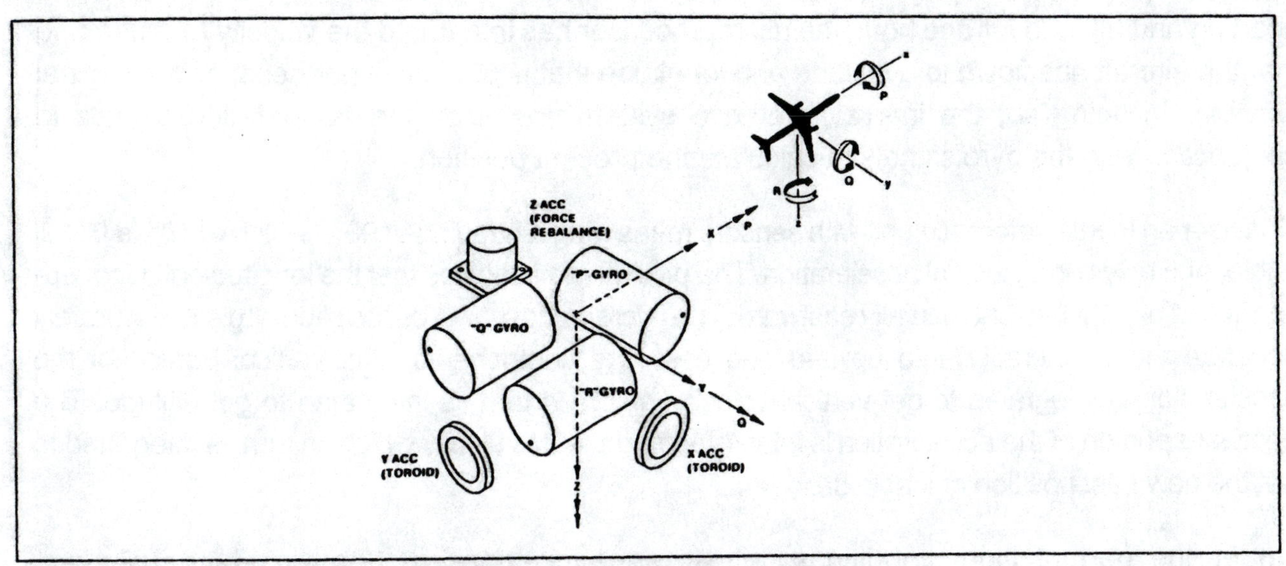

Figure 11-7. Strapdown Mechanization. *(Courtesy of Honeywell Commercial Flight Systems)*

The microprocessor also integrates the calculated velocity to determine position. For example, a vehicle traveling at a velocity of 30 feet per second for 10 seconds will have changed position by 300 feet. Velocity was simply multiplied by time to determine the position.

Although it is used to calculate velocity and position, acceleration is meaningless to the system without additional information. For example, consider an accelerometer strapped down to the longitudinal axis of the aircraft and measuring a forward acceleration. Is the aircraft acceleration north, south, east, west, up or down? In order to navigate over the surface of the earth, the system must know how this aircraft acceleration is related to the earth's surface. Because accelerations are measured by accelerometers that are mounted to the lateral, longitudinal, and vertical axes of the aircraft, the IRS must know the relationship of each of these axes to the surface of the earth.

The gyros in a strapdown system make the measurements necessary to describe this relationship in terms of pitch, roll, and heading angles. These angles are calculated from the angular rates measured by the gyros through an integration - similar to the manner in which velocity is calculated from measured acceleration. For example, suppose a gyro measures a yaw rate of three degrees per second for 30 seconds. Through integration, the microprocessor calculates that the heading has changed by 90° after 30 seconds.

Given the knowledge of pitch, roll, and heading that the gyros provide, the microprocessor resolves the acceleration signals into earth-related accelerations, and then performs the horizontal and vertical navigation calculations. This can be illustrated with the following two examples:

Suppose the gyro signals have been integrated to indicate that the aircraft's heading is 45° and the pitch and roll are both zero. The only acceleration measured has been in the longitudinal axis, and it has been integrated into a velocity of 500 miles per hour. After flying at a constant heading and attitude for one hour, the microprocessor has integrated the velocity to determine that the aircraft has flown to a latitude and longitude that is 500 miles northeast of the original location. In doing so, the inertial reference system has used the acceleration signals in conjunction with the gyro signals to calculate the present position.

Consider an inertial reference unit with sensors measuring a heading of 90°, a pitch of 10°, a 0° roll angle, and only a longitudinal acceleration. The pitch angle indicates that the longitudinal acceleration is partially upward and partially eastward. The microprocessor uses the pitch angle to accurately separate the acceleration into upward and eastward components. The vertical portion of the acceleration is integrated to get vertical velocity which, in turn, is integrated to get altitude. The eastward portion of the acceleration is integrated to get east velocity which, in turn, is integrated to get the new east position or longitude.

Under the normal flight conditions, all six sensors sense motion simultaneously and continuously, thereby entailing calculations that are substantially more complex than shown in the previous examples. A powerful, high-speed microprocessor is required in the IRS in order to rapidly and accurately handle this additional complexity.

In addition to the basic strapdown concepts that have been discussed, there are some additional details that must be considered in order to navigate with respect to the earth's surface. These special considerations are necessitated by the earth's gravity, rotation, and shape. A strapdown IRS compensates for these special effects with the microprocessor's software.

OPERATION OF THE HONEYWELL AHZ-600 ATTITUDE AND HEADING REFERENCE SYSTEM (AHRS)

The Honeywell AHZ-600 Attitude and Heading Reference System is an inertial sensor installation which provides aircraft attitude, heading, and flight dynamics information to cockpit displays, flight controls, weather radar antenna platform, and other aircraft systems and instruments. The AHZ-600 differs from conventional vertical and directional gyro systems in that the gyroscopic elements are rate gyros which are "strapped down" to the principal aircraft axes. A digital computer mathematically

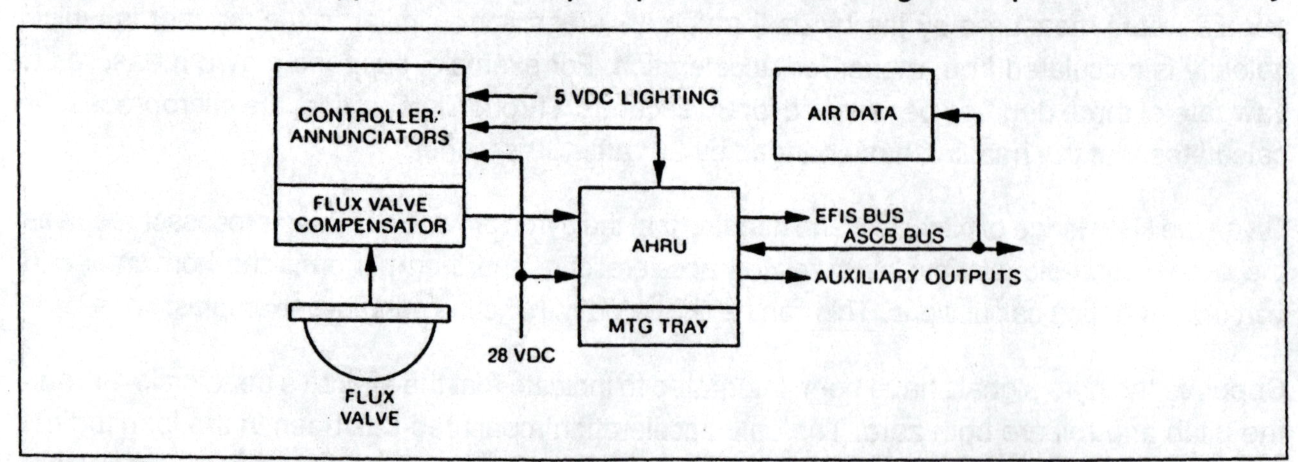

Figure 11-8. AHZ-600 System Components. *(Courtesy of Honeywell Commercial Flight Systems)*

integrates the rate data to obtain heading, pitch, and roll information. A flux valve and three accelerometers provide long-term references for the system. As shown in Figure 11-8, three Line Replaceable Units (LRUs) make up each AHRS installation: the flux valve, the compensator/controller, and the Attitude and Heading Reference Unit (AHRU).

The flux valve detects the relative bearing of the earth's magnetic field and is usually located in the wing or tail section away from disturbing magnetic fields. The compensator/controller, shown in Figure 11-9, provides the single cycle error correction for the flux valve. The compensator/controller is mounted in the cockpit area. The AHRU is the major component of the system and is composed of four major subsystems, as shown in Figure 11-10.

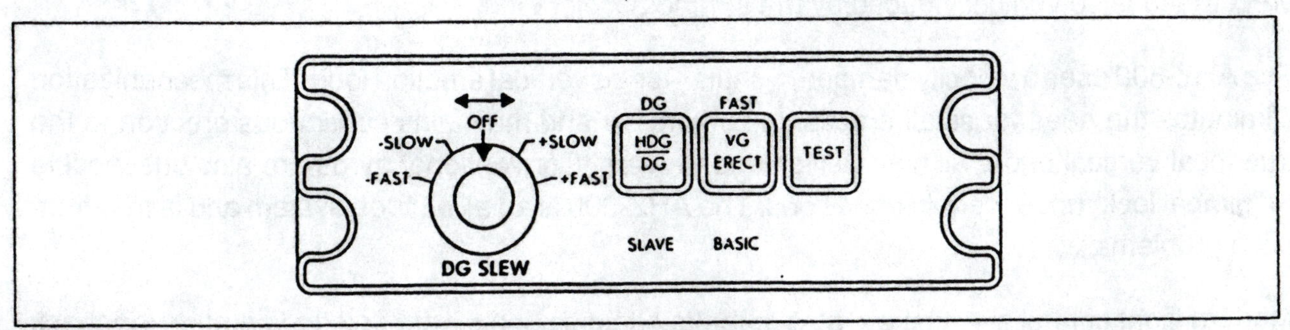

Figure 11-9. ***AC-800 AHRS Controller.*** *(Courtesy of Honeywell Commercial Flight Systems)*

The Inertial Measurement Unit (IMU) senses the aircraft's body dynamics. It contains the rate gyros, accelerometers, and support electronics. The Central Processor Unit (CPU) performs the numerical computations necessary to extract the attitude and heading information. It contains the microprocessor, arithmetic logic unit, program memory, and scratchpad memory. In addition to its computational activities, the CPU controls and monitors the operation of the entire system. The Input/Output (I/O) unit supervises the analog-to-digital and digital-to-analog conversions. The flux valve is

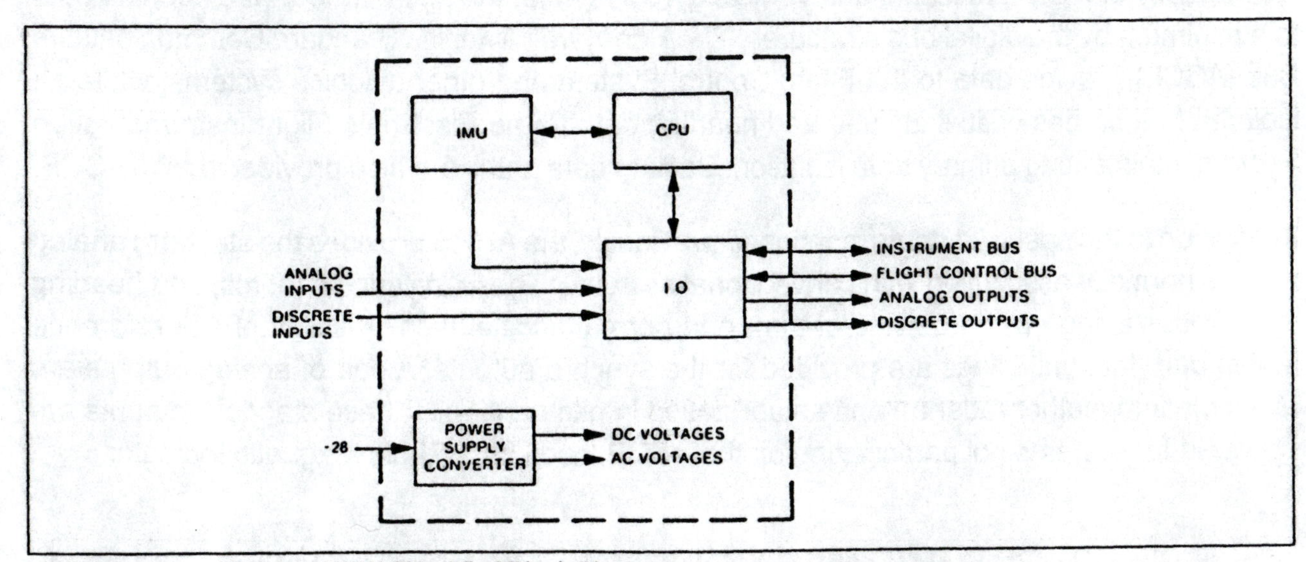

Figure 11-10. ***AHZ-600 AHRU Simplified Block Diagram.*** *(Courtesy of Honeywell Commercial Flight Systems)*

connected to the I/O unit through the compensator using the current servo approach. The power supply converts aircraft power to the regulated DC voltages and DC power signals required by the system. In order to maximize the effectiveness of the flight control system, it is recommended that the AHRU be mounted at the center of gravity of the aircraft.

The AHZ-600 offers several advantages over existing vertical and directional gyros. Principal among these is in the area of performance. In a conventional vertical gyro, automatic vertical erection is "cut off" when the roll angle, or in some cases the pitch angle as well, exceeds a certain value, typically 5° to 10°. This causes the gyro to produce a vertical error proportional to its free drift during extended large bank angle maneuvers. A related problem concerns shallow bank angles just below "cutoff." In this case, the automatic erection loop causes the gyro to erect to the false vertical induced by the turning acceleration.

The AHZ-600 uses a velocity damped, Schuler-tuned vertical erection loop. This mechanization eliminates the need for small erection cutoff angles and maintains continuous erection to the true local vertical under all normal flight maneuvers. Conventional gyros are also susceptible to "gimbal lock" under certain conditions. The AHZ-600 is an all attitude system and is free from such problems.

Modern flight control systems are able to make effective use of rate and acceleration feedback terms. These terms, which are not directly available from simple vertical or directional gyros, have had to be derived from position data or obtained from extra rate gyros and accelerometers. The AHRS, however, provides direct measurements of these quantities and supplies all required data for the Flight Control System in both earth-based and aircraft body axis coordinate reference frames. Angle-of-attack is computed for the flight control system as a function of true airspeed, altitude rate, and pitch angle.

While the number and variety of data outputs and formats available from the AHZ-600 is beyond the capacity of most directional and vertical gyro systems, the interconnect wire count is held to a minimum by the digital bus structures. The Honeywell Avionics Standard Communications Bus (ASCB) provide data to the Flight Control System and other avionics systems, while an isolated bus carries critical attitude and heading data to the Electronic Flight Instrumentation System, maintaining primary data isolation. Backup data paths are also provided by the ASCB.

In addition to the special outputs mentioned previously, the AHRS provides the standard analog signals normally associated with conventional systems. These include pitch, roll, and heading synchros, slaving error, true rate-of-turn, and normal acceleration. A set of phase reference inputs and data valid flags are provided for the synchro outputs. A pair of analog outputs are available for weather radar antenna stabilization in pitch and roll. These standard outputs are provided for systems not participating on the ASCB, such as a Radio Magnetic Indicator.

The AHZ-600 provides continuous system monitoring. The central processor in the AHRU performs self-checking of data and computations. A preflight test provides pilot verification of system operation through special sensor and signal path tests.

While mounting any AHRS system requires care in physical alignment, the AHZ-600 minimizes the problem of field replacement. The AHRU is installed on a tray which has previously been aligned to the aircraft axes using a special tool. The tray provides positive indexing of the IMU to the aircraft, eliminating any requirement for releveling should field maintenance or replacement of the AHRU be required. Longitudinal alignment of the tray to the aircraft may be at any 90° increment, permitting installation in an increased variety of previously unacceptable locations.

Two modes are provided for routine operation: the "normal" mode in the attitude channel, and the "slaved" mode in the heading channel. The "normal" mode uses true airspeed from the air data computer to compensate for acceleration induced attitude errors. The AHZ-600 also computes true airspeed as a monitor function to verify the reasonableness of the data received. The loss of TAS is recovered by using the computed value or, if available (as in the optional dual air data installation), the opposite side's TAS data. Use of other than the normal true airspeed channel is automatic and is not annunciated. The "slaved" mode uses the flux valve to align the heading outputs. To prevent errors due to the "North turning" phenomena, the flux valve slaving is cut off during maneuvers which would induce such errors.

Two reversionary modes are provided to maintain performance in the event of certain types of system failures: "Basic" and "DG." The "Basic" mode operates without TAS compensation and results in a simple first-order vertical erection scheme similar to that of a conventional vertical gyro with roll cut-off. Since this kind of operation is susceptible to stand-off errors following certain types of maneuvers, a fast-erect command input from the controller is provided. Entry into the "basic" mode is automatic and is annunciated on the controller. If all true airspeed data is invalid, but indicated airspeed remains valid, the AHZ-600 will use indicated airspeed for compensation and will annunciate the "basic" mode.

The "DG" mode is entered by a flight crew command on the controller and is provided to disable the automatic slaving of the heading outputs. Operation in this mode is similar to that of a conventional directional gyro and is annunciated on the controller. Although the "DG" mode may be entered at any time, the mode is usually reserved for operation in the event of a slaving loop failure. Such failures are indicated by the "slaving error" output or the "SLAVE FAIL" annunciator on the controller. A two-speed manual slaving input switch is provided on the controller to manually slew the heading output while operating in the "DG" mode. Use of the slew control in the "slaved" mode is inhibited, and no single failure of the slew command inputs will result in a runaway heading output. The free drift of the internal AHRU heading reference is less than 24° per hour while operating as a directional gyro. An additional feature of the AHZ-600 provides fast slaving when changing from the "DG" mode to the "slaved" mode. This reduces the time spent at the end of the runway should the system be inadvertently left in the "DG" mode prior to flight.

Several test modes are incorporated into the AHZ-600. The "preflight" test is performed automatically upon application of power to the AHRU or by flight crew command. These tests extend the normal AHRU self-monitoring activities. All flags are dropped into view on the flight instruments upon entry into this mode to protect the aircraft from accidental activation during flight. In addition, the AHRS annunciates the test mode selection with a separate output discrete so that user equipment, such as the Electronic Flight Instrumentation System, may make appropriate use of the test data. Once the AHRU is on-line, the flight crew may initiate a display interface test which causes displacement slewing of the various displays and exercises the valid flag indications. The ASCB interface is verified through the test feature on the Flight Control System. A lamp test is performed during all tests to verify operation of the controller lamps. Since all valid flags are dropped during the various tests, the test modes are not locked out during flight. If this is desired, the test commands may be interlocked with the "weight on wheels" switch.

OPERATION OF THE ROCKWELL COLLINS AHS-86 ATTITUDE AND HEADING REFERENCE SYSTEM

The Collins AHS-86 is a fully aerobatic, strapped-down, attitude/heading reference system designed for trouble-free installation in high-performance fixed-wing aircraft. The AHS-86 has replaced traditional vertical, directional, and rate gyros and accelerometers with state-of-the-art, piezoelectric multi-sensor technology. This enables the system to achieve outstanding performance levels in its measurements of angular and linear acceleration rates about the body of an aircraft up to 10g. Using this digitally processed data, highly accurate heading and attitude information is provided for display to the pilot. Digital implementation allows more sophisticated leveling/slaving techniques, and the monitoring architecture provides a high-coverage check of output validity.

The standard AHS-86 system is composed of the AHC-86 Attitude/Heading Computer, an FDU-70 Flux Detector Unit, and a compass compensator, such as the ECU-86 External Compensation Unit or the RCP-65 Remote Compensation Panel. A Control and Compensation Unit, the CCU-65, which allows for unslaved heading operation while providing for compass compensation functions, is also available for use with the system.

Figure 11-11 is a simplified system block diagram of the interface capabilities of the AHS-86 system. The AH-86 system provides three standard X-Y-Z synchro outputs capable of supplying data to conventional heading, pitch and roll indicators. To complement this capability, separate high-impedance outputs are also provided as a buffered source of attitude information to the autopilot. Analog angular pitch and roll turn rates, as well as linear acceleration outputs are also provided. Additionally, the system provides analog two-wire pitch and roll stabilization outputs for weather radar. The AHC-86 provides three completely independent, high-speed ARINC 429 bus outputs, each supplying three Euler angles (earth-referenced heading, pitch, and roll), three body rates, three body accelerations, and three level-axis accelerations.

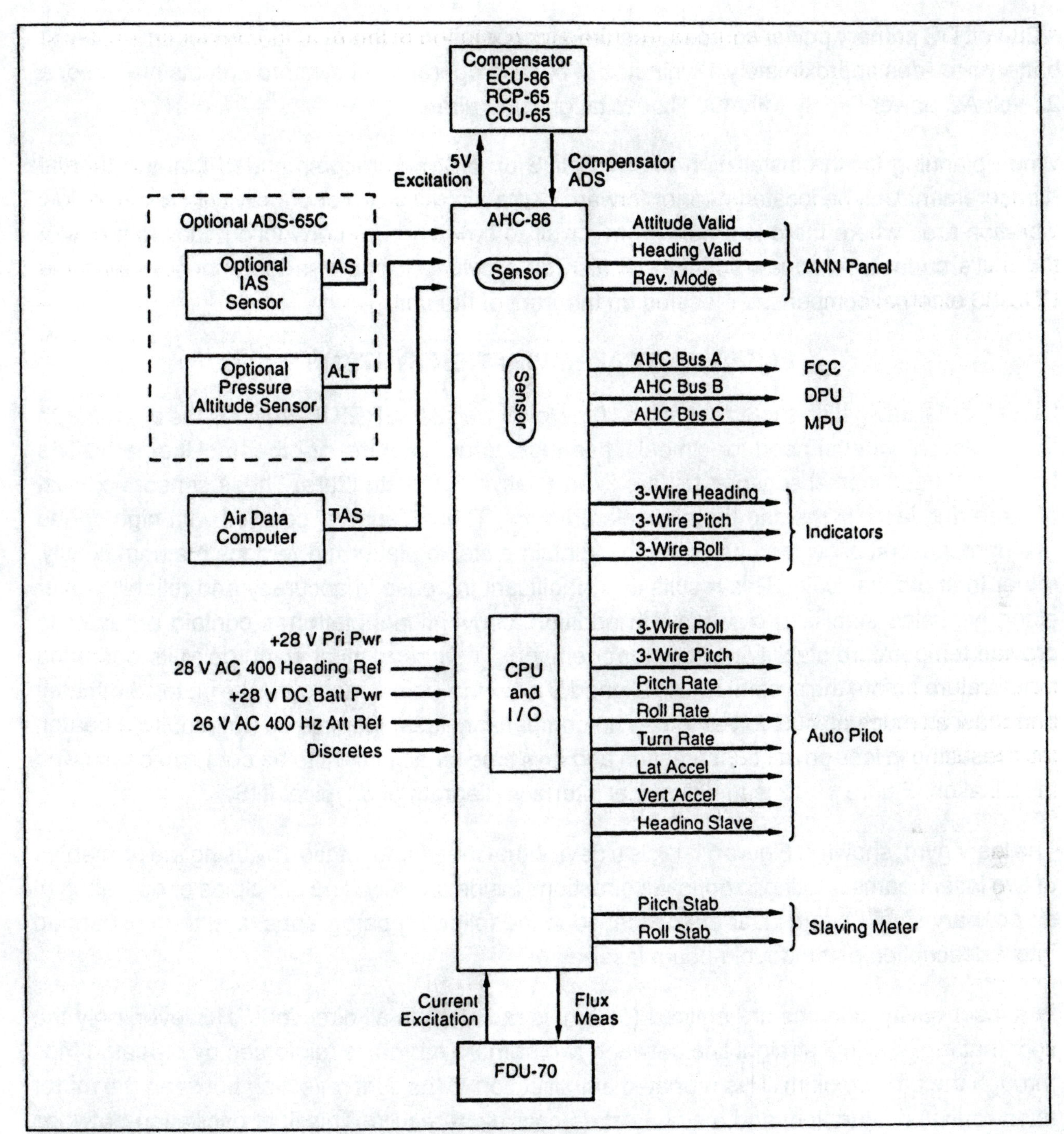

Figure 11-11. Rockwell Collins AHS-86 System Block Diagram.

The AHC-86 can alternately monitor true airspeed or pseudo-true airspeed in order to enhance leveling during in-flight maneuvers. TAS and altitude rate data are gathered directly from an air data computer that provides ARINC 429 low-speed digital bus output. Pseudo-TAS (pressure altitude/indicated airspeed), and vertical speed can be derived from conventional analog air data sensors with only slight degradation in performance.

A 28-volt DC primary power source is required for operation of the AHC-86. An external 28-volt battery provides approximately 11 minutes of backup operation. If synchro outputs are used, a 26-volt AC power supply with 400-Hz excitation is required.

When planning for the installation of any AHRS or INS, it is recommended that the Inertial Measurement Unit be located near or forward of the aircraft's center of gravity in a stable, low vibration area where there is ample room for air to flow over the convection cooling areas of the unit's surface. Adequate room must also be provided for access to connectors and the ECU-86 external compensator located on the front of the unit.

LASER INERTIAL NAVIGATION SYSTEMS

Laser inertial navigation systems, such as the Honeywell LASEREF II, provides the aircraft with flight data, without the need for gimbals, bearings, torque motors, or other moving parts. The Laser INS uses inertial sensors that are fixed relative to the structure. These sensors consist of three ring laser gyros and three accelerometers. These sensors, coupled with high-speed microprocessors, allow the Laser INS to maintain a stable platform reference mathematically, rather than mechanically. This results in a significant increase in accuracy and reliability over older, gimbaled stabilized platforms. In addition, conventional platforms contain a heater to provide temperature stability needed for operation. The heater must warm up to its operating temperature before the system is fully aligned. The warm-up process can take up to 15 minutes and draw as much as 400 volt-amperes. In comparison, laser inertials do not require a heater, thus resulting in less power consumption and less time for alignment to be completed following initialization. Figure 11-12 is an I/O signal interface diagram of a typical INS.

The laser gyro, shown in Figure 11-13, is a device that measures rotation by using the properties of two laser beams rotating in opposite directions inside a cavity. The principles of operation of an ordinary single-beam laser are described in the following paragraphs, and then expanded into a description of the double-beam laser gyro.

In a laser cavity, photons are emitted (or light is radiated) in all directions. However, only the light that radiates in a straight line between two or more mirrors is reinforced by repeated trips through the gain medium. This repeated amplification of the light reflecting between the mirror soon reaches saturation, and a steady-state oscillation results. This light oscillating between the mirrors is typically called a laser beam. To obtain useful laser light outside the laser cavity, a small percentage of the laser beam is allowed to pass through one of the mirrors.

A laser gyro operates much like an ordinary laser, but rather than just two mirrors it contains at least three so that the laser beams can travel around an enclosed area. Such a configuration allows the generation of two distinct laser beams occupying the same space. One beam travels in a clockwise direction and the other travels in a counterclockwise direction. The operation of a laser gyro is founded on the effects rotation motion has on the two laser beams.

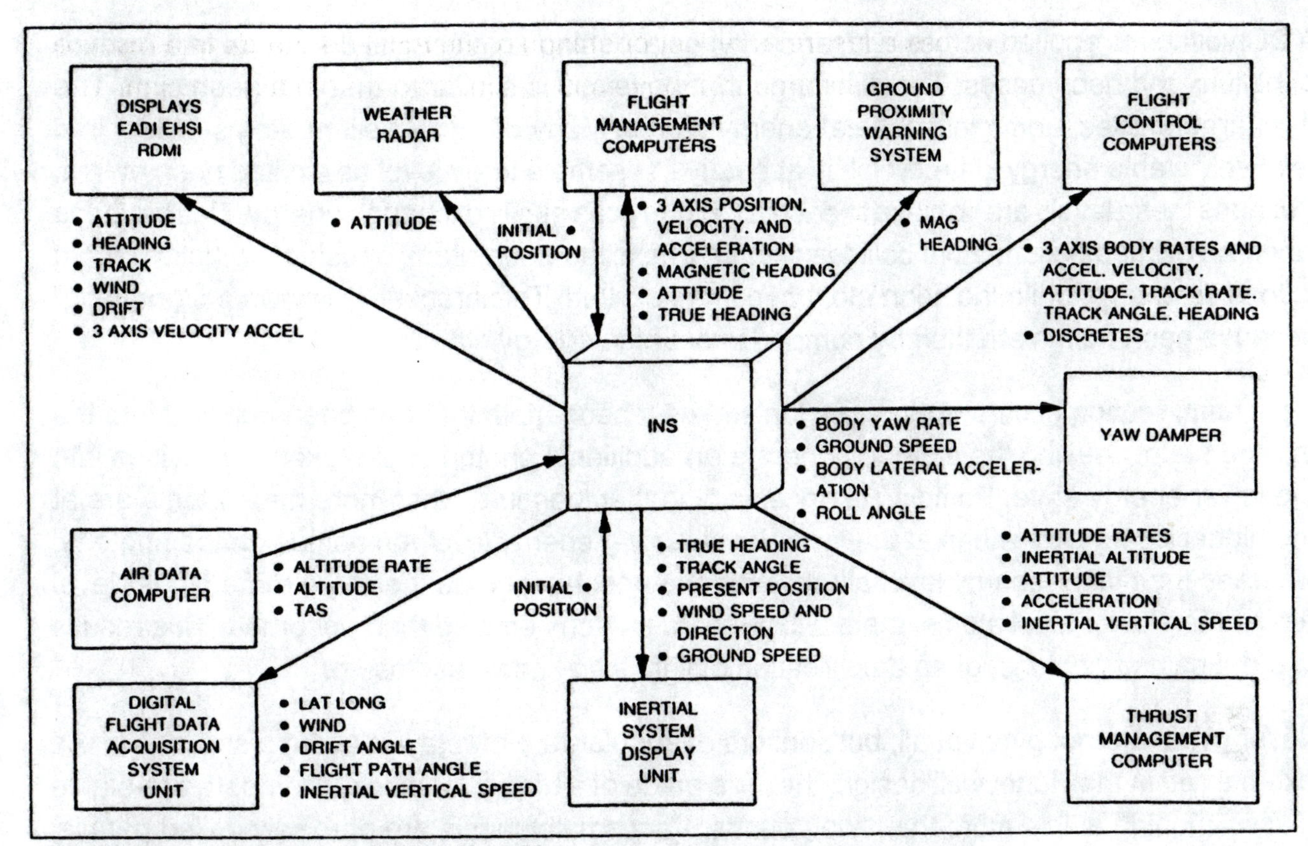

Figure 11-12. Typical Laser INS Interface Diagram. *(Courtesy of Honeywell Commercial Flight Systems)*

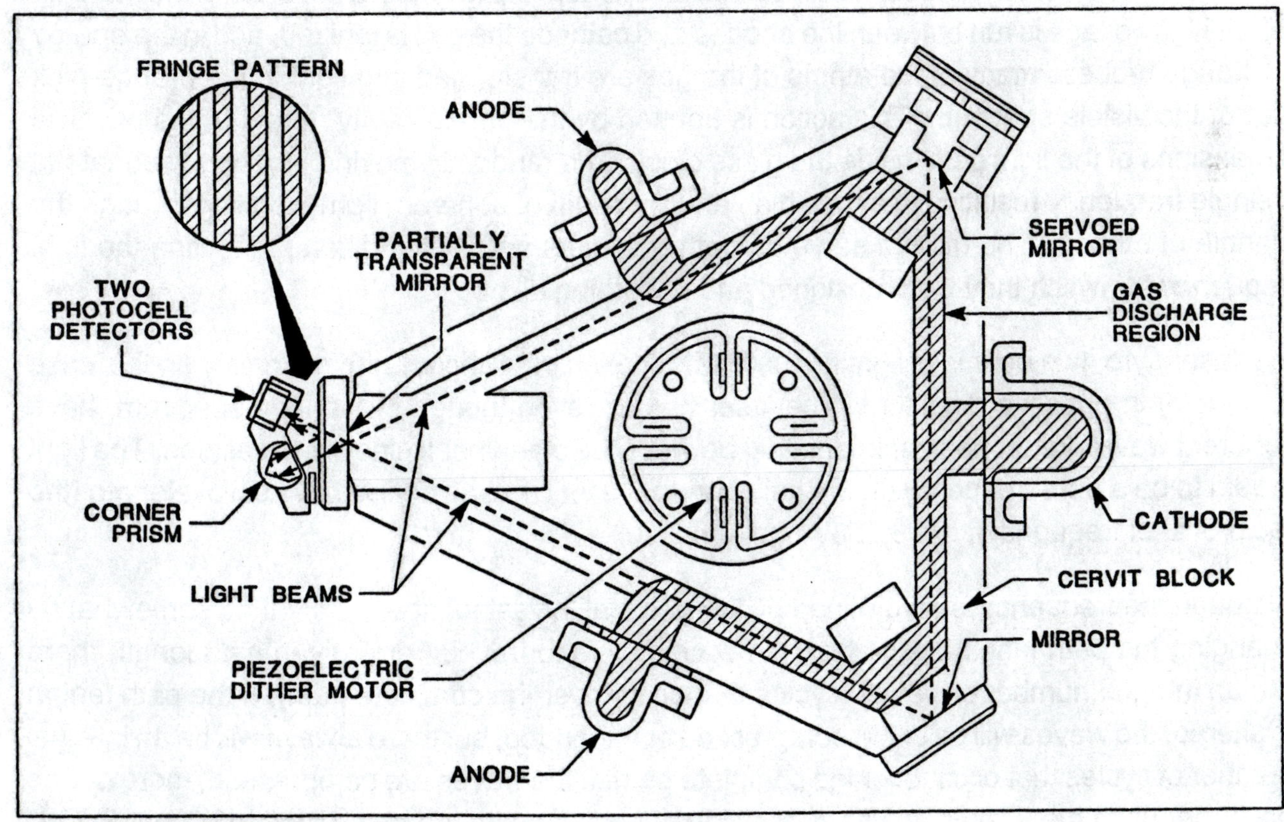

Figure 11-13. Laser Gyro Diagram.

A DC voltage is applied across a laser cavity, establishing an electrical discharge in a mixture of helium and neon gases. The discharge that develops is similar to that in a neon sign. The discharge excites atoms into several energy states. Many of these helium atoms collect in a relatively stable energy state, which is at nearly the same energy level as excited neon atoms. Because these levels are so close to each other, they can easily exchange energy. This happens when an excited helium atom collides with an unexcited neon atom, causing the helium atom to lose its energy while the neon atom becomes excited. This process is known as "pumping" because neon atoms are thereby pumped into a high-energy state.

Light amplification occurs when a photon strikes a neon atom that has been pumped into the excited state, causing the atom to generate an additional photon as it makes a transition into the lower energy state. Pumping is necessary in that it ensures that more neon atoms are at the higher energy level than at the lower level. Lower energy level atoms use up the photons, whereas the higher energy level atoms emit photons. As long as there are more higher level atoms than lower level atoms, there will be more photons emitted than absorbed. This results in a net gain of photons, or an amplification of light also known as "lasing."

Laser gyros are not gyros at all, but sensors of angular rate of rotation about a single axis. As exemplified in the Honeywell design, they are made of a triangular block of temperature-stable glass weighing a little more than two pounds. Very small tunnels are precisely drilled parallel to the perimeter of the triangle, and reflecting mirrors are placed in each corner. A small charge of helium-neon gas is inserted and sealed into an aperture in the glass at the base of the triangle. When high voltage is run between the anodes and cathode the gas is ionized, and in the energy exchange process many of the atoms of the gas are transformed into light in the orange-pink part of the visible spectrum. This action is abetted by the "tuned cavity" effect of the physical dimensions of the light path inside the glass block. The randomly moving particles resonate at a single frequency resulting in a thin, high energy beam of coherent light traveling through the triangle of tunnels. The mirrors serve as both reflectors and optical filters, reflecting the light frequency for which they were designed and absorbing all others.

In a laser gyro, two beams of light are generated, each traveling around the cavity (in this case a triangle) in opposite directions. The laser beams, even though in the light spectrum, have coherent wave-like properties, undulating between zero and peak sine-wave fashion. The light is said to be a pure frequency. In the helium-neon laser gyro, as defined by its wavelength (the reciprocal of frequency), it is 6,328 Angstroms.

Although the frequency is determined by the gas that is "lasing", it can be varied somewhat by changing the path length over which the waves have to travel. For a given path length there are an integral number of waves (cycles that occur over the complete path). If the path length is altered, the waves will be either compressed or expanded, but there always will be an integral number of cycles that occur over the complete path. If the waves are compressed, more cycles occur per unit time. Hence, the frequency increases. If expanded, the opposite is true.

Since both contrarotating beams travel at the same constant speed (speed of light), it takes each the same exact time to complete its circuit. However, if the gyro were rotated on its axis, the path length of one beam would be shortened, while that for the other would be lengthened. Since, as explained, the laser beam adjusts its wavelength for the length of the path, the beam that traveled the shorter distance would rise in frequency (wavelength decreases), while the beam that traveled the longer distance to complete the circuit would encounter a frequency decrease. This frequency difference between the two beams is directly proportional to the angular rate of turn about the gyro's axis. Simply stated, that is the principle of the laser gyro. Thus, frequency difference becomes a measure of rotation rate. If the gyro doesn't move about its axis, both frequencies remain equal (since the path lengths of both beams are equal) and the angular rate is zero.

For those who might find it hard to understand that the laser gyro turning about its axis shortens the path length for one beam and lengthens it for the other, here is another way to explain the phenomenon: Consider a particle of light, a photon, just leaving the cathode and traveling toward the mirror on the right-hand corner (see Figure 11-13). If the gyro turns clockwise on its axis, the mirror would move closer to the photon that was on its way toward it. Hence the photon's path length is shortened in the distance from cathode to mirror, and in the entire distance around the triangular race. Remember the photon is traveling in inertial space; it is not fixed to the gyro. Thus, this one photon and the millions of its traveling companions move around the circuit in a shorter time, and in doing so, they compress the waves in the laser beam and raise its frequency. Of course, the opposite happens to the photon traveling clockwise from the emitter, because when the gyro turned clockwise about its axis, its mirror moved away from the clockwise-traveling photon, forcing it to travel farther to reach the mirror and to complete the circuit.

The difference in frequency in the laser gyro is measured by an optical detector that counts the fringes of the pattern generated by the interference of the two light waves. Since the fringes are seen as pulses by the photocell, the detected frequency difference appears at the output of the detector in digital form, ready for immediate processing by the system's associated digital electronics. Note that there are two photocells. The function of one is to tell the direction in which the fringes are moving, which is an indication of whether the gyro is rotating to the left or right.

As indicated in Figure 11-13, the three corner mirrors are not identical. One is servoed so that it can make micro-adjustments to keep the physical path always the same. Another (the one at the apex of the triangle) permits a small amount light to pass through so as to impinge on the photocell detectors. The prism, as can be seen, flips one beam around causing it to meet and interfere with the beam aimed directly at the photocells. The interfering beams alternately cancel and reinforce each other, thus generating the fringe pattern.

To start the lasing action, 3,000 volts are applied across the anodes to the cathode. Although one can't see the laser beams in the laser gyro, a plasma is formed between the cathode and the two anodes that glows an orange pink that is the same part of the visible spectrum as the

6,328 Angstrom beams. This plasma can be seen. In the center of the Cervit glass block is a device called a dither motor. The motor, which vibrates at 319 Hz, eliminates "laser lock," a hangup that sometimes occurs in the deadband around the zero-rate point.

Accuracy of a laser gyro is influenced by the length of its optical path. The longer the path, the higher the accuracy. The relationship is not linear. For example, a small increase in path length makes for a larger increase in accuracy. As with spinning wheel gyros, the major source of error in a laser gyro is random drift. While in spinning wheel gyros, the root cause is imperfect bearings and mass imbalances, in the laser types it is noise, due almost exclusively to imperfect mirrors including mirror coatings.

As shown in Figure 11-14, the readout corner cube prism, which translates and returns the laser output, is fixed to the gyro base. The gyro readout mirror, therefore, moves with respect to this prism. The path length in the readout system is increased and decreased as the two elements move with respect to each other. By dithering the gyro about a point slightly removed from the center of the gyro, this path difference in the readout is made equal and opposite to the fringe motion created by the gyro oscillator phase changes. The cancellation is instant by instant with the readout detectors thus "seeing" a fringe motion equal to the base motion of the gyro with a small residual dither amplitude, usually about plus or minus one count. Changes in the dither amplitude are also corrected automatically.

LONG-RANGE RADIO NAVIGATION (LORAN)

In the early 1940's the first LORAN systems were placed into operation primarily for shipboard navigation. Twenty years later, an improved version, known as LORAN-C, was implemented

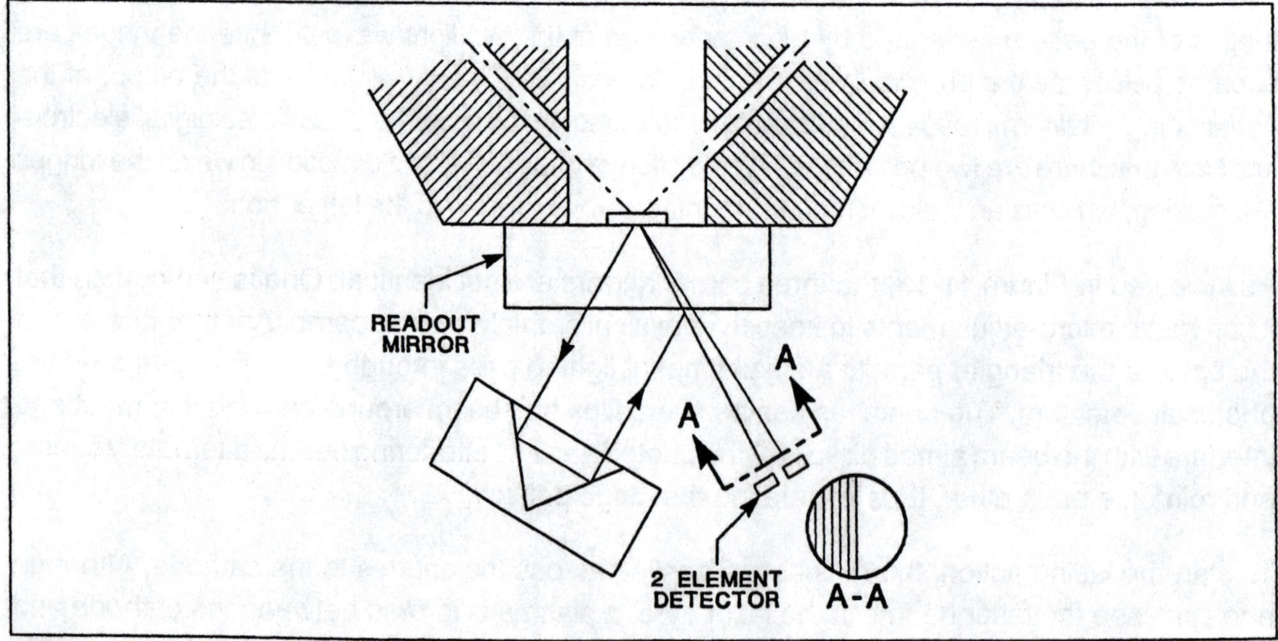

Figure 11-14. Readout Optics.

which is in use today. All LORAN systems operate on a frequency of 100 kHz. The principle of LORAN is based on hyperbolic navigation. A master ground station transmits a pulse which is received by a slave ground station. When the pulse is received, the slave station sends out another pulse. The aircraft LORAN receiver will detect both pulses and will compare the time delay between the pulses to determine a location on a hyperbola. When the airborne LORAN equipment receives a second set of pulses (at a different pulse repetition rate) from another pair of master/slave stations, the location on a second hyperbola is determined. The position of the aircraft can then be calculated at the intersection of these two hyperbolas. The major drawback of LORAN is the limited amount of station installations which are mostly located along the North American coastlines.

VERY LOW FREQUENCY (VLF)/OMEGA RADIO NAVIGATION

The VLF/Omega Radio Navigation System, which was first introduced in the 1960's, operates on the frequencies of 10.2 kHz, 11.05 kHz, 11.33 kHz, and 13.6 kHz during four of eight time slots. The four additional slots are used to transmit a unique frequency, as shown in Figure 11-15. This frequency range is free from ionospheric propagation disturbances, and with a transmit power of ten kilowatts versus four kilowatts for LORAN, VLF/Omega has even longer range.

Airborne VLF/Omega systems receive signals from up to eight Omega navigational facilities located worldwide, augmented by signals from nine U.S. Navy VLF communication systems.

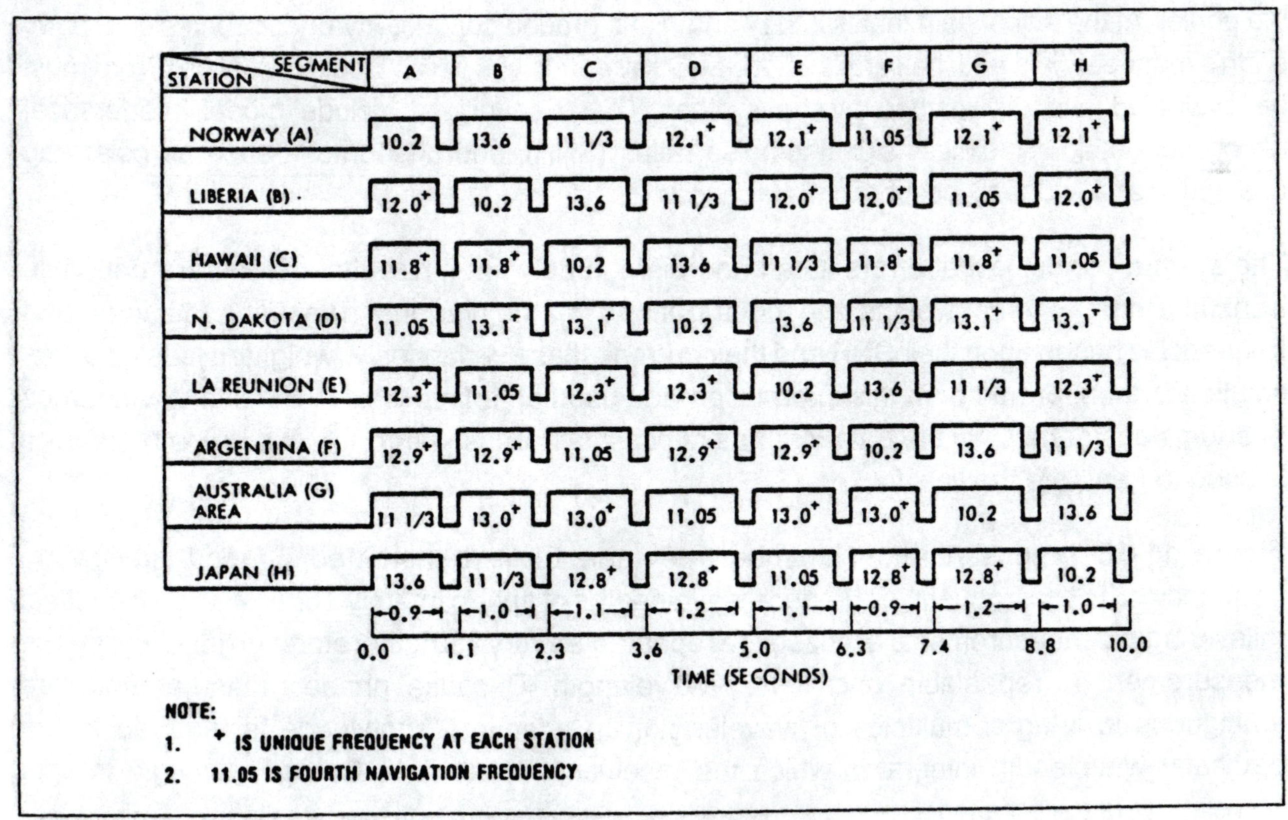

Figure 11-15. ***Omega Transmission Format.*** *(Courtesy of Litton Aero Products)*

The Omega stations are located in Argentina, Australia, Hawaii, Japan, La Reunion, Liberia, North Dakota, and Norway. The Navy VLF stations are located in Australia, Great Britian, Hawaii, Japan, Maine, Maryland, Norway, Puerto Rico, and Washington. They continually change frequency and pulse spacing, repeating the sequence every ten seconds. Reception of the VLF/Omega signals provides the pilot with position information accurate to within three nautical miles anywhere in the world.

The principle of VLF/Omega navigation is not based on measuring the time interval between transmissions, as is the case with LORAN-C, but instead depends on measuring the phase angle between the received pulses. These pulses produce "lanes" that are separated 16 to 144 nautical miles apart and are usable at a range of approximately 8,000 nautical miles (nm) from the transmitter.

The VLF/Omega system is first initialized on the ground with the current latitude and longitude the same as is done to initialize an INS. However, since the VLF/Omega system has no inertial sensors to compute distance and direction from the initial location, it must instead track the lanes that are passed along its course.

The VLF/Omega System automatically synchronizes itself to the Omega pattern and then automatically applies propagation corrections and phase shift through the receiver to the signals received. Propagation corrections computed are based upon input of Greenwich Mean Time (GMT), date, and initial aircraft position coordinates. To improve navigation accuracy, corrections take into account the portion of the signal path that is in day and night, ground conductivity under each signal path, earth's magnetic field and the earth's ellipticity. Other conditions which would introduce errors must be evaluated and disregarded by the system. These conditions include modal interference, directional ambiguity, excess signal-to-noise ratios (SNR), and phenomena such as polar cap absorption and sudden ionospheric disturbances.

The system corrects what errors it can and disregards what it has determined are unusable transmissions. It then weighs the contributions of the various remaining stations and frequencies based upon their SNR and their current phase variances. A weighted least squares solution is then formed from this information and used as a final filter to remove any sources of anomalous data. Only the data that has successfully passed through this entire procedure is used to generate the final fix.

Figure 11-16 graphs relationships between the three transmitted Omega navigation frequencies. These frequencies produce wavelengths of approximately 16, 14.4 and 12 nautical miles. Angular measurements of phase are repeatable every 360°; therefore, an identical phase measurement is repeatable once every wavelength. Because phase measurements are ambiguous in integral multiples of wavelength, an adequate knowledge is required of the particular wavelength integral in which the receiver is located. In Omega, a single integer wavelength is called a lane.

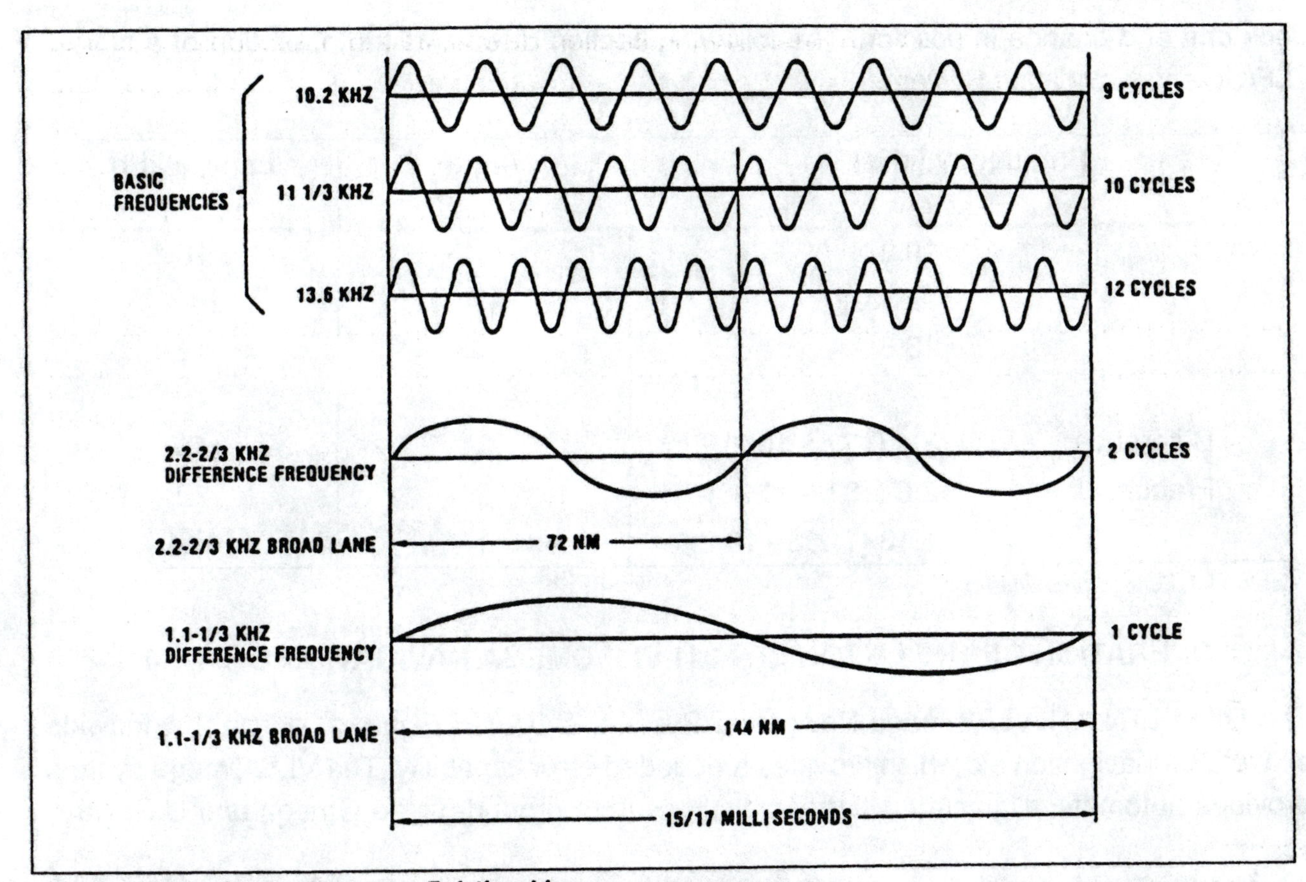

Figure 11-16. Omega Frequency Relationships. *(Courtesy of Litton Aero Products)*

The three Omega navigation frequencies are phase-locked to Universal Time Coordinates (UTC) such that all three cross zero phase with a positive slope at exactly six seconds in 1977 before 0000 hour UTC. The use of a frequency ratio of 9:10 and 9:12 between the 10.2 kHz, and the 11.333 kHz and 13.6 kHz causes this simultaneous crossover to occur every 15/17 milliseconds thereafter. This time interval corresponds to approximately 144 nm. The phase of this 144 nm wide lane signal can be artificially created by subtracting the phase of the 10.2 kHz from the phase of the 11.333-kHz signal. This is the so-called "difference frequency" of 1.133 kHz. Two additional phase values can be formed by subtracting the phase at the 11.333 signal from the phase of 13.6 kHz to produce a lane width of 72 miles, and by subtracting the phase at the 10.2 kHz from 13.6 kHz to produce a lane width of 48 miles. The resulting lane widths are summarized in Table 11-1.

The time slot allocated to the 11.05-kHz frequency is used by the system to receive VLF signals. Four VLF stations are selected on the basis of geometry and signal strength. Each station is sampled at 120 samples per second for an approximate one second burst every five seconds. The VLF stations transmit one of three possible frequencies depending on transmission format. All three must be sampled to determine the active mode on the basis of signal-to-noise ratio. The resulting VLF phase information is initialized at the position of the receiver when the signal comes on line with an acceptable SNR. Subsequent changes in phase are resolved through clock drift and change in position. The following section discusses the operation of a typical VLF/Omega Navigation System.

	Frequency (kHz)	Ratio	Lane Width (nm)
	10.2	9	16
	11.33	10	14.4
	13.6	12	12
Difference	3.4 (13.6 - 10.2)	3	48
Frequencies	2.6 (13.6 - 11.33)	2	72
	1.13 (11.33 - 10.2)	1	144

Table 11-1. Omega lane widths.

OPERATION OF THE LITTON LTN-311 VLF/OMEGA NAVIGATION SYSTEM

The Litton LTN-311 VLF/Omega Navigation System, shown in Figure 11-17, is a worldwide all-weather navigation aid which provides a bounded error capability. The VLF/Omega system provides automatic alignment with transmissions from ground-based Omega and U.S. Navy

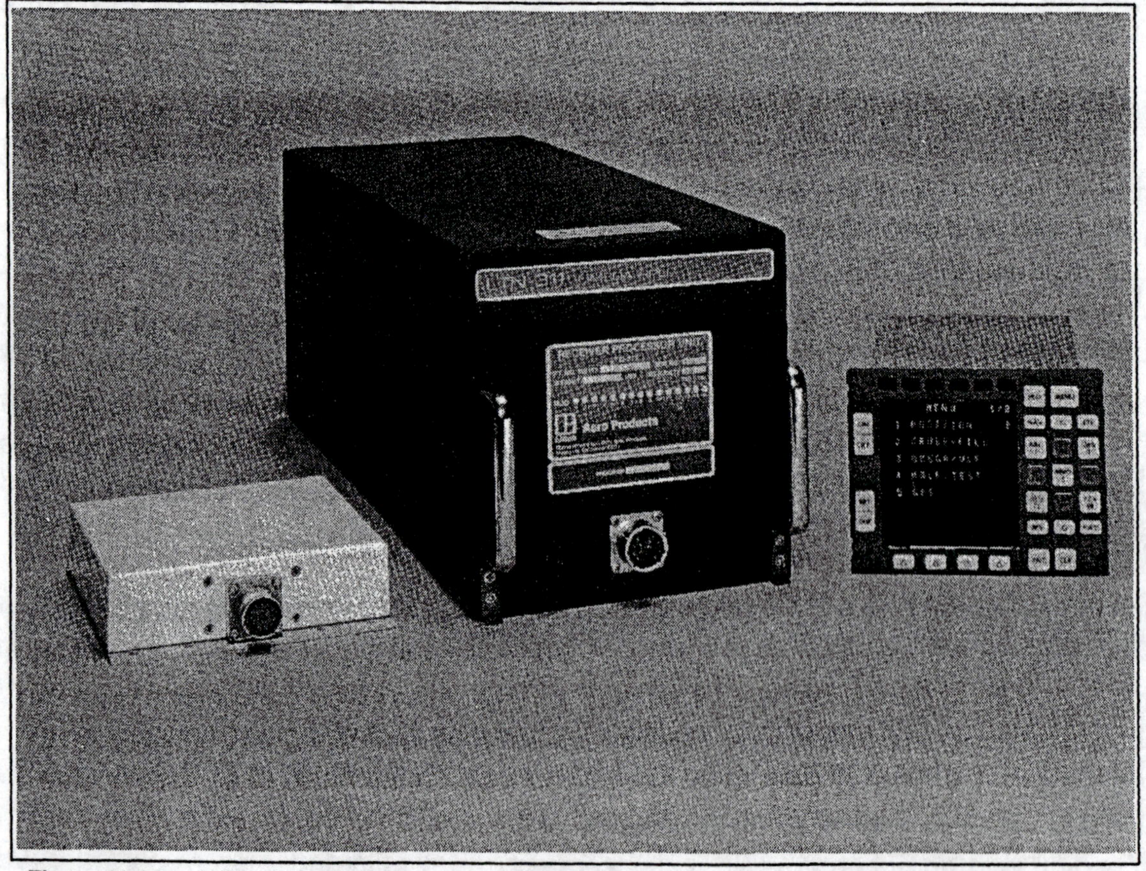

Figure 11-17. LTN-311 Omega Navigation System. *(Courtesy of Litton Aero Products)*

VLF transmitting stations. Using these signals, the system continuously displays all guidance parameters necessary for long-range great circle navigation. In addition to the primary VLF/Omega mode of operation, the system is configured with a backup dead reckoning (DR) mode based on the available aircraft velocity and heading. Switchover to the DR mode is automatically initiated when the number and quality of received signals is below that required for position tracking navigation.

The pilot initializes the system by entering time, date and a flight plan; initialization methods range from a simple verification of the internal time/date clock and a request for flight plan data from an external source to manual data entry via the Control Display Unit keyboard. Once initialized, the LTN-311 automatically navigates from waypoint to waypoint, providing continuous, accurate navigation data.

During operation, the system automatically selects the stations to be used for position and navigation based on measured signal-to-noise ratios of the received transmissions. This method yields the maximum position accuracy because it considers both the quality of the selected stations and the propagation stability, thereby using the stations least likely to be affected by diurnal transitions. The LTN-311 processes signals from the three standard Omega frequencies, 10.2 kHz, 11.3 kHz and 13.6 kHz, as well as the fourth frequency, 11.05 kHz. This fourth frequency is used in conjunction with a VLF converter to process signals from the U.S. Navy VLF transmitters.

The LTN-311 system hardware consists of a Receiver Processor Unit (RPU), a Control Display Unit (CDU), and an Antenna Coupler Unit (ACU), as shown in Figure 11-18.

The RPU contains the major system electronics:

1) Computer board assembly

2) Omega/VLF receiver board

3) Digital interface board

4) Analog interface board

5) Power supply

The computer/processor board is a complete integrated navigation computer subsystem on a removable printed circuit board. The assembly contains a Z8001 central processor, program memory, nonvolatile memory, sensor inputs for heading and speed, CDU communications transmitter and receiver, program pin sensing, discrete drivers, computer bus control, and communication with the other subsystems in the RPU.

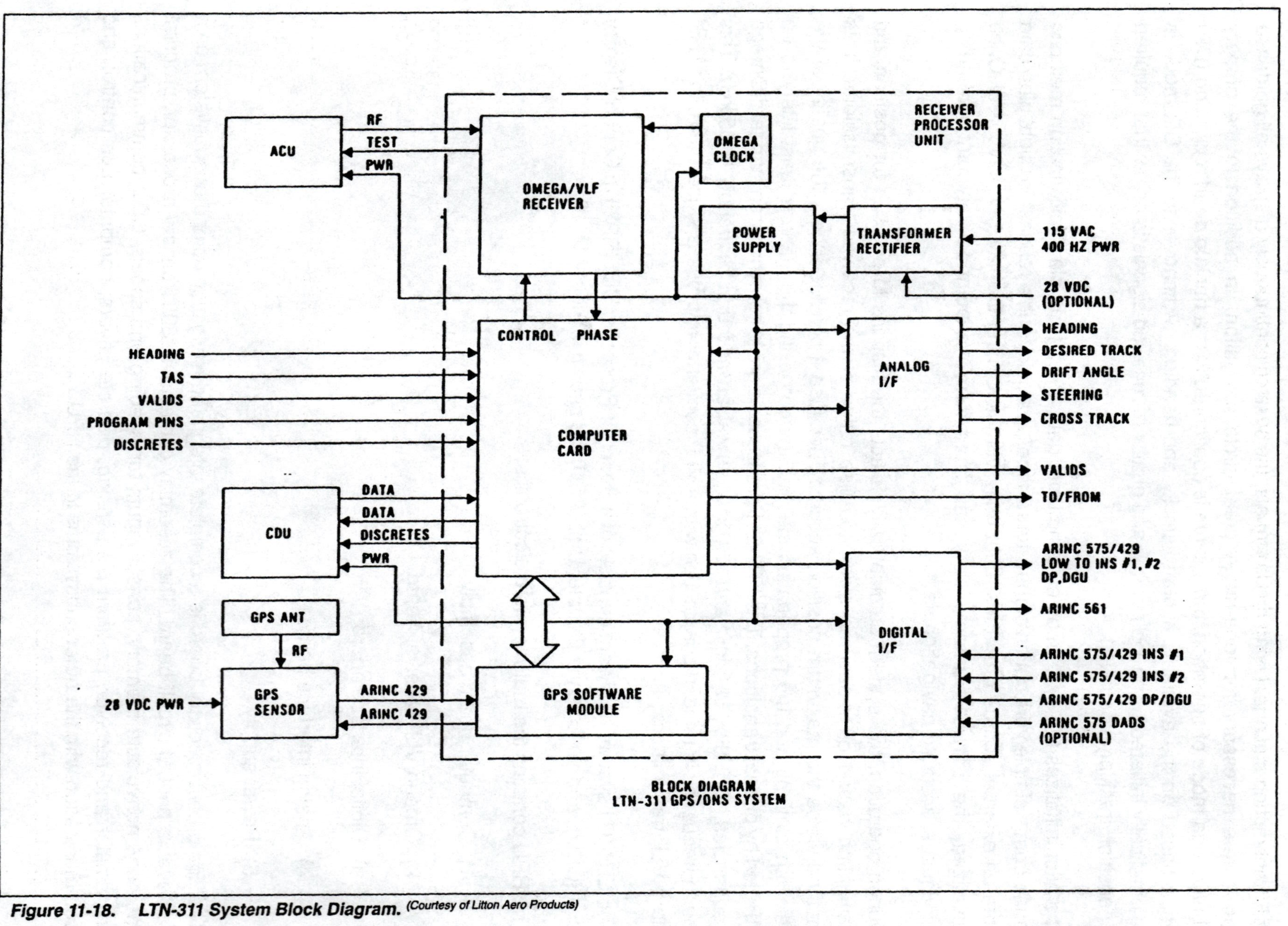

Figure 11-18. LTN-311 System Block Diagram. *(Courtesy of Litton Aero Products)*

The receiver board is a printed circuit board containing circuits for receiving, preprocessing and phase measurements of Omega signals. It contains the antenna loop-selection matrix, four Omega frequency amplifier/filter/limiter channels, reference frequency generators, and the phase measurement circuit.

A VLF converter subassembly configures one Omega channel to process VLF signals from selected U.S. Navy communication stations. The converter provides outputs at 11.05 kHz. The board is separated into a signal processing section and a digital phase measuring section. The signal processing section has three or four receiver channels -- one for each of the Omega frequencies, 10.2 kHz, 13.5 kHz, 11.3 kHz, and the fourth frequency, 11.05 kHz. The fourth frequency channel is used for VLF.

The digital interface board is the digital communication link between the Omega Navigation System and other aircraft systems. The card incorporates four ARINC 575 receivers, one ARINC 575 transmitter, and one ARINC 561 transmitter. The reception and transmission of data is controlled by the RPU computer.

The analog interface board provides analog signals associated with the steering functions of the autopilot and instruments. Four synchro parameters plus high- and low-level cross-track deviation signals are provided to the aircraft avionics. The board contains four digital-to-synchro converters, a digital-to-dc converter, a cross-track deviation, and a synchro self-test.

The CDU is the main interface between the pilot and the Omega Navigation System. Communication is via a transmitter/receiver port which is electrically compatible with ARINC 429. The CDU receivers power, display, and annunciator data from the RPU, and transmits the switch depressions to the RPU. The CDU consists of a front panel, LED display board, four circuit board assemblies, a power supply, and a chassis. The front panel assembly includes 23 keyboard pushbuttons, six annunciator lights, three indicator lights, two rocker switches, and an optical filter for the LED assembly. Behind the front panel is the LED Display Board consisting of a two-color 128 character display matrix of 32 quad chips. Providing signals to the Display Board is the Display Driver Board, mounted vertically on the inside front of the chassis.

The CDU computer is a Z8 microprocessor with support circuits mounted on a single card. It is mounted horizontally at the top of the CDU. Mounted horizontally beneath the computer board, the Display Refresh Board consists of a single printed circuit card containing circuitry for the Display Sequencer/Refresh Control, Watch Dog Timer, and Power On Reset function.

The Analog/IF Board and Power Supply are plugged together to form a single module. The module contains four power supplies, five lamp drivers, a power supply monitor, and an ARINC transmitter/receiver. Also contained on the module are auxiliary circuits consisting of the ON/OFF control circuit and an edge lighting transformer. The two units are mounted together on a heat sink at the bottom of the CDU chassis.

The LTN-311 uses an H-Field bi-directional loop antenna. Unlike E-Field antennas, this unit is insensitive to precipitation static. E-Field antennas characteristically will develop a negative charge on the surface of the antenna when flown through clouds which contain excess electrons. Omega signals generally have a field strength of only a few hundred microvolts; therefore, this P-static electron charge would cause severe signal degradation and frequent signal loss in poor weather conditions.

The antenna components are mounted on a single printed circuit board which is fastened together with the interface connector to an aluminum backing plate. The entire assembly is then encapsulated in lightweight syntactic foam. The antenna is attached to the aircraft with flat-head screws inserted through counterbores in the epoxy and fetching on the aluminum back plate. The exposed surface of the antenna is then coated with epoxy to provide the desired color and finish.

GLOBAL POSITIONING SYSTEM (GPS) NAVIGATION

The Navstar GPS is a space-based radio positioning and navigation system that provides fifteen meter accuracy of three-dimensional position data, velocity information, and system time to suitably equipped users anywhere on or near the earth. The testing and development of the GPS is currently under the Department of Defense (DoD). As development progresses, the system has the potential to replace most radio navigation systems currently in use.

As shown in Figure 11-19, the GPS consists of three major segments: space system segment, control system segment, and user system segment. The operational space system segment deploys

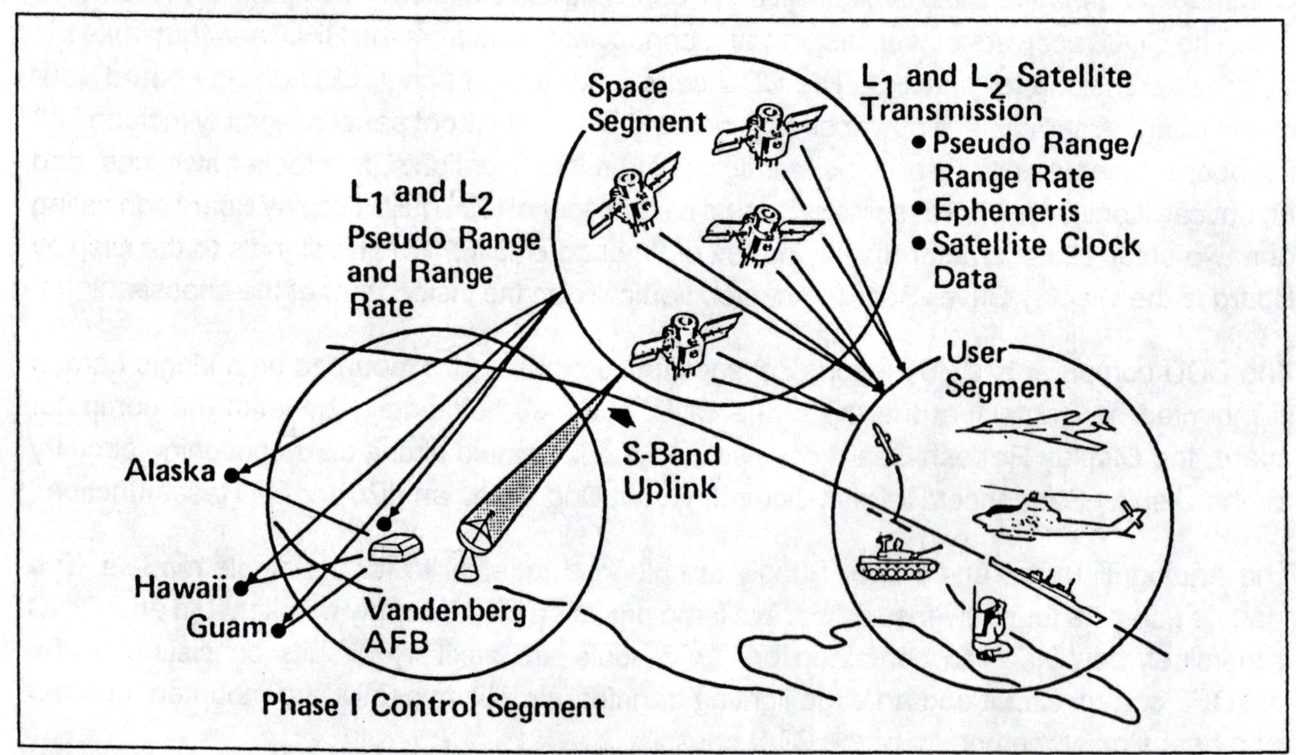

Figure 11-19. Navstar GPS Segment Operation. (Courtesy of Magnavox)

six planes of satellites in circular 10,898 nautical mile orbits. Each satellite has an orbital inclination of 55° and a 12-hour period. Each plane has four satellites. This deployment provides the satellite coverage for continuous three-dimensional positioning, navigation, and velocity determination. Each satellite transmits a composite signal at two L-band frequencies consisting of a precision (P) navigation signal and a coarse acquisition (C/A) navigation signal. The navigation signals contain satellite coordinates at a specific time known as ephemerides, atmospheric propagation correction data, and satellite clock bias information provided by a master control station. In addition, the second L-band navigation signal permits the user to correct for the ionospheric group delay or other electromagnetic disturbances in the atmosphere.

An airborne GPS receiver computes its distance from a satellite by measuring the travel time of the satellite's signal. Using the distances from at least three satellites, the receiver can triangulate the aircraft's current position. With measurements from four satellites, the GPS receiver can compute elevation as well.

In the current control system segment, four widely separated monitor stations, located in U.S. controlled territory, passively track all satellites in view and accumulate ranging data from the navigation signals. The ranging information is processed at a master control station located in the continental United States to use in satellite orbit determination and systematic error elimination.

The orbit determination process derives progressively refined information defining the gravitational field influencing spacecraft motion, solar pressure parameters, location, clock drifts, and electronic delay characteristics of the ground stations, and other observable system influences. An upload station located in the continental United States transmits the satellite ephemerides, clock drifts, and propagation delay data to the satellites as required.

Each of the satellites emits a carrier which is modulated with a pseudorandom noise code of very low repetition rate. The generation of this code is synchronized to the satellite clock time reference. The user receiver also maintains a time reference used to generate a replica of the code transmitted by the satellite. The amount of time skew that the receiver must apply to correlate the replica with the code received from the satellite provides a measure of the signal propagation time between the satellite and the receiver. This time of propagation is called the pseudorange measurement since it is in error by the amount of time synchronization error between the satellite and receiver clocks. The receiver also measures the Doppler shift of the carrier signals from the satellite. By measuring the accumulated phase difference in this Doppler signal over a fixed interval, the receiver can infer the range change increment. This measurement is called the delta pseudorange measurement and is in error by an amount proportional to the relative frequency error between the emitter and receiver clocks. Since the carrier wavelength is short, the delta pseudorange is a finely quantized measurement.

Measurements from four satellites provide the receiver with sufficient information to solve for three components of user position, velocity, and user clock errors. To accomplish the navigation function, pseudorange and delta pseudorange measurements are used to update a running estimate of user position.

The major issue in certifying a GPS Sensor in a multi-sensor environment is Integrity. Integrity is the ability of the equipment to identify and provide timely warnings when the system is providing navigation information outside its specified limits. Receiver Autonomous Integrity Monitor (RAIM) is the technique being developed in the aviation industry as a means of determining and annunciating failures of the navigation solution.

Using RAIM, a GPS sensor or a combination of sensors in a multi-sensor configuration can detect system failures so that the safety requirements for all phases of flight can be met. For each phase of flight, enroute, terminal and approach, the navigation performance requirements and time to alarm can be specified.

For a GPS-based system to provide the cost benefits demanded by the airline community, it needs to be certifiable as a Sole Means of Navigation. To qualify as a Sole Means of Navigation system, the GPS system must be able to detect and isolate a failed satellite from the navigation solution and continue to provide the navigation solution within specified limits. In contrast, Supplemental Means of Navigation requires only the detection and annunciation of the failure of the navigation solution.

Industry studies indicate that the Navstar GPS constellation of 21 satellites with three active spares will not provide sufficient availability of satellites and redundancy of signals to perform RAIM to a level required for Sole Means of Navigation for the primary phases of flight.

Therefore, to achieve the integrity required for Sole Means, a GPS sensor needs to be augmented with other navigation or position determining techniques. Wide-Area Augmentation Systems (WAAS) being considered include satellite based, radio navigation based and inertial navigation system based. Also being considered is GPS Integrity Channel (GIC), whereby a small set of geostationary satellites provide integrity information of the satellites in real time. These geostationary satellites can also provide differential corrections along with integrity information.

Differential GPS (D-GPS) is a technique used to overcome selective availability, whereby the Defense Department dithers the GPS signal, thus diluting the precision for civilian users from 15-meter to 100-meter horizontal accuracy. D-GPS not only overcomes signal degradation, but eliminates satellite clock errors, and ionospheric and atmospheric propagation errors. Differential GPS for instrument approaches is obtained by having a fixed GPS receiver at the runway measure the selective availability error and transmit an error correction message, via a data link, to incoming aircraft. Upon receipt of the error correction signal, the onboard GPS receiver will display the correct position of the aircraft.

The most convenient complement to the Navstar GPS satellite constellation is the Commonwealth of Independent States' (CIS) equivalent system, GLONASS. The combination of GPS and GLONASS satellites will provide enough redundancy and accuracy to achieve sole means of navigation for enroute, terminal and non-precision approach phases of flight. To realize precision approach capabilities, other means such as differential corrections will be required. In commercial aviation the concept being developed is called Global Navigation Satellite System (GNSS). GNSS is envisioned to be composed of GPS, GLONASS and possibly other geostationary satellite systems. These systems will allow GNSS to provide navigation, integrity and differential corrections so that precision navigation for all phases of flight including precision approaches, and eventually autoland capabilities can be achieved.

Commercial aviation will profit immensely from the capabilities GNSS will provide for all phases of flight. GNSS promises improved traffic flow, safety and economy. Whether flying transoceanic routes, remote non-aided areas, or in the terminal environment, GNSS-equipped aircraft will benefit from the extreme accuracy of the system. GNSS will have a major impact on commercial aviation, helping to fly increasing numbers of passengers safely worldwide.

OPERATION OF THE LITTON LTN-2001 GLOBAL POSITIONING SYSTEM SENSOR UNIT (GPSSU)

The LTN-2001 MK2 sensor is an eight channel C/A-code continuous tracking GPS receiver. When upgraded with the GNSS receiver card, the LTN-2001 becomes a MK3 sixteen channel receiver, with each of the channels capable of tracking GPS, GLONASS, GIC, Geostationary satellites or Pseudolite signals. This architecture provides for an all-in-view capability when both the GPS and GLONASS constellations are fully operational. It also allows for growth to all envisioned applications of satellite-based navigation systems for commercial air transports. These features include precision approach and autoland using differential corrections, enhanced integrity monitoring by tracking of GLONASS satellites and/or processing signals from GIC, and/or track to the available pseudolite signal near the airport. These features will allow the Litton GNSS to be certifiable for sole means of

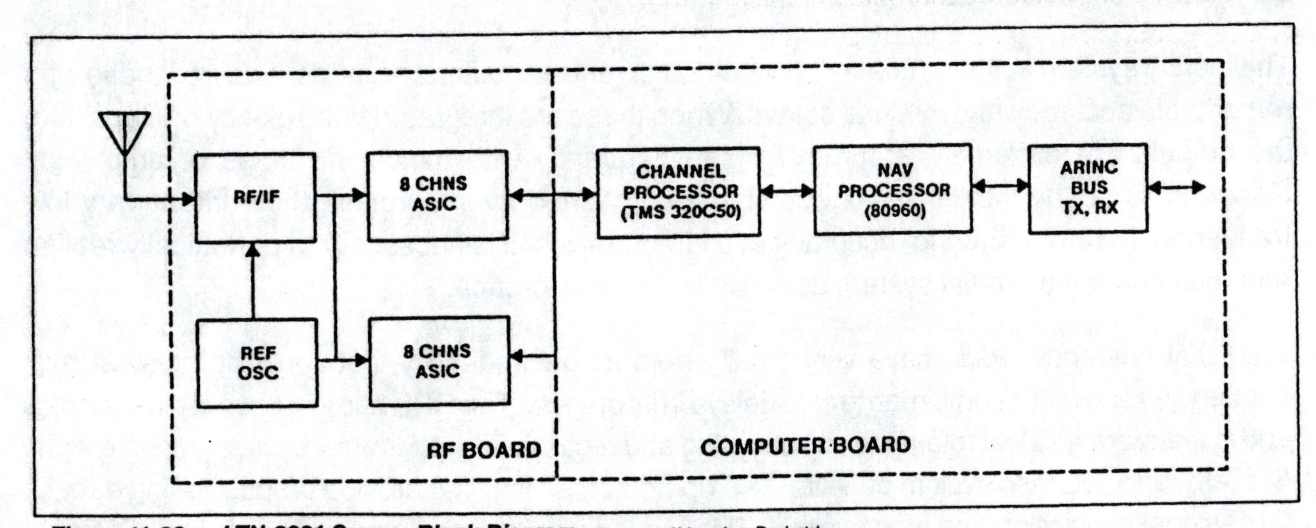

Figure 11-20. LTN-2001 Sensor Block Diagram. (Courtesy of Litton Aero Products)

navigation for all phases of flight, as well as future applications of GPS for commercial aviation, such as automatic dependent surveillance and curved path approaches.

A functional block diagram of the LTN-2001 MK2 sensor unit is shown in Figure 11-20. The sensor unit contains two hardware modules, the Receiver module and the Computer module. The Receiver module comprises RF/IF circuitry and the Baseband Application Specific Integrated Circuit (ASIC). The RF/IF circuitry consists of L band and intermediate frequency (IF) amplifiers, local oscillator, crystal oscillator, frequency synthesizer, and analog-to-digital (A/D) converter. Adequate filtering function is also provided to reject SATCOM and other interference. The Baseband ASIC is made of a high-speed CMOS standard cell in one micron semiconductor technology. The density of the chip is about 50,000 gates.

Eight channels are in each chip, so two chips will be used in the MK3 GNSS sensor unit. The computer module consists of TMS320C50 Digital Signal Processor (DSP), Intel 80960 Navigation processor, memory and ARINC I/O transceiver. The DSP performs the acquisition and tracking functions of the incoming digitized signal and generates the pseudorange, delta-range measurements and demodulates the satellite message for further processing in the navigation processor. The navigation processor receives the raw measurements from the DSP and computes the three-dimensional position by using an extended Kalman filter. (A Kalman filter is a linear system in which the mean squared error between the desired output and the actual output is minimized when the input is a random signal generated by white noise.)

INTEGRATED GPS/INS

Integrating a GPS receiver with an Inertial Navigation System can provide a navigation system which has superior performance to stand-alone installation of either or both systems. Since the position accuracy of GPS is so much better than that of an INS, it may first appear that the principle advantage in integrating the two is the bounding of inertial position errors by the GPS. However, an inertial system does not directly measure position as does the GPS, so that a comparison of position accuracies is misleading.

The inertial system actually measures acceleration and angular rates, and measures these with remarkable accuracy and minimal delays. When these are integrated into velocity and position, the outputs are extremely accurate at high frequency with almost no noise or time lags. However, since the inertial system outputs are obtained by integration, they drift at very low frequency. To obtain superior accuracy at all frequencies, it is necessary to periodically realign and recalibrate the inertial system using an external reference.

The ideal reference would have very small errors at low frequency, but could be noisy at high frequencies and could contain large time delays in its outputs. The GPS has just such characteristics and is therefore an ideal reference for realigning and recalibrating the inertial system during a flight. Similarly, an integrated system benefits GPS operation. An INS can provide prepositioning data for GPS signal acquisition and reacquisition. This reduces the search domains in space and time for the GPS receiver and reduces the time to the first satellite fix. Additionally, although GPS is very accurate under ideal conditions, such conditions will not always exist.

Independent GPS navigation requires four satellites in good geometry. GPS accuracy will degrade due to bad satellite geometries, periodic satellite failures, dynamic lags during maneuvers, and jamming. The INS provides an ideal means for minimizing the degradation of accuracy under sub-optimal conditions. The reason is that the inertial errors build up very slowly during the initial period following loss of GPS update data. This build-up of errors is much smaller than generally realized. Instead of errors building up at the nominal rate of a pure inertial system, the error growth rate is determined by the accuracy with which the inertial system was calibrated and aligned just before the GPS was interrupted. Since this calibration occurred at stabilized temperatures in the existing environment and without turning off the system, it is extremely accurate. The error build-up is almost entirely due to uncompensated gravity anomalies, with an additional small contribution due to gyro bias random walk.

GPS failures must not be allowed to contaminate the calibrated pure inertial solution. Such contamination can be avoided by exploiting the low noise characteristics and slow error build-up of the inertial system. The incorporation of the calibration and alignment corrections can be delayed until the integrity of the GPS signals used to determine the corrections is established. This is done in the integrated solution by propagating inertial solutions forward without corrections, or "open loop" for short periods, while the closed loop corrected solutions are propagated in parallel.

A major advantage of a strapdown system over a gimballed platform system is that such parallel solutions are possible. This is true because, in a strapdown system, the reference frames are computed, and therefore correspond to "imaginary" platforms of which there may be more than one, rather than a single physical gimballed platform. Once the GPS integrity is verified, the closed-loop solution is accepted and becomes the starting point for the next cycle of open and closed loop propagation.

In addition to the improved long-term accuracy obtained by the use of GPS updates to periodically recalibrate and realign the INS, the integration of GPS with the inertial system makes in-flight alignments and alignment while taxiing or takeoff possible. Thus the INS and GPS systems form a truly synergistic relationship, with the best features of each being combined to form an ideal navigation system.

OPERATION OF THE LITTON GPS INTEGRATION MODULE

The GPS Integration Module (GPSIM) is a single card computer. When the GPSIM is inserted into its reserved slot in the updated LTN-92 INS, the integrated system (with appropriate software) becomes capable of interfacing with one or two GPS Sensor Units (GPSSU). A block diagram illustrating the interface of the GPSIM to the LTN-92 system is presented in Figure 11-21. The GPSIM is shown resident in the LTN-92 with connections to two GPSSU's. Interfaces

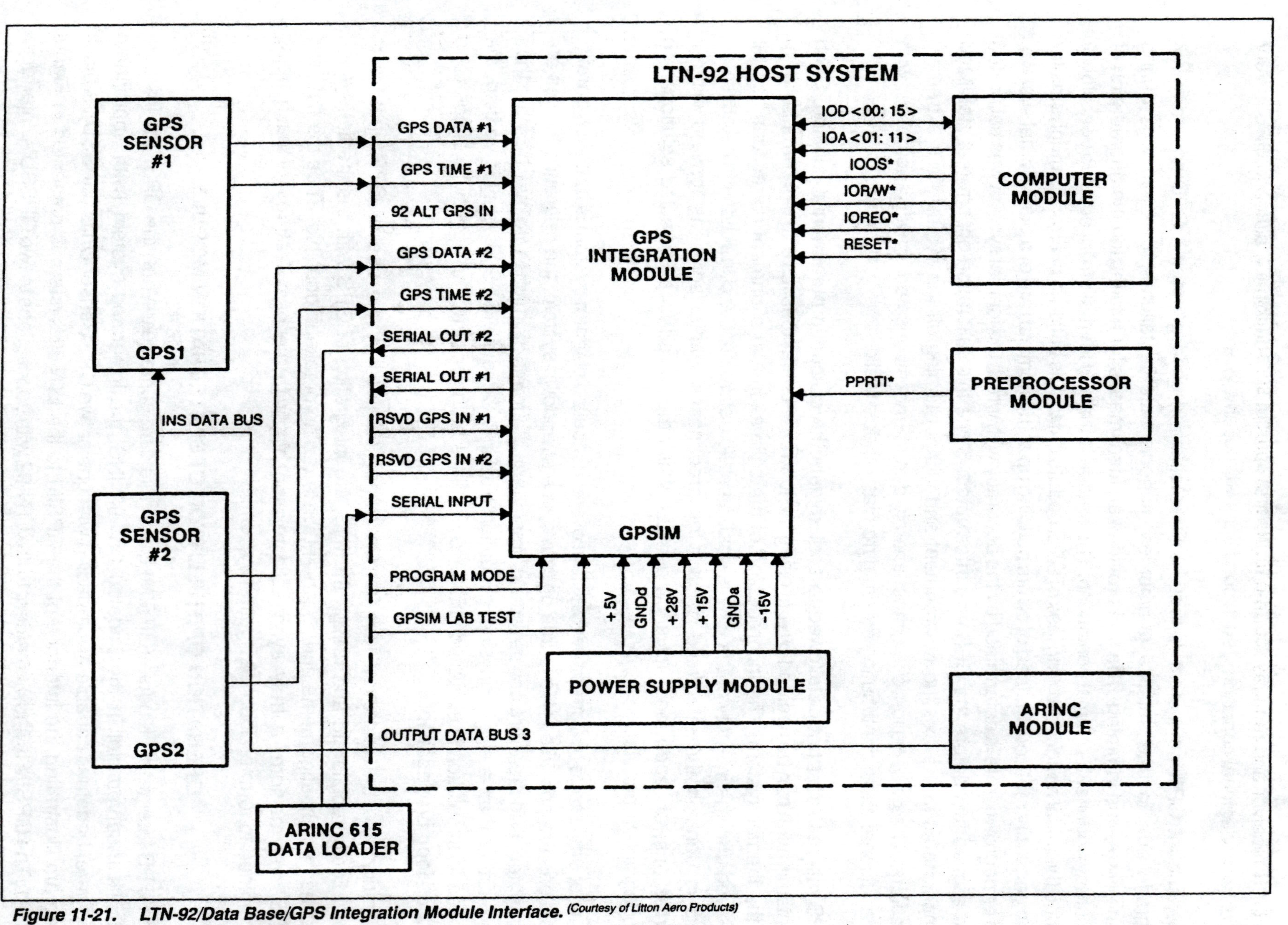

Figure 11-21. LTN-92/Data Base/GPS Integration Module Interface. (Courtesy of Litton Aero Products)

to the LTN-92 power supply, computer, and preprocessor modules are routed through the LTN-92's motherboard.

The interfaces to the remote GPSSU's are routed through the LTN-92 rear panel connector via the system's motherboard and flex cables. The GPSIM receives GPS data from the GPSSU's and inertial navigation data from the LTN-92's computer. The GPSIM software automatically determines which of the three types of GPSSU's it is being interfaced to by analyzing the incoming data. Upon making this determination, the GPSIM software sets the appropriate data receiver parameters and enables the proper integration algorithms. The GPSIM then processes the GPS and inertial data and outputs a hybrid INS/GPS solution, as well as autonomous GPS information, to the LTN-92's computer. The LTN-92 system transmits this information along with the unaided inertial navigation data on a standard serial output bus for use by other aircraft systems.

When configured with optional application memory chips, the GPSIM will also support a worldwide data base option for the LTN-92. This card will contain two megabytes of memory that will have the capacity to store: high and low altitude airways, high and low intersections, airports over 4,000 feet, non-directional beacons, co-located navaids (VOR/DME, VORTAC, TACAN), and up to 1,000 custom routes of up to 98 waypoints each.

CONCLUSION

This chapter began with a discussion of the Gyrosyn Compass System, which was the predecessor of modern day Inertial Navigation Systems. A gyrocompass is more accurate than a magnetic compass, which relies on the earth's magnetic field to indicate magnetic north. A gyrocompass points to true north, and is not affected by magnetic forces or by the pitch or roll of the aircraft. We learned the principals of inertial navigation is based on measuring and integrating accelerations from a known fixed position. Discussed was the gyro-stabilized platform and the strapdown platform. Recent advances in laser and microcomputer technology has brought us the Laser INS, which requires no moving parts and consumes much less power than traditional inertial systems. Instead of using accelerometers, the laser inertial systems measures the change in frequency due to changes in the light path length.

Finally, we discovered the principals of operation of radio long-range navigation systems used for transoceanic flight, such as LORAN, VLF/Omega, and the very latest in onboard long-range navigation, the Global Positioning System (GPS), proven invaluable as a navigation aid in Operation Desert Storm and currently being certified by the FAA for commercial Category III operations. In the near future, it is predicted that the primary means of long-range navigation on commercial transport aircraft will be GPS integrated with INS.

The last chapter will present how this abundance of navigational information is integrated through the Flight Management System and presented to the pilot on Electronic Multi-Function Displays. Also to be presented is how the Automatic Flight Control System uses this information to fly the aircraft safely to its intended destination.

REVIEW QUESTIONS

1. The Gyrosyn Compass combines the functions of what two instruments?
2. What component in the Gyrosyn Compass System slaves the directional gyro to provide accurate and stable magnetic heading information?
3. What is the purpose of the Remote Magnetic Compensator?
4. What type of information is supplied from Inertial Navigation Systems?
5. How does the INS convert acceleration into distance?
6. Why are the accelerometers mounted on a stabilized platform and how is the platform stabilized?
7. How is an Inertial Navigation System initialized before takeoff?
8. How are angular rates derived in a Strapdown IRS?
9. Explain the operation of the Honeywell AHZ-800 AHRS in the "normal" and "slaved" modes.
10. Define the installation procedures for an AHRS or INS.
11. What are the advantages of a Laser INS over conventional gimbaled stabilized platform inertial systems?
12. Explain the principles of operation of a laser gyro.
13. Define what is meant by "pumping" a laser.
14. What happens to the frequency as the path length of the laser beam is shortened?
15. Compare the major source of errors in a laser gyro to a spinning wheel gyro.
16. Explain the theory of hyperbolic navigation used in LORAN-C.
17. How does the VLF/Omega Navigation System determine position by measuring phase angles between received pulses?
18. Explain the operation of the three GPS segments and the GNSS concept.
19. What function does the DSP perform in the Litton LTN-2001 GPS Receiver?
20. What are the benefits of integrating GPS with INS?

Chapter 12
Flight Management and Electronic Display/Control Systems

INTRODUCTION

The proliferation of navigation systems onboard aircraft has led to a cockpit information explosion. Hence, Flight Management Systems (FMS) were developed, first as RNAV systems which provided phantom waypoints independent of VORTAC station locations, then as integrated navigation information control and display systems. These signals are coupled to a Flight Director System to provide steering cues and/or to an Autopilot to relieve the pilot of manually controlling the aircraft. Electromechanical displays are slowly being replaced with color CRT displays to present a composite picture of the navigational solution to aid the pilot's situational awareness and reduce his/her workload. This chapter presents Electronic Flight Instrumentation Systems and Flight Management and Control Systems in use today on modern commercial and corporate jet aircraft.

AUTOMATIC FLIGHT CONTROL SYSTEMS (AFCS)

Early automatic pilots were primarily pilot relief devices, which did little more than hold the aircraft straight and level. The introduction of transistorized electronics permitted dramatic changes in the size, weight, and power requirements of automatic pilots. However, these advantages were directed toward adding more functions and features. The "autopilot" has grown to become a system that is used in all phases of flight and has, as such, acquired its more modern identification as an Automatic Flight Control System.

Automatic Flight Control Systems used in modern jet transports are all uniquely tailored to the specific aircraft, but all share common features. For example, the flight aerodynamics of a MD-11 are different from those of a Boeing 767; however, both aircraft would most likely require an "attitude hold" mode of operation. In this case, the attitude hold feature is common to both autopilot designs, but gains in the two autopilots will differ to accommodate the differences in the aerodynamics of each aircraft. Each AFCS receives attitude and heading signals from a vertical and directional gyro or Long-Range Navigation System and has its own rate gyro/accelerometer system to develop attitude and flight path stabilization signals. The AFCS computers comprise an electronic "brain" that receives signals from its "senses" to compute the proper responses and provides outputs to electric and/or hydraulic actuators that are then "muscles" which move the aircraft's control surfaces.

As complex as some of today's autopilot systems have become, they all can be narrowed down to providing at least one main function. That function is stability. Today's modern autopilots are designed to provide pitch, roll, and yaw axis stabilization around the pilot's desired reference attitude. To do this, the autopilot system must detect changes in aircraft attitude and respond to those changes more quickly and smoothly than its human counterpart. For an autopilot to maintain this stability, it must know what the pilot's desired aircraft attitude is, know what the actual aircraft attitude is, compare the two, correct for the difference or error, and control the speed of the correction.

The human pilot controls the aircraft by detecting a change in aircraft attitude by one of his senses. His/her brain then computes the necessary corrective action required and transmits a signal to his/her muscles to move the flight controls. Again his/her senses will detect that corrective action has taken place and he/she will move the flight controls back to where they started. A typical autopilot, as shown in Figure 12-1, would have to do all that the human pilot does, but would do it through electronic or electrohydraulic devices. The autopilot is divided into four parts: sensors, computer, controls, and loads.

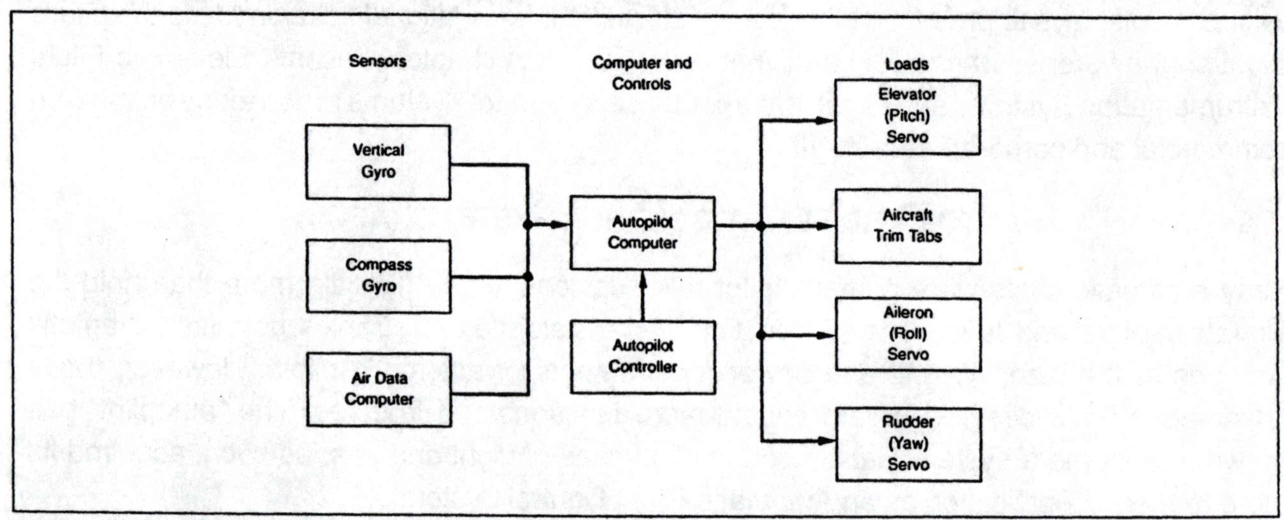

Figure 12-1. ***Typical Autopilot System.*** *(Courtesy of Honeywell Commercial Flight Systems)*

The sensors take the place of our pilot's "senses" to detect various changes in aircraft attitude. This information is fed to the computer, which calculates the size of its output signal and which axis to send it on. The controller turns the autopilot on and off and provides other system inputs not discussed here. Finally we come to the loads which are the "muscle" of our system and move the aircraft's flight control surfaces in response to the output signal of the computer. As the aircraft responds to these signals, the sensors, through aerodynamic feedback, detect the attitude change and tell the computer when the aircraft is back where it should be.

ROCKWELL COLLINS APS-85/86 AFCS SYSTEM DESCRIPTION

A typical dual flight guidance system consists of three servos, two mode select panels, a control panel and a flight control computer. The Collins APS-85/86, shown in Figure 12-2, is a fully digital 3-axis fail-passive autopilot system. The digital flight control computer contains four microprocessors installed on seven circuit boards. This computation power provides for flight precision and operational flexibility on two completely independent flight guidance channels.

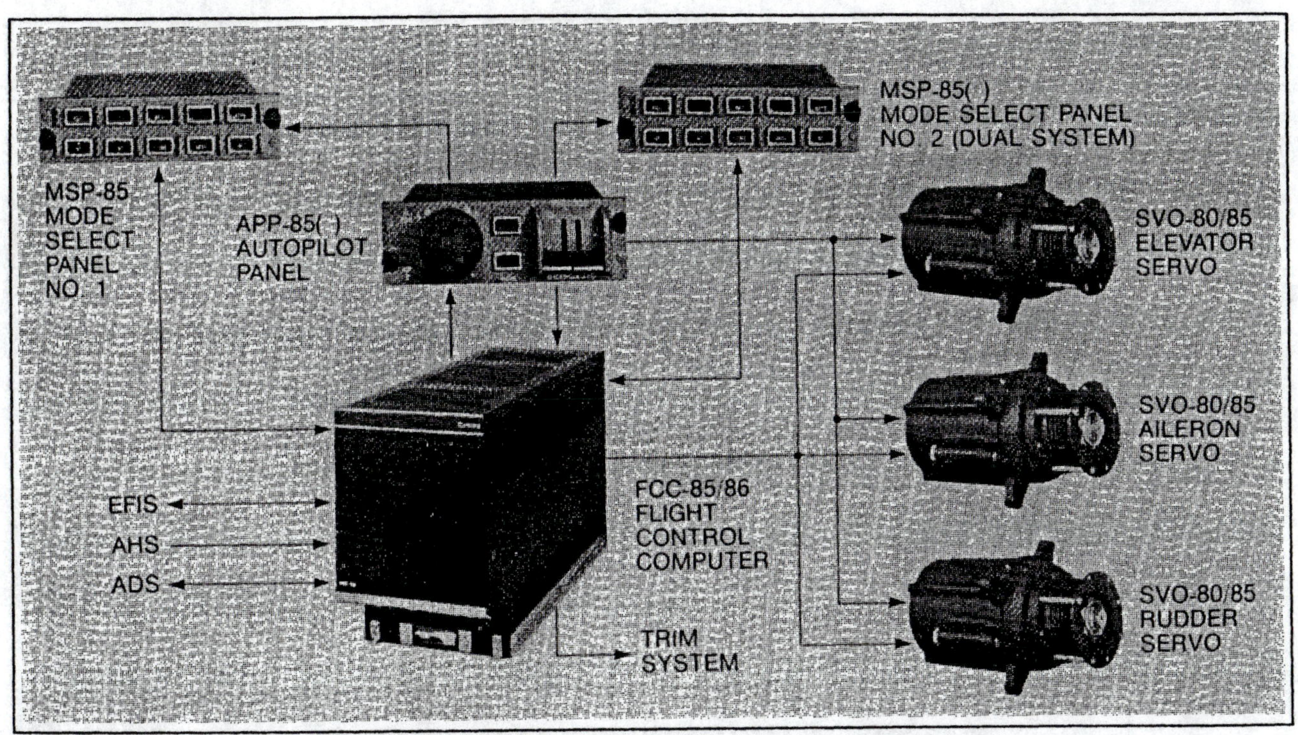

Figure 12-2. Collins APS-85/86 AFCS System Block Diagram. *(Courtesy of Rockwell Collins Avionics)*

The completely digital design of the Collins APS-85/86 makes possible fail-passive autopilot loops, which limit malfunction responses to no more severe than a normal heading change from the heading bug. The configuration of the control panel offers fully independent lateral and vertical mode capability on both sides of the cockpit.

FLIGHT DIRECTOR SYSTEMS (FDS)

After World War II, the advent of high performance aircraft produced a gap between the capabilities of the aircraft and the capabilities of the flight instruments. It was apparent that both display and scale ranges were inadequate to keep pace with the increased performance and complex systems of the aircraft. In 1950, an entirely new idea in flight instrumentation, the original flight director, was introduced.

This system took information from the cockpit instruments and gyros, processed it through an electronic computer, and presented it as "How to Fly" information on a single indicator. Thus, instead of having to calculate from five or six different instruments, a pilot could navigate and make landing approaches simply by "zeroing," or centering two cross-pointers or command bars, on the Attitude Director Indicator.

A flight director system, as shown in Figure 12-3, in simple form is designed to provide computed steering commands to the command bars of the Attitude Director Indicator (ADI) and/or to an autopilot system. The system uses various signal inputs such as: VOR/ILS/DME/ADF radio, air data, course information, heading information, and attitude information to generate the

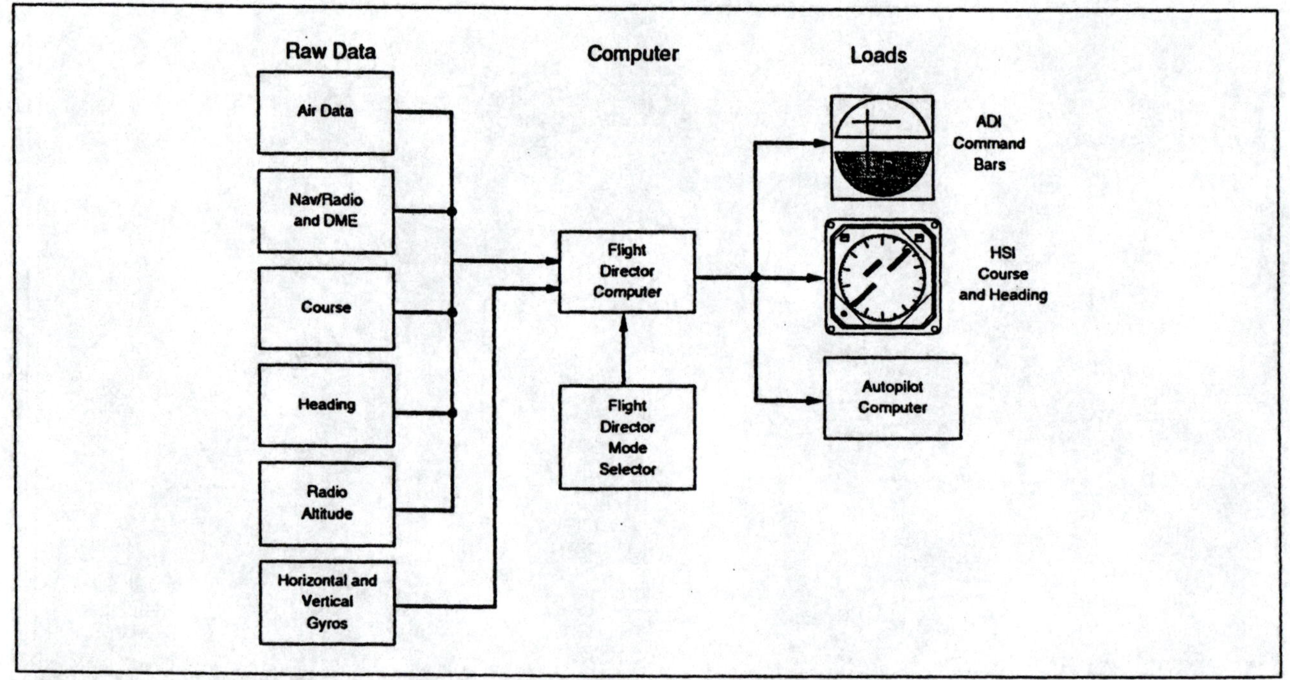

Figure 12-3. ***Typical Flight Director System.*** *(Courtesy of Honeywell Commercial Flight Systems)*

computed steering commands. A typical flight director system, like an autopilot system, can be divided into four parts: sensors, computer, controls, and loads.

The sensors provide the raw data to be processed by the computer. The flight director mode selector tells the computer which raw data to use, depending on pilot mode preference. The computer processes the raw data and scales the information to be displayed on the ADI command bars and/or to the autopilot computer. An illustration of a typical ADI is shown in Figure 12-4. Course, heading and navigation radio information is presented on a Horizontal Situation Indicator (HSI), as shown in Figure 12-5.

ELECTRONIC FLIGHT INSTRUMENT SYSTEM (EFIS)

The EFIS, shown in Figure 12-6, is comprised of three subsystems, the Pilot's Display System, the Copilot's Display System and the Weather Radar System. The pilot and copilot systems are identical, with both providing ADI and HSI information to the flight crews. The weather radar system provides weather information for display on both the pilot's and copilot's display systems.

Each display system consists of two Electronic Cathode Ray (CRT) Tube Displays, one Symbol Generator, one Display Controller, and one Source Controller. The Electronic Displays are identical and interchangeable. Both Electronic Displays use a combination of manual and photoelectric dimming for various light conditions. (A photograph of a typical electronic display is shown in the Preface.)

The Symbol Generator is the heart of the system and receives all the aircraft sensor inputs. The sensor information is processed and transmitted to the Electronic Displays. The Display

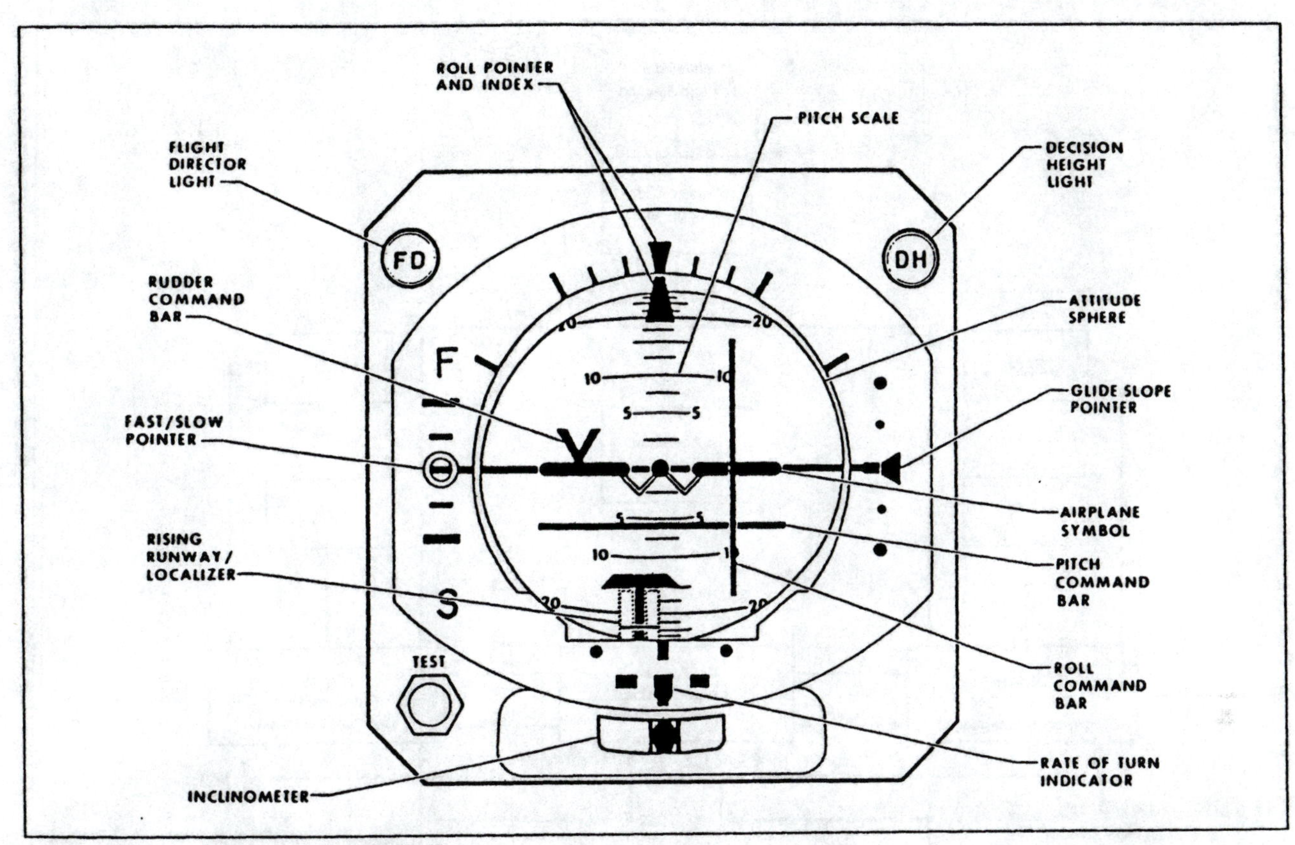

Figure 12-4. Attitude Director Indicator. *(Courtesy of McDonnell Douglas Aerospace Corporation)*

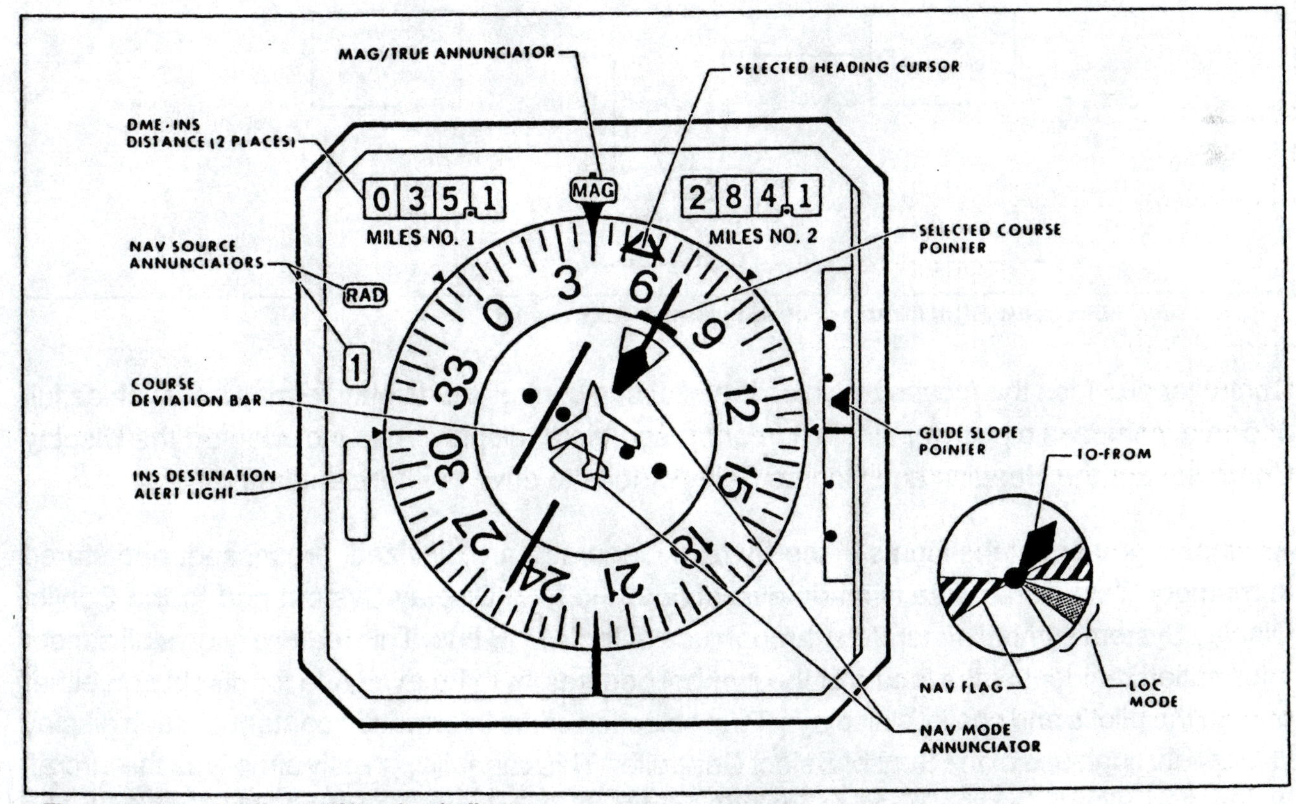

Figure 12-5. Horizontal Situation Indicator. *(Courtesy of McDonnell Douglas Aerospace Corporation)*

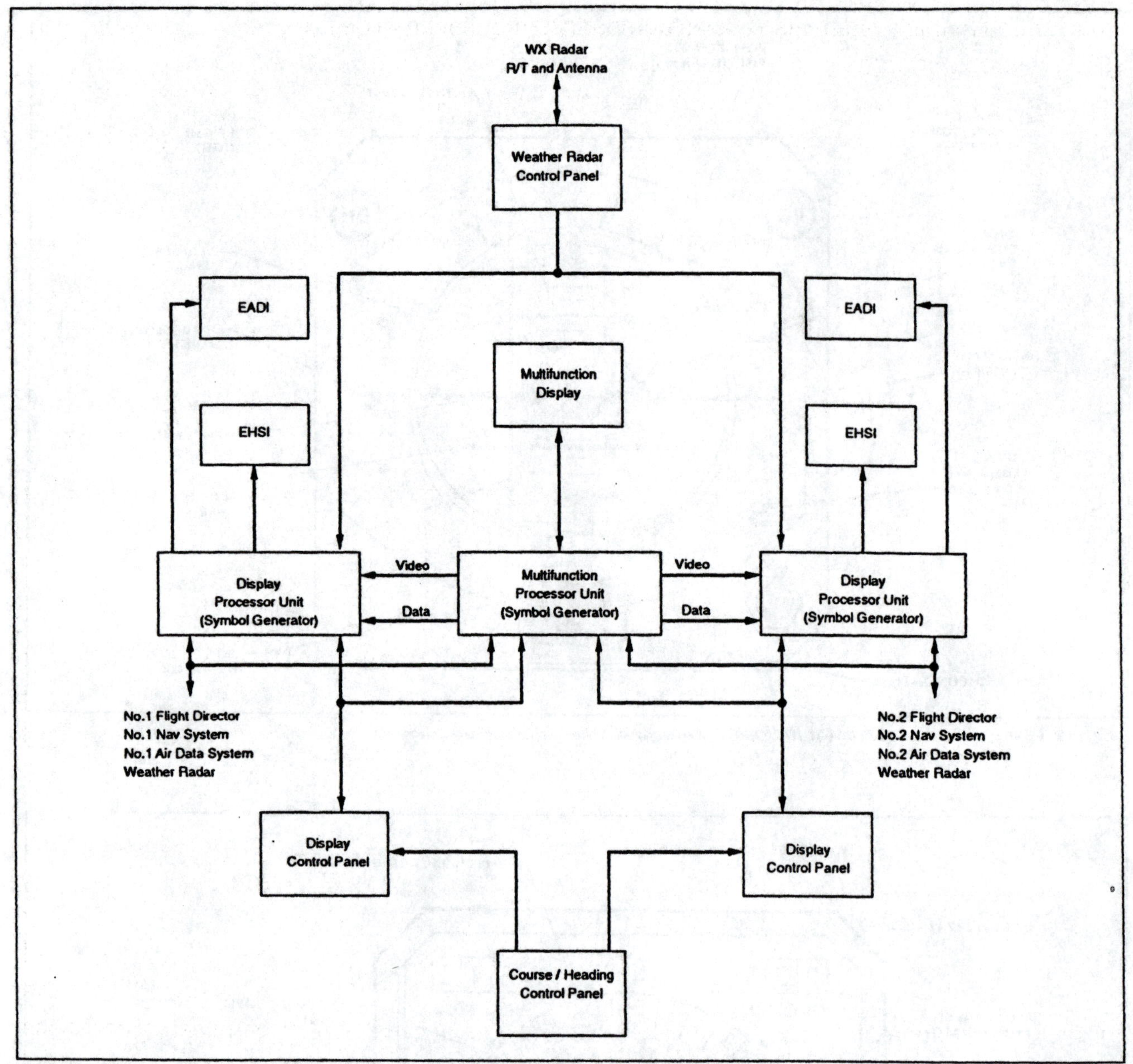

Figure 12-6. Electronic Flight Instrumentation System Block Diagram.

Controller provides the means by which the pilot can control the display formatting, such as full or partial compass display or single cue or cross-pointer display. Also included on the Display Controller are the Heading and Course Select knobs to drive pointers on the HSI.

All signals present at the inputs of the Symbol Generator are digitized, processed, and stored in memory. This stored data is made available to the Pilot Display System and to the Copilot Display System symbol generators through use of the ASCB bus. This feature allows all sensor information resident at the inputs of the symbol generators to be available for display on either or both the pilot's and copilot's displays. Pilot selection of the information content of each display is made through use of the Sensor Select Controller. This capability greatly simplifies the aircraft wiring and eliminates the need to include complex switches to perform this function. The organization of this system also allows comparison monitoring to be performed continuously in both symbol generators, eliminating the need for a separate Comparison Monitor. (A Comparison Monitor is a separate unit that monitors the outputs of a dual Flight Director System and provides an annunciation to the pilot on the Caution/Warning panel if an appreciable deviation is detected between the two flight director computers.)

EFIS Display Controller Operation

A Display Controller, as shown in Figure 12-7, is included in both the Pilot Display System and the Copilot Display System. This controller is the pilot interface by which the display functions may be selected or adjusted.

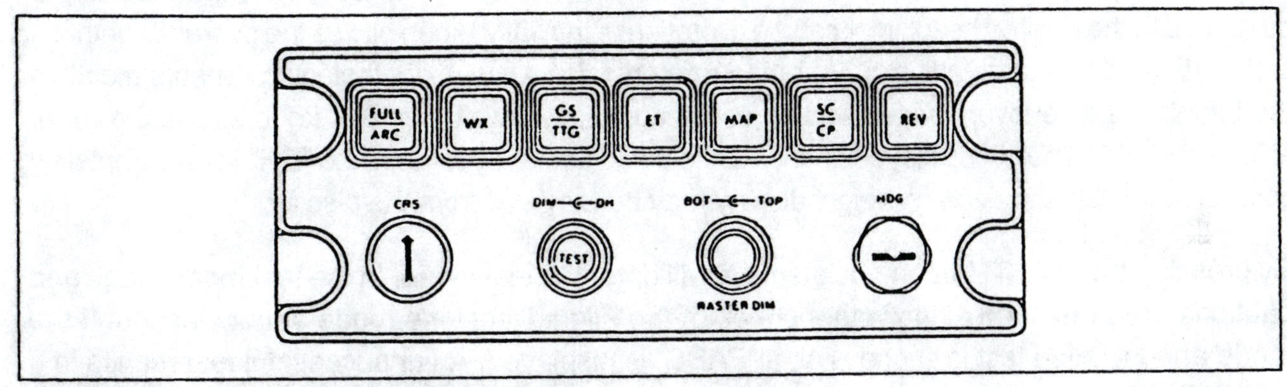

Figure 12-7. EFIS Display Controller. *(Courtesy of Honeywell Commercial Flight Systems)*

The "FULL/ARC" button is used to change the HSI display from full compass to partial compass format. In the full compass mode, 360° of heading are displayed. In the partial compass mode, 90° of heading are displayed. Successive pushes of the "FULL/ARC" button change the display back and forth from the full compass to the partial compass display. The "WX" button is used to call up weather radar returns on the partial compass display of the HSI. If the HSI is in the full compass mode initially, pressing the "WX" button changes the display to the partial compass mode and displays weather radar returns from the weather radar. A second push of the "WX" button will remove the weather information.

By pressing the "GS/TTG" button, ground-speed or time-to-go will alternately be displayed in the lower right corner of the HSI. By pressing the "ET" button, elapsed time is displayed. If elapsed time is being displayed, pushing "ET" will initialize the displayed time to zero. The toggling sequence of the "ET" button is initialize, start, stop. By pressing the "MAP" button, the full compass display is changed to the partial compass format allowing the active waypoint(s) and VOR/DME ground station positions to be displayed. By pressing the "SC/CP" button, the flight director command cue(s) is toggled back and forth from single cue configuration to the cross-pointer configuration.

In the event of an ADI/HSI display failure, the "REV" button may also be used to display a composite format on the remaining good display. The first push of the button blanks the HSI and puts a composite display on the ADI. The second push of the button blanks the ADI and puts a composite display on the HSI. The third push of the button returns the display to normal operation.

In the event of a Symbol Generator failure, pushing the "REV" button three times enables cross-side display information to be transferred to the onside ADI/HSI display. The reversionary mode is annunciated on the pilot and copilot ADI's as "SG1" or "SG2", depending on whether the source is the number 1 (pilot) or number 2 (copilot) symbol generator. Pushing the "REV" button a fourth time reverts the ADI and HSI displays back to normal operation.

Rotation of the course or heading select knobs allows the course or heading pointer on the HSI to be rotated to the desired heading or course. Rotation of the "DIM" knob allows the overall brightness of the ADI and HSI to be adjusted. After the reference level is set, photoelectric sensors maintain the brightness level over various lighting conditions. Rotation of the "DH" knob allows the radar altimeter decision height displayed on the ADI to be adjusted. By rotating the "DH" knob completely counterclockwise, the decision height display may be removed from the display.

By pressing the "TEST" button, the displays will enter the test mode. In the test mode, flags and cautions are presented along with a check of the Flight Director's mode annunciations. If the mode annunciation test is successful, a "PASS" is displayed. An unsuccessful test results in a "FD FAIL." Rotation of the bottom "RASTER DIM" knob allows the raster display on the HSI to be adjusted. Rotation of the top knob allows the raster display on the ADI to be varied. Rotation of the heading select knob allows the heading select bug to be rotated to the desired heading.

The Source Controller, shown in Figure 12-8, is used to select the available sources of heading, attitude, bearing, and navigation for display. Since each aircraft has different equipment, the Source Controller is tailored to fit each need. The flight manual supplement should be used as a reference for any unique function of the source controller in a particular aircraft. The Source Controller can contain up to seven select buttons and two bearing knobs.

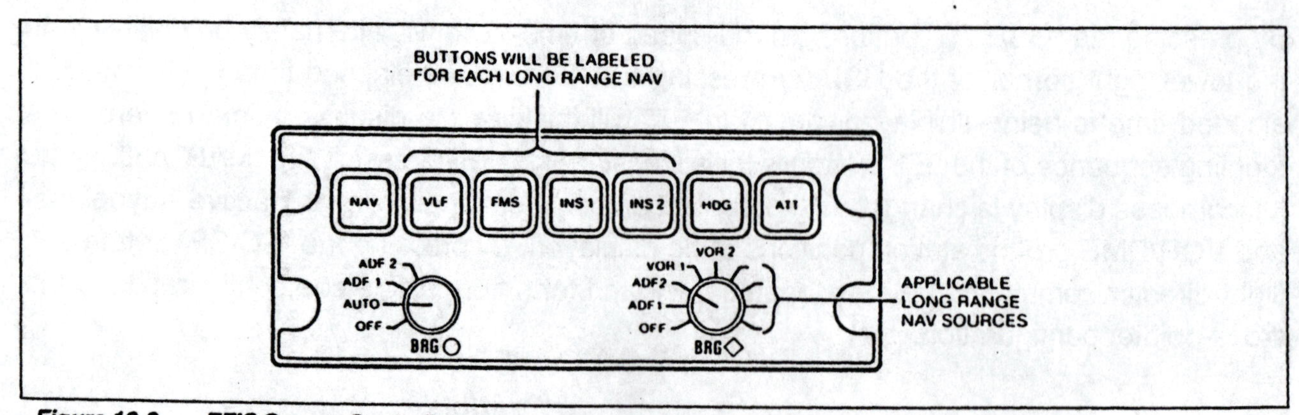

Figure 12-8. EFIS Source Controller. *(Courtesy of Honeywell Commercial Flight Systems)*

The "NAV" button is used to control the source of VHF NAV display information. Each push of the button will toggle the source between pilot's and copilot's NAV information. Display data being switched include DME distance, TO-FROM, NAV deviation, localizer deviation, glideslope deviation, and selected course. Further course selection is made with the course select knob from the side supplying NAV information. The Source Controller contains a button for each long-range NAV, up to four. Each button is labeled for the long-range NAV it controls. Pushing the button will display the applicable long-range NAV information. For INS and VLF Navigation System selections, the system will automatically switch heading source to the appropriate heading for the INS or VLF.

The "HDG" button selects the source of heading information. Each push of the button will select a different source for display. The "ATT" button selects the source of attitude information. Each push of the button will select a different source for display. The left Bearing ("BRG") knob has up to four positions. The "OFF" position removes the bearing pointer from the display. In the "AUTO" position, the bearing pointer displays bearing from the NAV source selected on the select buttons. In the "ADF1"/"ADF2" position, the bearing pointer displays the applicable ADF information. The right Bearing knob has an "OFF" position and a position for each available bearing source. In the "OFF" position, the bearing pointer is removed from the display. In each of the other positions, the applicable bearing information is displayed.

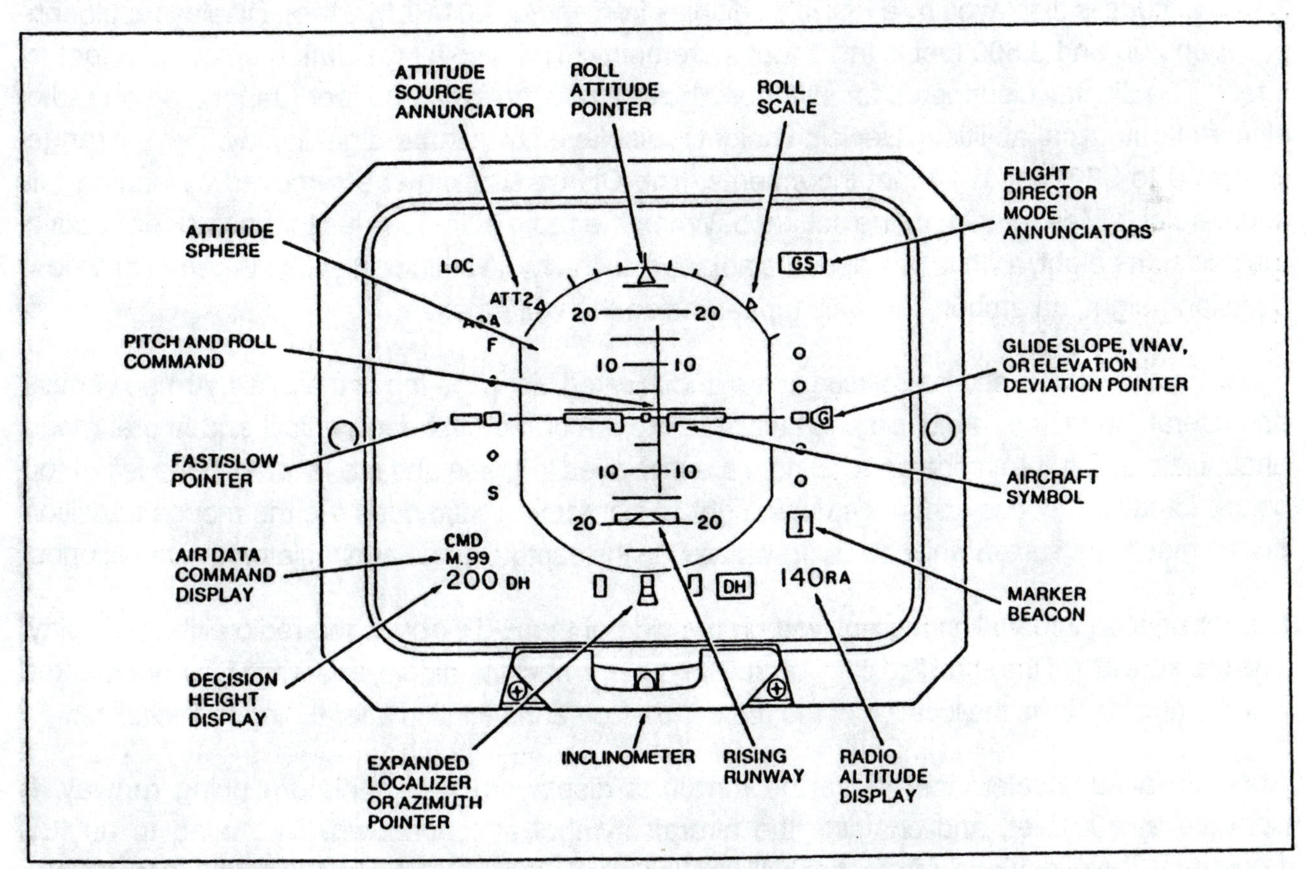

Figure 12-9. ***Electronic ADI Display.*** *(Courtesy Honeywell Commercial Flight Systems)*

Electronic ADI (EADI) Display Operation

As shown in Figure 12-9, the attitude sphere moves with respect to the aircraft symbol to display actual pitch and roll attitude. The pitch attitude display has white scale reference marks at 5°, 10°, 15°, 20°, 30°, 40°, 60°, and 80° on a blue and brown sphere. The EADI displays actual roll attitude through a movable index and fixed-scale reference marks at 0°, 10°, 20°, 30°, 45°, 60°, and 90°.

The aircraft symbol serves as a stationary symbol of the aircraft. Aircraft pitch and roll attitudes are displayed by the relationship between the fixed miniature aircraft and the movable sphere. The flight director command cue(s) displays computed commands to capture and maintain a desired flight path. The commands are satisfied by flying the aircraft symbol to the command cue. The cue(s) will bias from view should an invalid condition occur in either the flight director pitch or roll channel.

The fast/slow pointer indicates fast or slow error provided by an angle-of-attack, airspeed or alternate reference system. The EADI uses a conventional inclinometer which provides the pilot with a display of aircraft slip or skid, and is used as an aid for coordinated maneuvers. The selected attitude source is not annunciated if it is the normal source for that indicator. As other attitude sources are selected, they are annunciated in white at the top left side of the EADI. When the pilot and copilot sources are the same, the annunciation is amber.

Radar altitude is displayed by a four-digit display from minus 20 to 2,500 feet. Display resolution between 200 and 2,500 feet is in 10-foot increments. The display resolution below 200 feet is 5 feet. The display disappears for altitudes above 2,000 feet or 2,500 feet (depending on radio altimeter output capabilities). Decision height is displayed by a three-digit display. The set range is from 0 to 990 feet in 10-foot increments. The DH display may be removed by setting full counterclock-wise rotation of the set knob. When the radio altitude is less than 100 feet above the decision height, a white box appears adjacent to the radio altitude display. When at or below decision height, an amber "DH" will appear inside the white box.

Flight director vertical and lateral modes are annunicated along the top of the EADI. Armed vertical and lateral modes are annunciated in white to the left of the captured vertical and lateral mode annunciators. Capture mode annunciators are displayed in green and are located to the left of top center for lateral modes and in the upper right corner for vertical modes. As the modes transition from armed to capture, a white box is drawn around the capture mode annunciator for five seconds.

Marker beacon information is displayed on the side of the EADI above the radio altitude display. The markers are of the specified color of blue for outer, amber for middle, and white for inner marker. A white box identifies the location of the marker beacon annunciation after tuning to a localizer.

Absolute altitude reference above the terrain is displayed by a miniature rising runway. It appears at 200 feet, and contacts the aircraft symbol at touchdown. By tuning to an ILS frequency, the glideslope information will be displayed on the EADI. Aircraft displacement from glideslope beam center is then indicated by the relationship of the aircraft to the glideslope pointer. The letter "G" inside the vertical scale pointer identifies the information as glideslope deviation. When tuning to other than an ILS frequency, the glideslope display is removed.

By tuning to an ILS frequency, the rate-of-turn display is replaced by the expanded localizer display. Raw localizer displacement data from the navigation receiver are amplified 7.5 times to permit the localizer pointer to be used as a sensitive reference indicator of the aircraft position with respect to the center of the localizer. When tuning to other than an ILS frequency, the expanded localizer display is replaced by the rate-of-turn display. The rate-of-turn is displayed by a pointer at the bottom of the EADI.

The Vertical Navigation (VNAV) deviation pointer indicates the VNAV's computed path center to which the aircraft is to be flown. In this mode, the letter "V" inside the vertical scale pointer identifies the information as VNAV deviation. This scale and pointer is displayed on the EADI and Electronic HSI (EHSI) if VNAV is selected without the multiple waypoint EHSI selection. When the multiple waypoint MAP mode is selected on the EHSI and the VNAV mode is selected on the flight director, the vertical navigation scale is presented on the EADI, but not on the EHSI.

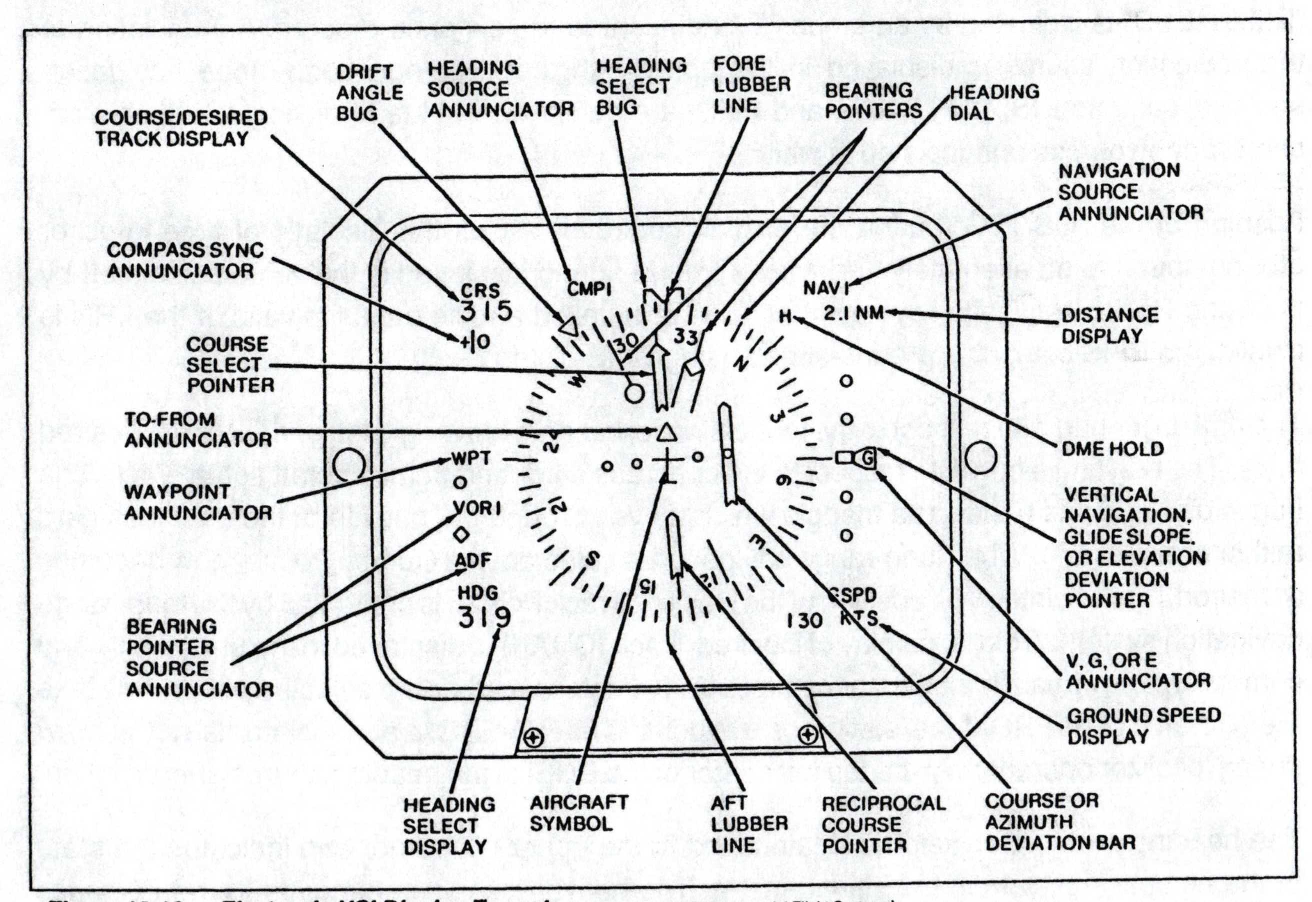

Figure 12-10. Electronic HSI Display Format. *(Courtesy Honeywell Commercial Flight Systems)*

Electronic HSI (EHSI) Display Operation

As shown in Figure 12-10, the aircraft symbol provides a quick visual cue to the aircraft position in relation to the selected course and heading, or actual heading. The heading dial displays heading information on a rotating heading dial graduated in 5° increments. Fixed heading indexes are located at each 45° position. The notched blue heading bug is positioned around the rotating heading dial by the remote heading select knob on the Display Controller. A digital heading select readout is also provided for convenience in setting the heading bug. Heading select error information from the heading bug is used to fly to the bug.

The course deviation bar represents the centerline of the selected navigation or localizer course. The aircraft symbol pictorially shows aircraft position in relation to the displayed deviation. A digital course select readout is also provided for convenience in setting the select course pointer. Course error information from the course select pointer is used to fly the selected navigation path.

The distance display indicates the nautical miles to the selected DME station or Long-Range Navigation (LRN) waypoint. Depending on whether DME or LRN is used, the distance will be displayed in a zero to 399.9 nautical miles or a zero to 3,999 nautical miles format respectively. "DME HOLD" is indicated by an amber "H" adjacent to the distance readout. Annunciation of the navigation source is displayed in the upper right hand corner. Long-range navigation sources, such as INS, VLF, RNAV, and FMS, are displayed in blue to distinguish them from short-range sources annunciated in white.

Pushing of the "GS/TTG" button the display controller allows the displays of time-to-go or ground-speed to be alternately displayed. Ground-speed displayed is the value calculated by the Long-Range Navigation system if an LRN is installed and its output is valid. If the LRN is invalid, the EFIS computes ground-speed using DME distance.

The drift angle bug with respect to the lubber line represents drift angle left or right of the desired track. The drift angle bug with respect to the compass card represents aircraft actual track. The bug is displayed as a magenta triangle which moves around the outside of the compass card (either partial or full). When long-range navigation is selected, the Course Pointer now becomes a Desired Track Pointer. The position of the Desired Track Pointer is controlled by the long-range navigation system. A digital display of Desired Track (DRAK) is displayed in the upper left-hand corner. An arrowhead in the center of the EHSI indicates whether the selected course will take the aircraft TO or FROM the station or waypoint. The TO-FROM annunciator is not in view during localizer operation. At the top left center of the EHSI is the heading source annunciation.

The heading sync annunciator is located next to the upper left corner and indicates the state of the compass system in the slaved mode. The bar represents commands to the compass gyro to slew to the indicated direction ("+" for increased heading and "o" for decreased heading). Heading sync is removed during compass "FREE" mode and for LRN derived heading displays. If a cross-side compass display is selected, the sync display is removed.

The vertical navigation display comes into view when the VNAV mode on the flight director is selected. The deviation pointer then indicates the VNAV's computer path center to which the aircraft is to be flown. In this mode, the deviation display is identified by a letter "V" inside the scale pointer. The Microwave Landing System elevation display comes into view when the MLS mode on the source controller is selected. The deviation display is identified by a letter "E" inside the pointer. The pointer indicates the MLS elevation beam center to which the aircraft is to be flown. The glideslope display comes into view when a VHF NAV source is selected on the EHSI and that NAV source is tuned to a localizer frequency. The deviation pointer then indicates the glideslope beam center to which the aircraft is to be flown. The deviation display is identified by a letter "G" inside the scale pointer.

The bearing pointers indicate relative bearing to the selected navaid. Two bearing pointers are available and can be tuned to navaids or selected off from the source controller. When the "OFF" position is selected, the bearing pointer and annunciator are removed from the EHSI display. The bearing source annunciations are symbol and color coded with the bearing pointers. When the bearing pointer navigation source is invalid or a localizer frequency is chosen, the respective bearing pointer is removed. When in the Elapsed Time (ET) mode, the ET display can read minutes and seconds or hours and minutes. The hour/minute mode will be distinguishable from the minute/second mode by an "H" on the left of the digital display.

The partial compass mode displays a 90° arc of compass coordinates. The following features are available during partial compass operations: Wing Vector, Range Rings, Weather, and Navaid Position. Wind information is displayed in any partial compass format. The wind information can be shown as magnitude and direction or as head/tail component and cross-wind component. In both cases, the arrow shows the direction and the number indicates velocity of the wind. The type of display is determined during installation. The wind vectors are available from long-range NAV systems, such as VLF and INS. Range rings are displayed to aid in determining the position of radar returns and navaids. The range ring is the compass card boundary and represents the select range on the radar. Weather information from the radar can be displayed in partial compass mode. Navaid position can be selected during MAP mode. The source of the navaid position markers is selected and annunciated in conjunction with the associated bearing source and is color coded.

In the event of a display unit failure, the "REV" button on the Display Controller is used to display a composite attitude and NAV format on the remaining good display, as shown in Figure 12-11. As in normal EADI and EHSI presentations, all elements are not displayed at the same time. The presence or absence of each display element is determined by flight phase, NAV radio tuning, selected flight director mode, absolute altitude, etc. The failure, caution and warning annunciations function is much the same as for the normal display mode.

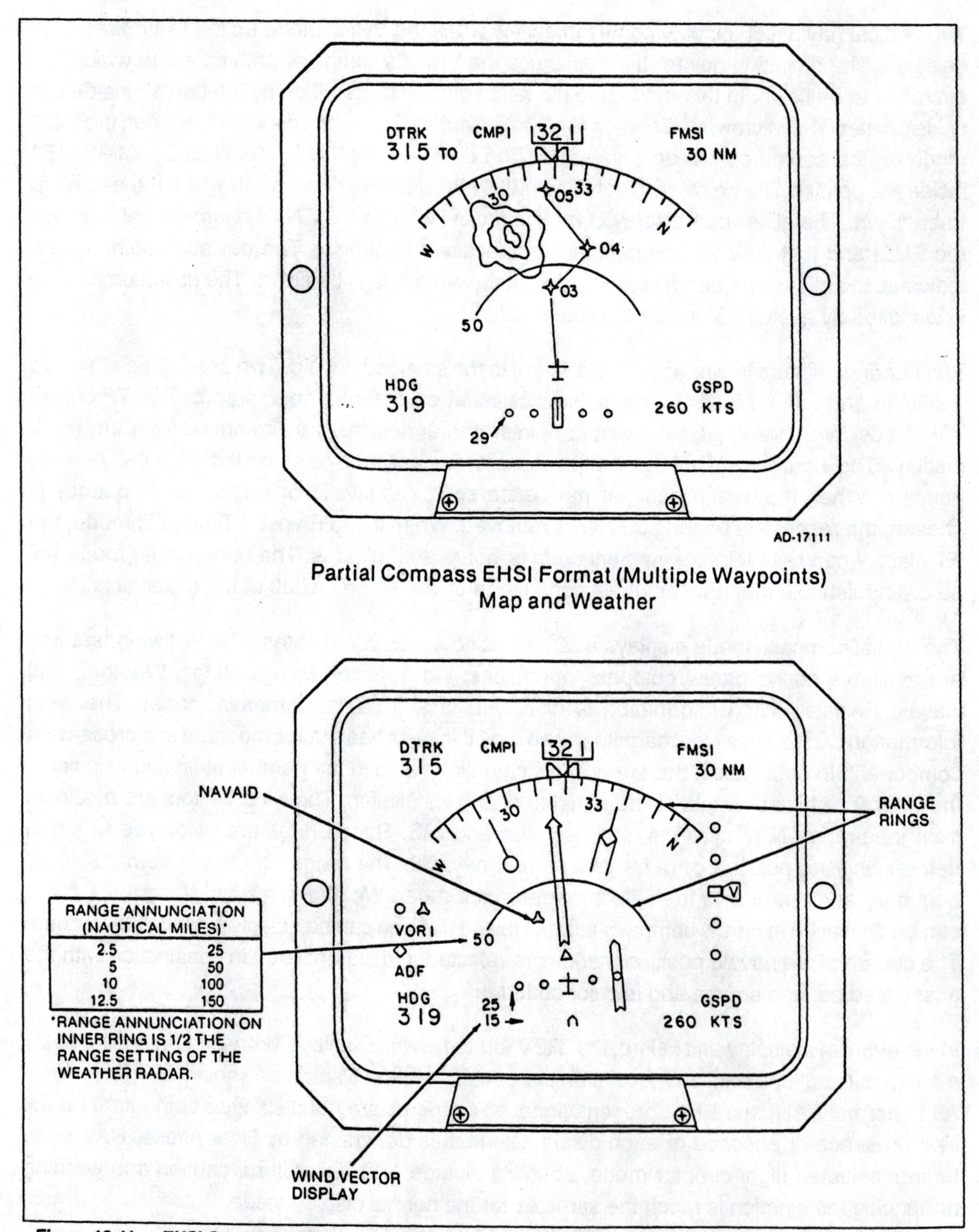

Figure 12-11. EHSI Composite Display Formats. (Courtesy Honeywell Commercial Flight Systems)

ROCKWELL COLLINS EFIS-85 FUNCTIONAL OPERATION

The Rockwell Collins EFIS-85/86 Electronic Flight Instrument System, illustrated in Figure 12-12, uses color cathode ray tubes driven by digital computers to provide the cockpit crew with system level capabilities far beyond conventional, electromechanical flight instruments. The electronic ADI display features a full screen sky/ground representation, flight control mode annunciation, and a runway symbol which expands as it rises for more accurate final approaches.

The electronic HSI display includes the conventional full compass rose format with time-to-go, ground-speed, dual course arrows, and wind vector. By pressing a button on the display control panel, the crew can select expanded compass formats with NAV waypoints and weather radar information superimposed. A rotary knob located on the display control panel allows the crew to select the NAV source to be connected to the course arrows and bearing pointer via internal, solid-state switches.

The Multifunction Display (MFD), located in the center panel, provides the crew with a systems information window which currently provides: Radar display; expanded navigation display, which includes VOR/LRN waypoint strings; up to 100 pages of checklist data; display of alpha/numeric data from up to four remote systems; and a joystick control for defining new waypoints. Additional functions can be easily added to the display system via software as the new generation of electronic displays evolves.

The Collins EFD-85/86 and MFD-85A contains a CRT, deflection yoke, video and deflection amplifiers, and high-voltage power supply. The CRT has an in-line gun with shadow mask, using a black matrix surrounding pigmented phosphor dot trios; this provides high contrast along with high resolution. A multi-bandpass optical filter is used to provide enhanced contrast. Both stroke and raster scanning techniques are used to provide large area color background for weather and distinct alphanumerics and symbols with high brightness. Mode select and display control functions are located on the display bezel (for the MFD-85A).

The EADI display provides an ADI presentation which can include the following: pitch and roll attitude, steering commands, vertical deviation (glideslope), localizer deviation, autopilot mode annunciation, radio altitude, decision height, marker beacon annunciation, fast-slow deviation, indicated airspeed, trend vector, VNAV, rising runway, and MLS. Figure 12-13 illustrates the EFD-85 EADI display.

The EHSI display has two basic modes: full compass rose and compass sector. The full compass mode presents the compass rose similar to existing electromechanical instruments. Information available in this mode includes the following: true/magnetic heading, selected heading, selected course, navigation data source annunciation, lateral deviation, to/from, distance-to-station waypoint, ground speed, time-to-go, elapsed time, bearing pointer (VOR, ADF, etc.), vertical deviation (VNAV or GS), weather radar return information, wind vector, and map or sector mode.

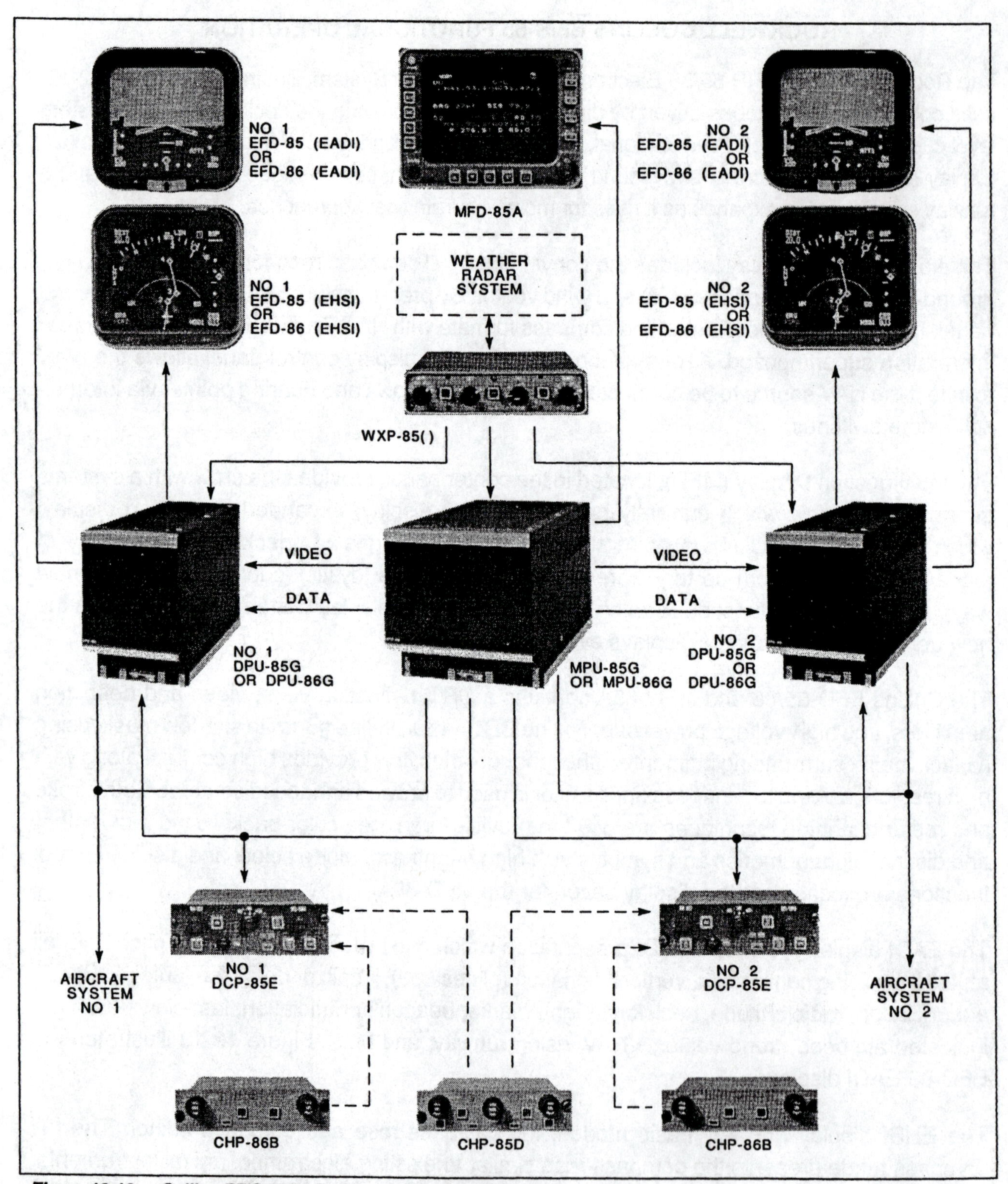

Figure 12-12. ***Collins EFIS-85/86 System.*** *(Courtesy Rockwell Collins Avionics)*

Figure 12-13. ***Collins EFIS-85 ADI.*** *(Courtesy Rockwell Collins Avionics)*

The compass sector mode presents a portion of the compass card (approximately 80°) across the top of the display, which gives the pilot basic horizontal situation information in an enlarged, easily readable format. Other information presented in this mode is similar to that for the full compass mode.

As shown in Figure 12-14, radar information from the Collins WXT-250B Weather Radar Receiver-Transmitter (discussed in Chapter 8) can be added to the sector EHSI display. Radar modes are controlled by the WXP-85C/85D Weather Control Panel. WXT-250A and appropriate WXP-85A/85B may be used if three-color radar display is desired.

The MFD-85A is a multicolor CRT display unit that mounts in the instrument panel in the space normally provided for the weather radar indicator. Mode select and display control functions are located on the display bezel. The memory and associated logic for up to 100 pages of checklist data are also contained in the MFD-85A. The MFD-85A can show a variety of navigation, checklist, and diagnostic information in addition to weather. The MFD-85A displays information that is supplied by the MPU-85G/86G Multi-Function Processor Unit.

The DPU-85 Display Processor Unit (DPU) generates the necessary deflection and video signals to draw the desired display on the Electronic Flight Display (EFD) units. The DPU

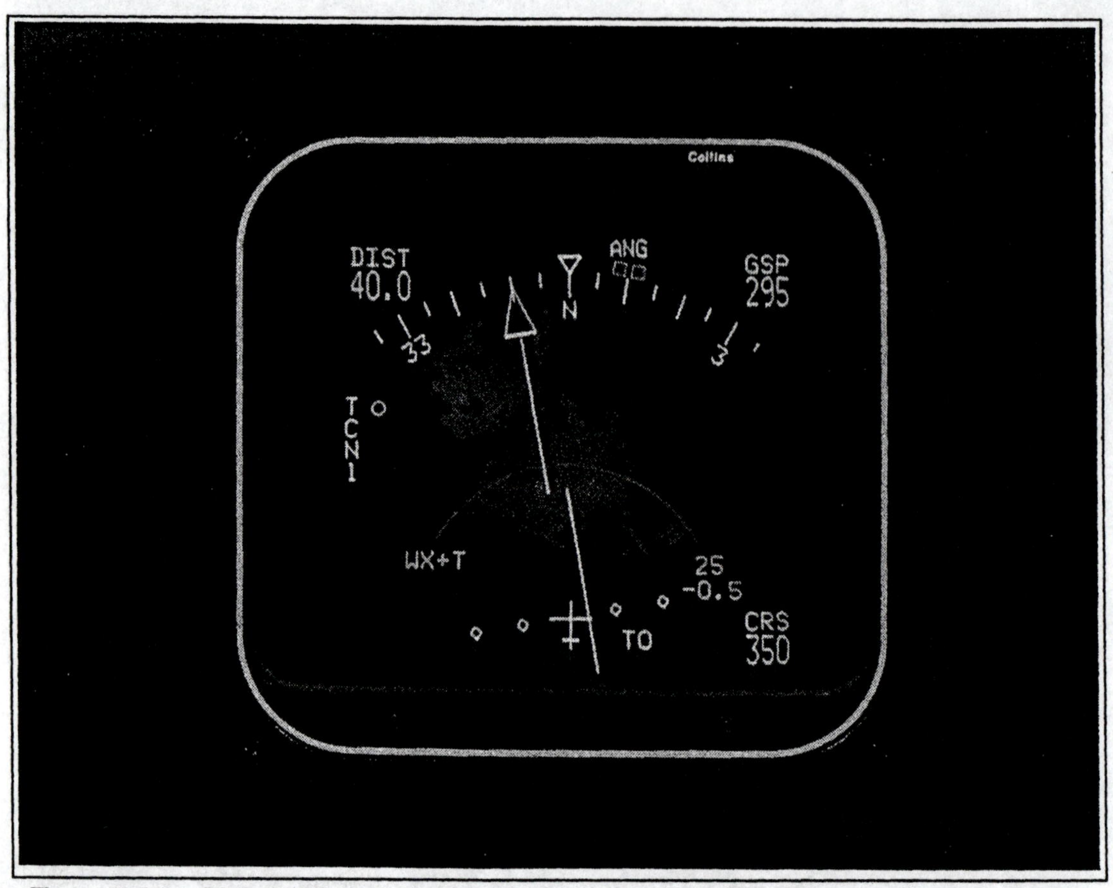

Figure 12-14. ***Collins EFIS-85 Navigation Display.*** *(Courtesy Rockwell Collins Avionics)*

contains an aircraft system's I/O interface, a display processor and a symbol generator to provide stroke-writing, an ADI sky raster generator, a multiplexer circuit, and a power supply.

The I/O section of the DPU-85 is responsible for receiving analog and digital data from aircraft systems and other EFIS-85/86 units, operating upon that data, and providing that data to the display processor. There are also a number of digital intro-EFIS-85/86 communication channels which are supervised by the DPU-85/86 I/O System. The I/O section is capable of receiving data from approximately 17 different aircraft sources. In addition, there are several intra-EFIS-85/86 data sources which must be accommodated. All together, five general data formats are received, digitized, filtered, and formatted by the I/O logic. These five formats, and the data found in that form, are as shown in Table 12-1.

AC Analog Inputs	DC Analog Inputs
ADF bearing	Radio altitude
Roll angle	Localizer deviation
Pitch angle	Glideslope deviation
Heading	Roll steering command
Desired track	Pitch steering command
Drift angle/bearing to waypoint	Crosstrack deviation
VOR bearing	Fast/slow
Serial Digital Inputs	
To/from	Air data computer information
Instrument flags	DME distance
Instrument modes	Long-range navigation information
EFIS and I/O straps	DCP data
FCS modes	MPU-85 cross-side data

Table 12-1. DPU-85 display processor inputs.

Along with receiving data, the DPU-85 must also provide several outputs to other EFIS-85 units and other aircraft systems. The I/O section provides these outputs again with a variety of formats as listed in Table 12-2.

The Symbol Generator creates signals required to draw characters and lines on EFD units. The Sky/Ground Raster Generator produces the deflection and video blanking signals needed for the sky and ground portions of the EADI display. The DPU is capable of driving two EFD display units with different deflection and video signals. The DPU can, for example, draw an ADI on one display while drawing an HSI on another.

The DCP-85B Display Control Panel provides EHSI display sensor selection control, EHSI display format control, EADI and EHSI display dimming, and EADI radio altitude decision height set capability. Inclusion of this panel in the system provides the basics for display modes switching capability.

AC Analog	DC Analog
Heading error Course datum	Crosstrack deviation
DC Discrete	**Serial Digital**
Attitude warn (comparator output) Heading warn (comparator output) Glideslope warn (comparator output) Localizer warn (comparator output) Radio altitude warn (comparator output) Master warn (comparator output) To/from Back localizer Modes/status	DCP control FMS-90 Cross-side DPU/MPU data

Table 12-2. DPU-85 Display processor outputs.

PRINCIPLES OF AREA NAVIGATION (RNAV)

Area navigation is a method of point-to-point navigation along any desired course within the service area of a VOR/DME station, without the need for flight over the station. This course is plotted with "waypoints." A waypoint is defined as a geographic position located within the service area of a VORTAC station. It may be used for route definition and/or progress reporting. A waypoint is often called a "phantom" station because it provides the RNAV user with the same navigation information that a real VORTAC station at that location would provide.

A waypoint is described by its radial and distance from the selected VORTAC station. The waypoint, shown in Figure 12-15, is located on the 255.0° radial at a distance of 20.0 nautical miles from the ANX VORTAC.

From basic trigonometry we know that if two sides of a triangle and the angle they form are given, the other sides and angles can be calculated. This is what the RNAV computer does. Calculations

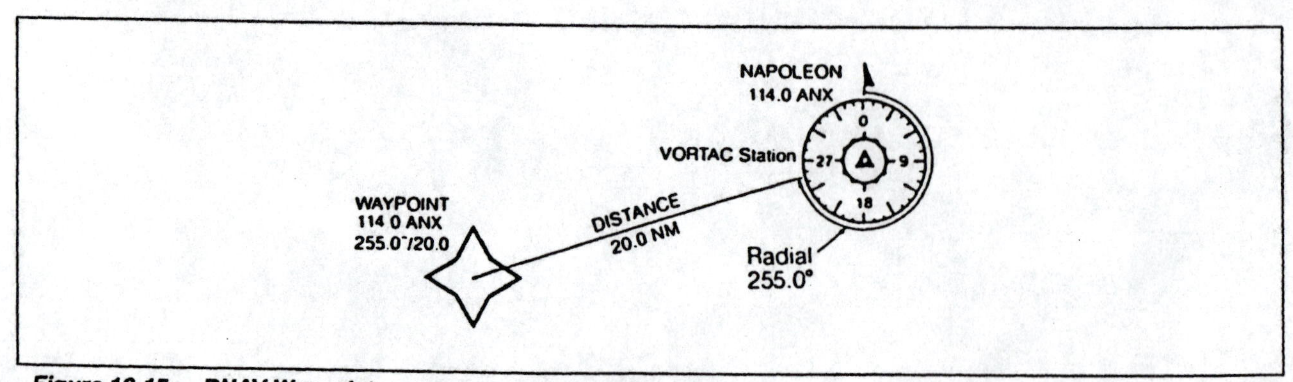

Figure 12-15. RNAV Waypoint. (Courtesy Allied-Signal Aerospace Company)

based on waypoint radial, waypoint distance, and cross-track deviation are supplied by the RNAV computer as solutions to continuously changing trigonometric equations during flight.

Known inputs to the RNAV computer are:

1) The distance from the VORTAC station to the waypoint.

2) The radial from the VORTAC station that the aircraft is on at any particular point in time.

3) The DME distance from the VORTAC station to the aircraft.

4) The selected course (OBS) to the waypoint.

As shown in Figure 12-16, the RNAV computer continuously processes this information to supply the aircraft distance from the waypoint and crosstrack deviation of the aircraft from the selected course in nautical miles (linear deviation instead of the angular deviation used in conventional VOR navigation).

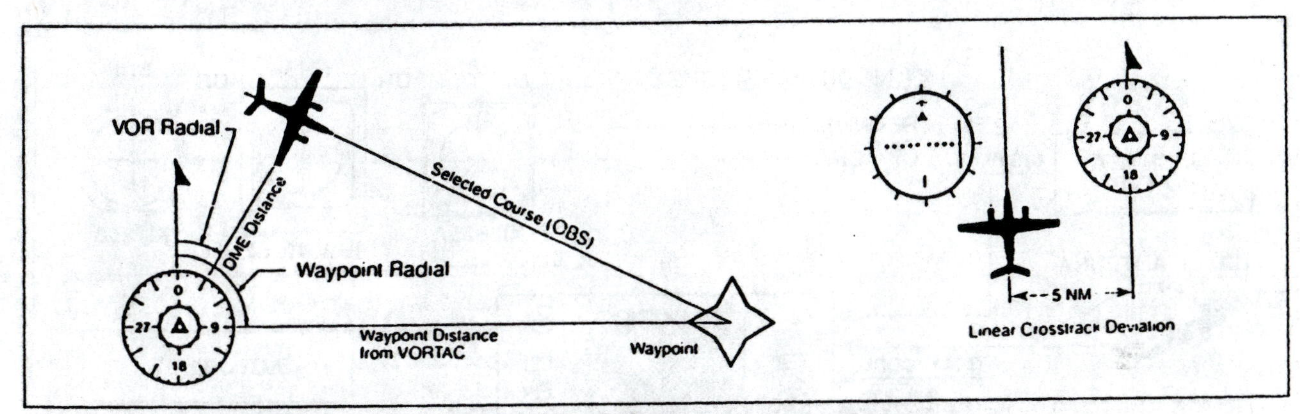

Figure 12-16. RNAV Distance and Crosstrack Deviation. *(Courtesy Allied-Signal Aerospace Company)*

Linear crosstrack deviation on a CDI or HSI permits flying parallel to a selected course by maintaining appropriate needle deflection. In an RNAV enroute mode, each dot represents 1.0 nautical mile off course. In an RNAV approach mode, each dot represents 0.25 nautical mile off course. In a VOR mode, the RNAV computer is by-passed so that deviation from the selected course is conventional angular crosstrack deviation expressed in degrees off course (one dot equals 2°). In a VOR/PAR (precision approach RADAR) mode, you have in effect put the waypoint over the VORTAC station and will, therefore, be provided linear crosstrack deviation as in an RNAV enroute mode. VOR and DME information must be provided to the RNAV computer to operate in this mode.

Under certain conditions, VOR scalloping can contribute significant variation in smoothness of the along track distance while operating in the RNAV modes. VOR scalloping is defined as an imperfection or deviation in the received VOR signal that causes radials to deviate from their standard track. VOR scalloping is generally the result of reflection from buildings or terrain. This deviation or scalloping effect causes the CDI needle to slowly or rapidly shift from side to side. Factors contributing to VOR scalloping include: VORTAC location in mountainous terrain when not a Doppler VOR, and snow cover on the ground around a VORTAC station. The accuracy of the RNAV system is poorest when the waypoint offset distance and the aircraft distance from the VORTAC station is large. On the other hand, the RNAV accuracy is greatest when the waypoint offset distance and the distance from the VORTAC station is small.

BENDIX/KING KLN-90 GPS RNAV FUNCTIONAL OPERATION

As shown in Figure 12-17, a basic Bendix/King KLN-90 system consists of a panel-mounted GPS sensor/navigation computer, a data base cartridge, and an antenna which is mounted on the top of the fuselage. An altitude input is required to obtain full navigation and operational capabilities. The following additional system components may be interfaced to the KLN-90 to increase its capabilities: HSI, RMI, fuel management system, air data system, emergency locator transmitter, autopilot, and external annunciators.

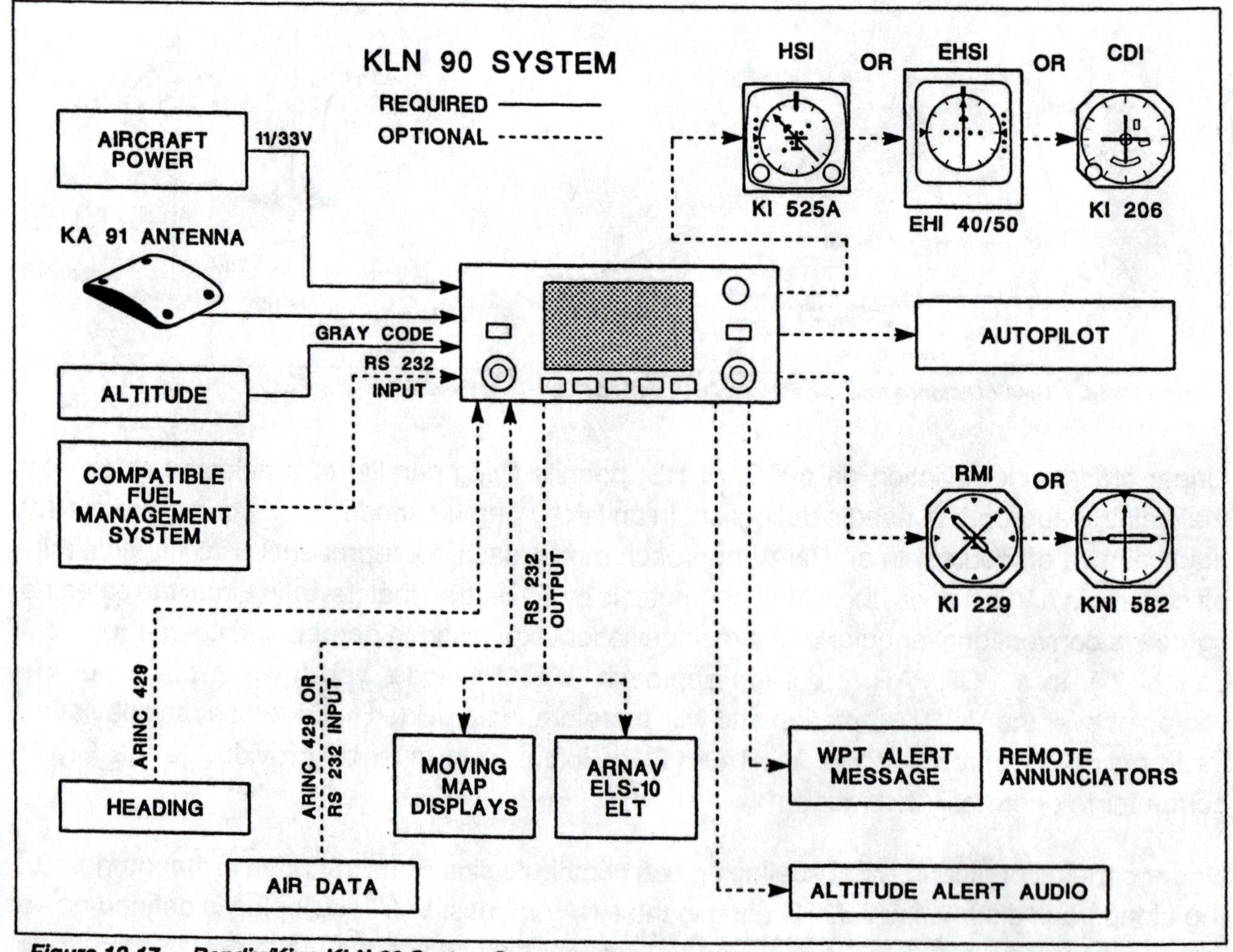

Figure 12-17. Bendix/King KLN-90 System Components.

The KLN-90 panel-mounted unit contains the GPS sensor, the navigation computer, a CRT display, and all controls required to operate the unit. It also houses the data base cartridge which plugs directly into the back of the unit. The data base cartridge is an electronic memory containing a vast amount of information on airports, navaids, intersections, special use airspace, and other items of value to the pilot. The data base is designed to be easily updated by the user by using a laptop computer and Bendix/King-furnished 3.5-inch diskettes. The data base may also be updated by removing the obsolete cartridge and replacing it with a current one.

The data base provides two primary functions. First, it makes pilot interface with the GPS sensor much easier. Rather than having to manually look up and then enter the latitude and longitude for a specific waypoint, it allows the pilot to merely enter a simple waypoint identifier. The data base automatically looks up and displays the latitude and longitude associated with the identifier. The data base saves a lot of tedious latitude/longitude entry and also greatly reduces the potential for data input mistakes.

The KLN-90 has analog outputs to drive the left-right deviation bar of most mechanical CDIs and HSIs. In addition, it has digital outputs to automatically drive the course pointer and display flight plan waypoints on the electronic HSIs. RMIs may be interfaced to the KLN-90 to provide a display of magnetic bearing to the waypoint. The NAV mode of the Flight Control Systems may be coupled to the KLN-90. Fuel management interfaces allow the pilot to view fuel related parameters calculated by the KLN-90, such as how much fuel will be remaining when the aircraft lands at the destination.

Some installations may require remote annunciator lights to be mounted in the aircraft panel in order to indicate the status of certain KLN-90 functions. Specifically, the KLN-90 has outputs to provide annunciation for waypoint alert and message.

The primary purpose of the KLN-90 is to provide the pilot with present position information and to display guidance information with respect to a flight plan defined by the pilot. Flight plan information is entered by the pilot via various knobs and buttons on the front panel. The display consists of a right and left page and a status line at the bottom of the display. The information on the left page is controlled by the concentric knobs on the left side of the unit and the information on the right page is controlled by the concentric knobs on the right side of the unit. Present position information is displayed on the unit CRT and can also be displayed on an EFIS or MFD.

The unit can use its present position information to determine crosstrack error, distance-to-waypoint, ground speed, track angle, time-to-waypoint, bearing-to-waypoint, and advisory Vertical Navigation guidance. The internal data base of the KLN-90 contains information concerning airports, VORs, NDBs, intersections, and outer markers throughout the world. Waypoints are stored in the data base by their International Civil Aviation Organization (ICAO) identifiers. The ICAO identifiers are in most cases taken directly from Jeppesen Sanderson or Government aeronautical charts.

The information stored in the data base eventually becomes out of date, therefore, to provide a means of updating the information, the database is housed in a cartridge which plugs into the back of the KLN-90. It is designed so that the user may easily remove the old database cartridge and install a current database cartridge. A secondary method of updating the database is by loading the information via a laptop computer.

The GPS Receiver Board, operating in conjunction with the KA-91 antenna provides the host computer (on the Main Board) with position, velocity, and time information through the RS-232 interface module. The host computer provides control information to the GPS Receiver through the RS-232 interface module, as shown in Figure 12-18.

The Main Board integrates a variety of functions onto a single board. The KLN-90 has three microprocessors; two reside on the GPS Receiver Board and the other, the host computer, resides on the Main Board. The host computer controls all RNAV functions associated with the KLN-90. In addition, it controls ARINC 429 and RS-232 communications, CDI activity data base communications, RTC (real time clock) activity, and the CRT video bit map. Most of the analog circuitry on the Main Board supports the CRT display. The Analog circuitry includes a CRT controller, deflection amplifiers, video driver, and high voltage power supply.

The host computer is responsible for displaying present position, defining flight path, navigation data, providing trip planning, providing calculator functions and providing user interface. It uses an Intel 80C186 operating at 10 MHz. The 10-MHz clock, driving the Microprocessor, is divided down to 2 MHz to drive the video timing circuit and 429 LSI. Three types of memory devices are used: Erasable-Programmable-Ready-Only-Memory (EPROM), Static-Read-Write-Memory (SRAM) and Electrically-Erasable-Programmable-Ready-Only-Memory (EEPROM). The KLN-90 has 64 Kbytes of volatile RAM and 64 Kbytes of non-volatile RAM and 8 Kbytes of video RAM.

The host computer communicates with the GPS sensor, fuel flow sensor, and ELT through the RS-232 communication controllers (CMOS Dual channel UART). Channel A is used for the GPS sensor and Channel B is used for fuel sensor/ELT communication. The host computer communicates with the data loader or remote RS-232 source through a single channel UART. An ARINC 429 LSI provides the KLN-90 with three 429 receivers and one 429 transmitter.

The low-voltage power supply (LVPS) uses a flyback converter circuit to transform the aircraft power bus voltage (11 to 33 volts) into the required voltages for the KLN-90's internal circuitry. The power supply incorporates overvoltage and short circuit protection to minimize the possibility of damage occurring in the power supply or other circuitry, resulting from abnormal circuit behavior within the KLN-90. The power supply is also protected by a fuse on the Main Board.

The KCA-167A Non-Volatile Memory Offloader System is to be used with KLN-90. The system consists of three items: KCA-167A 429 Card with hardware for PC kit, KUTIL90 program diskette, and the offloader harness. The KUTIL90 is an IBM 386/AT/XT-based tool for transferring the contents of the KLN-90 user memory (non-volatile RAM) to and from disk

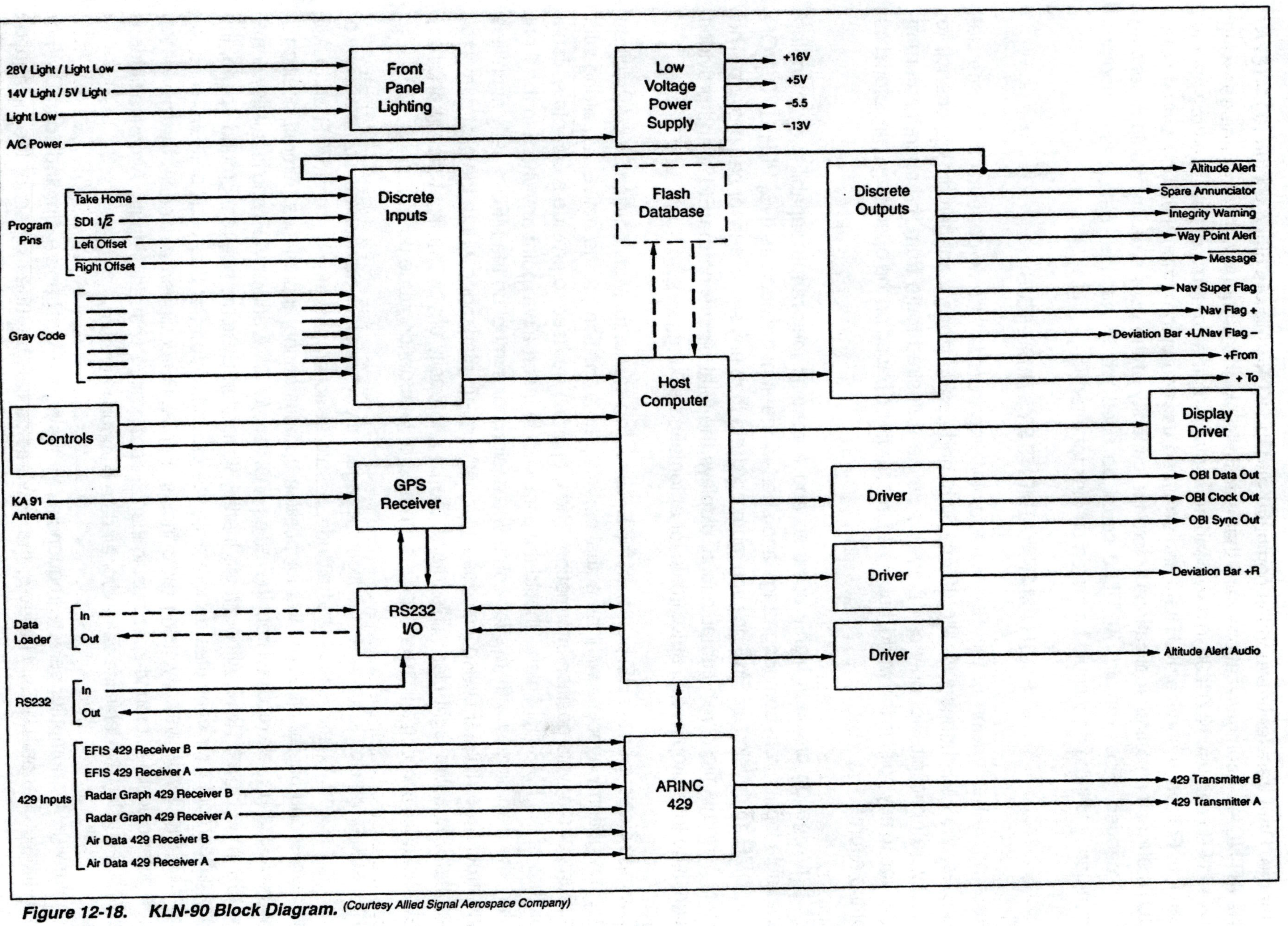

Figure 12-18. KLN-90 Block Diagram. *(Courtesy Allied Signal Aerospace Company)*

storage. Data transfer between the computer and the KLN-90 takes place via the KCA-167A, the ARINC 429 communications card that is installed in the computer. Technicians servicing the KLN-90 can use KUTIL90 in downloading and storing the contents of user memory prior to service. Following service, KUTIL90 can again be used to upload the previously saved data. KUTIL90 may also find applications where the capacity of the user memory would otherwise be exceeded. For example, KUTIL90 could be used when the user data base becomes completely full; the data could be stored on disk for later use.

FLIGHT MANAGEMENT SYSTEMS (FMS)

As a natural extension of the RNAV systems flying in today's new-generation commercial airlines, the FMS brings a wealth of total flight and performance management experience to the aircraft cockpit. Not only can it quickly define a desired route from the aircraft's current position to any point in the world, but the definition will be based on the operating characteristics of the aircraft.

FMS navigation provides highly accurate and automatic long-range capability by blending available inputs from both long-range and short-range sensors such as INS, VLF, GPS, VOR and DME, to develop an FMS position more accurate than any single sensor can provide. The FMS performs flight plan computations, displays the total picture on the EFIS, and provides signals to the autopilot and autothrottle for automatic tracking.

The map displays on the HSI portion of the EFIS and on the center Multifunction Display include the navigation waypoints with the actual curved path transition for the active leg, along with airports and navaid position information. The maps can be used to provide a complete visual picture of the FMS flight plan. Should the pilot wish to insert a waypoint anywhere in the flight plan, the map display with joystick-controlled designator symbol on the MFD greatly simplifies the procedure. The MFD also includes a unique North-up plan mode where the pilot can scroll through the entire flight plan. This allows the pilot to easily check the flight plan graphically rather than reviewing a list of waypoints, names, or latitude/longitude positions.

The radio navigation functions of the flight management system include accurate computations of aircraft position, ground-speed, and altitude, navaid selection and tuning of the VOR and DME receivers, and use of MLS and ILS for position updates during approaches. Aircraft position is computed using data received from the radios (VOR and/or DME and/or VLF) and inertial systems, as shown in Figure 12-19. Although the Inertial Navigation Systems output a ground-speed, the navigation computer computes a more accurate ground-speed (for INS equipped airplanes) by conditioning inertial velocity using radio inputs. For AHRS-equipped airplanes, the navigation computer calculates ground-speed using inertial acceleration and radio inputs. Aircraft altitude is computed using available inertial and/or air data computer sensors.

The navigation computer selects frequencies for tuning the VOR and a directed scan DME or two single channel DMEs. However, the NAV radios can be tuned by either the navigation computer supplied frequencies or a frequency transmitted by the VHF NAV control head. When in autotune, the VOR/DME is "listening" to the navigation computer for selection of the VOR frequency and as many as three DME frequencies. In the manual position, the VOR/DME is "listening" to the VHF NAV control head which can select a VOR and/or DME station for display on the RMI. Radio tuning is discussed in detail later in this chapter.

There are four position update modes that are a function of navaid and sensor availability. These are: radio/inertial, inertial only, radio only, and dead reckoning. These modes indicate the primary method used by the navigation computer to compute aircraft position and ground speed. For INS-equipped airplanes, the difference between the radio and inertial positions is

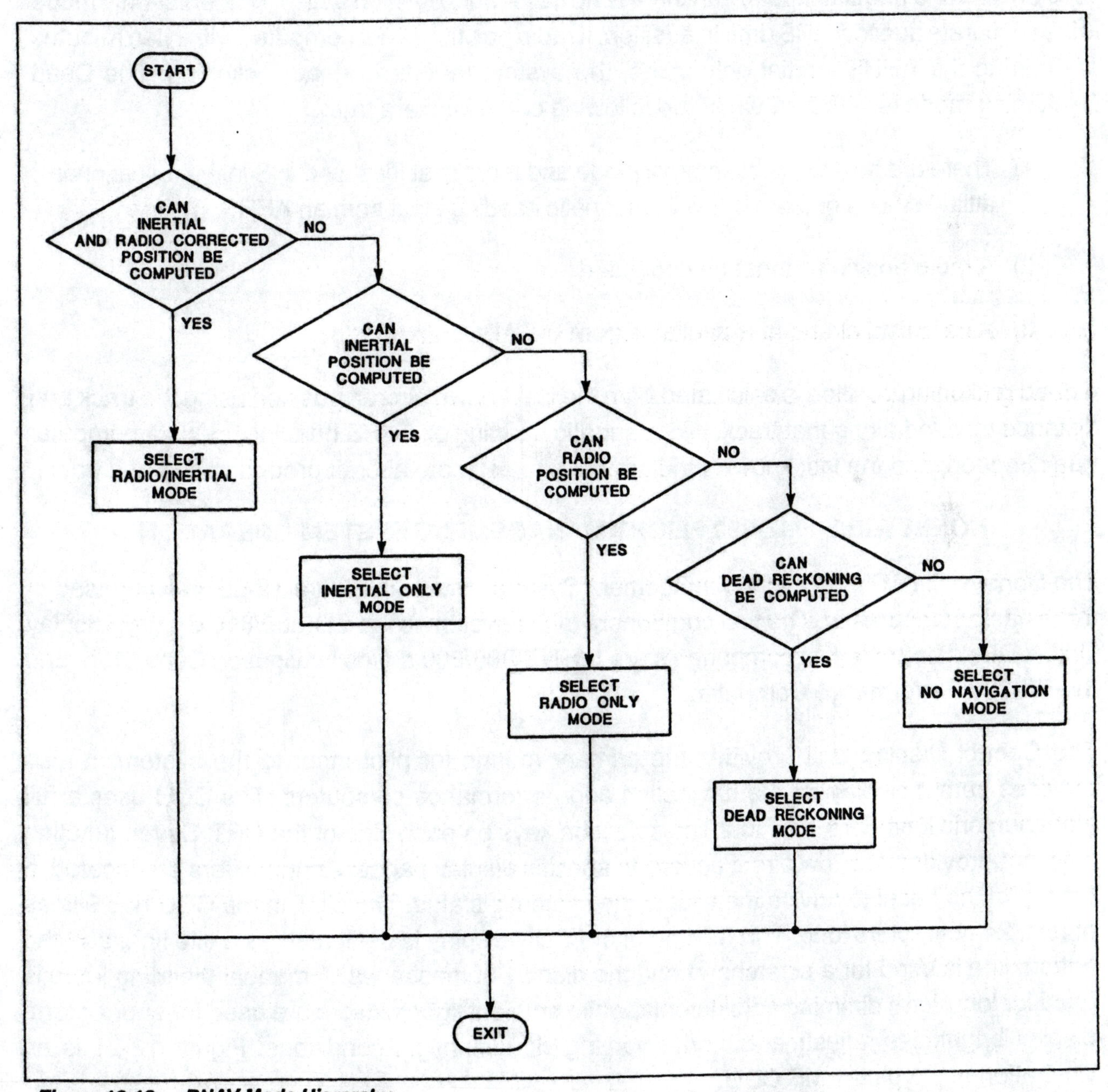

Figure 12-19. RNAV Mode Hierarchy.

filtered to compute a position error. This position error is then combined with inertial position and inertial velocities for computing radio/inertial position. For AHRS-equipped airplanes, the radio/inertial position is computed by using radio position in place of INS position and position error. The radio position is then combined with the calculated inertial velocity for computing radio/inertial position. The radio position and inertial acceleration are combined in the AHRS velocity filter to compute the north and east inertial ground speed components. These components are then combined with radio position in a complementary filter to compute the radio/inertial position.

With AHRS, the inertial-only NAV mode is active when at least one AHRS is providing valid acceleration and angular inputs, and there is no valid radio position data. The inertial-only mode is less accurate due to AHRS drift. In addition, if radio position is not computed within ten minutes of entering the AHRS inertial only mode, the system reverts to dead reckoning. The Dead Reckoning mode is activated when the following conditions are true:

1) There are no INSs in the normal mode and there is at least one INS in the reversionary attitude mode or there is a valid magnetic heading input from an AHRS.

2) A radio position cannot be computed.

3) A valid true airspeed is available from the ADC.

A dead reckoning position is calculated from the last known aircraft position using the track and distance traveled along that track. INS magnetic heading or AHRS heading, air data computer true airspeed, and the last known wind are used to estimate aircraft ground speed and track.

HONEYWELL FMZ-800 FLIGHT MANAGEMENT SYSTEM OPERATION

The Honeywell FMZ-800 Flight Management System, shown in Figure 12-20, is comprised of three basic components. The first component of the system is the CD-800/810 Control Display Unit (CDU). The other two components are the NZ-800/900 series Navigation Computers and the PZ-800 Performance Computer.

The Control Display Unit provides the primary means for pilot input to the system. It also provides output display for the navigation and performance computers. The CDU uses a full alphanumeric keyboard, with four-line selection keys on each side of the CRT. Seven function keys are provided to allow direct access to specific display pages. Annunciators are located in the top of the bezel to advise the pilot of the system's status. The CRT in the CDU has 9 lines of text, 24 characters long. The top line of the CDU display is dedicated as a title line, and the bottom line is used for a scratchpad and the display of messages. A manual dimming knob is used for long-term dimming adjustments, while ambient light sensors are used for shorter term display brightness adjustments under varying cloud/sunlight conditions. Figure 12-21 is an illustration of a typical FMS CDU.

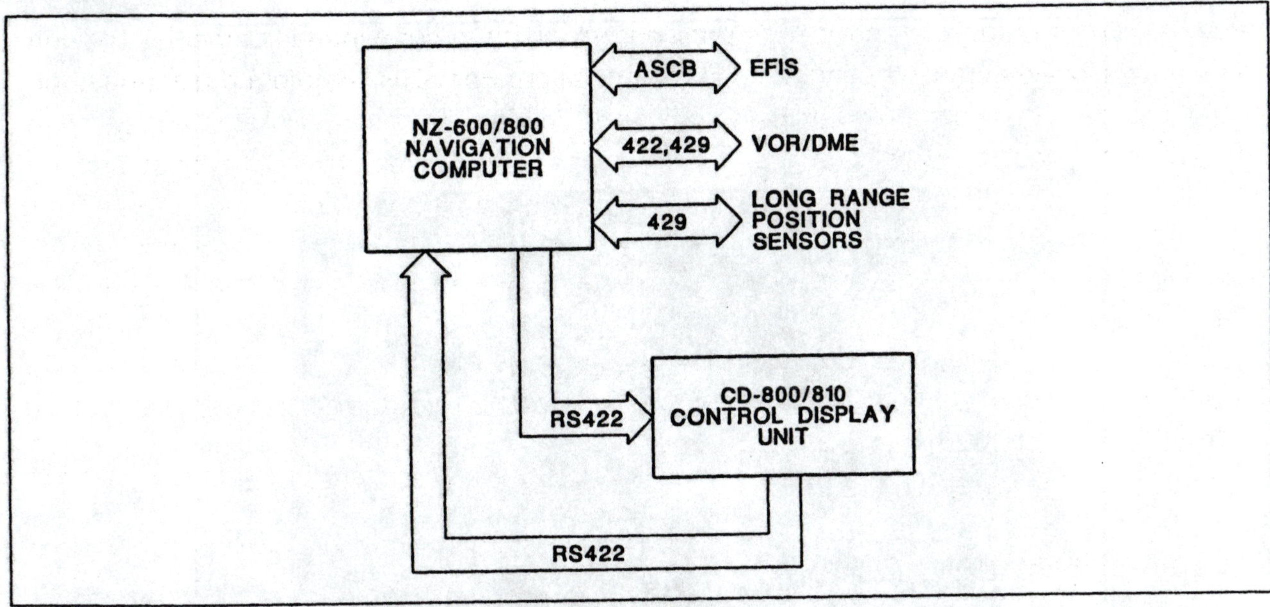

Figure 12-20. ***Honeywell FMZ-800 Flight Management System.***

The NRZ-800/900 Navigation Computer is the component in the FMS which provides both lateral and vertical navigation guidance. It has two different sizes of internal navigation data bases which are used for storage of waypoints, navaids, routes, airports, and other NAV data. Both units can interface with five long-range sensors; three via ARINC 429 buses and two by ASCB buses. Each navigation computer can also connect to DME and VOR receivers. The interface to the attitude heading reference, air data, MFD, EFIS, and autopilot is over the Avionics Standards Communications Bus. Flight Plans are also transferred between Navigation computers over the ASCB while the link to the CDU is over an RS-422 "private-line" interface. To provide high accuracy long-range navigation, the navigation computer is designed to connect to INS, VLF Sensors, plus VOR/DME. With links to the onboard navigation sensors, the navigation computer develops an FMS position based on a blend or mix of the sensors. The FMS does not directly display navigation maps on the CDU; however, the FMS is the source of map data for other cockpit displays, such as EFIS.

The lateral navigation function of the FMS is to provide navigation information relative to selected geographical points. Navigation management allows the pilot to define a route from the aircraft present position to any point in the world. The system will output advisory information and steering signals to allow the pilot or Automatic Flight Control System to guide the aircraft along the desired route. Routes are defined from the aircraft present position to a destination waypoint via a great circle route or via a series of great circle legs defined by intermediate waypoints.

The navigation computer and CDU provide many varied navigation functions; however, the primary function is to provide high accuracy long-range lateral and vertical navigation. To accomplish this function, the navigation computer connects to a variety of sensors. These sensors include VOR, DME, AHRS, INS, VLF, LORAN-C, and GPS receivers. The sensor inputs

Figure 12-21. Honeywell CD-800 (Mono)/810 (Color) Flight Management System Control Display Unit.
(Courtesy of Honeywell Commercial Flight Systems)

complement each other. For example, the Inertial Reference System sensor has very good short term accuracy but has long term drift. That error characteristic is complemented by the DME inputs which are not subject to drift error. By using a weighted average of the sensor inputs, the navigation computer can develop an FMS position which is as accurate as any single sensor under any given condition.

Another function of the navigation computer is the ability to store and retrieve information on navaids, earth reference points such as airports, intersections, runways, and routes. This storage area is referred to as the Navigation data base. The Navaid, earth reference points and published routes are subject to change and require updating every 28 days. The data base memory is also used to store pilot-defined waypoints and pilot defined routes; however, these are not updated every 28 days. They reside in memory until changed by the operator.

Other functions of the navigation computer are to provide guidance outputs to the autopilot and map display outputs to the EFIS. To provide high accuracy navigation and guidance, the navigation computer must determine the optimum DME stations to tune based on the geometry between the stations, and other characteristics, such as distance from the aircraft. A by-product of the data base is the capability to furnish map reference points near the aircraft's present position. The reference points can be both navaids and airports. Since the navigation computer has outputs to automatically tune the NAV radios, these outputs have been extended to provide keyboard control of electronic control heads, such as the ADF, communications radios, and ATC transponders.

The guidance outputs vary with the axis and the flight profile. The roll axis is controlled directly by the navigation computer through a lateral steering signal. The pitch axis is controlled by sending airspeed, vertical speed and altitude targets to the AFCS. The level offs and transitions are controlled by the normal AFCS trip points. The turn anticipation is calculated within the navigation computer to prevent overshooting or undershooting the next course.

The prime radio navigation inputs are VOR bearing and DME distance. This may be broken down into two categories - VOR/DME and multi-DME. Some DME receivers can be commanded to scan two or three channels, while others supply distance from a single channel (station). The navigation normally optimizes the present position calculation accuracy by using DME distance data from at least two stations versus VOR bearing and DME distance data from a single station. The VOR bearing information is subject to errors which are proportional to distance from the station. Using multi-DME, a much more accurate position may be derived.

Two types of radio (VOR/DME) configurations are supported. First, the FMS will support a system with single channel DMEs. In this system, both the on-side VOR/DME and the cross-side DME are brought into the navigation computer. The second system configuration supported by the FMS is a directed scanning DME. In this configuration, multiple distances can be received from a single DME receiver. In this configuration, distances from three stations can be used for a precise position fix.

Long-range sensors are used in addition to VOR/DME or DME/DME inputs for overland flight. The INS and VLF inputs are the only navigation source inputs used when VOR/DME signals are not receivable. The navigation computer will automatically choose the best navigation combination (VOR/DME, INS, VLF, AHRS) based on predefined priority. When using VOR/DME inputs, a blending of these inputs and INS information occurs. This blending is done via complementary filtering. Filtering lessens the effects of error and noise in both the VOR/DME and INS inputs, and thus provides a smooth and accurate position derivation. Airspeed information is used for blending in the event INS inputs are not available. The filtering also is done on the VLF positioning within the VLF sensor.

The navigation computer provides automatic tuning of the aircraft VOR and DME receivers. Calculation of aircraft present position from VOR/DME information requires input of bearing and distance and knowledge of the station coordinates. The data base is periodically used by the navigation computer to find the coordinates and frequencies of the VORTAC and VOR/DME stations in the aircraft vicinity. When the desired VOR and DME stations are chosen, the frequency is output to the navigation receivers. Automatic receiver tuning is operationally transparent to the pilot other than a periodic change in the receiver's frequency display. Provision is included for remote tuning of receivers via the CDU or manual tuning through the radio control heads. For remote tuning via the CDU, the pilot can choose to enter the VHF navaid identifier or enter the frequency. The frequency of the entered station is found in the data base, and output to the navigation control heads. For manual tuning via the radio control head, the navigation computer uses the frequency code from the receiver and compares it to the frequencies of stations in the aircraft vicinity. Frequency comparison allows the navigation computer to deduce what station is being tuned. A comparison of calculated bearing and distance to received bearing and distance is used to verify received information.

An important part of the navigation computer is the navigation data base, which contains information on navaids, airports and airways. The data base is integral to the navigation computer to allow quick access of the stored information. The data base can be loaded with one of five regionalized data bases. The data includes VOR, VORTAC, VOR/DME, ILS data, airport reference points, runway thresholds, high altitude airway intersections, airway routes and other procedures as defined as follows.

Flight Plan. The data base is the source of waypoints used to create a flight plan. Complete flight plans or segments of routes such as SIDS and STARS, are also accessible in the data base for selection as active flight plan or for modification of the currently selected active flight plan.

Positioning Fixing. Navaid position data from the data base is used in the computation of radio position.

Local Navaids. The navaid portion of the data base is the source of navaids available for auto-tuning. A list of 50 closest navaids is updated at a rate of once every two minutes while the aircraft is airborne.

Custom Data. Pilot defined waypoints and flight plans are stored in the custom portion of the data base. A maximum of 100 flight plans and 200 waypoints may be entered.

Data Display. All navigation information related to navaids, waypoints, airports, runways and airways is available for display on the CDU. Items copied into the custom data area from the standard data base during the creation of a flight plan are not viewable in the custom data index, but appear only in the flight plan.

CONCLUSION

This final chapter discussed some of the more complex integrated systems found in commercial transport and corporate jet aircraft. The Automatic Flight Control System, which has it roots in simplistic "wing leveler" autopilots, consists of vertical and directional gyros and an air data computer as sensor inputs to a computer that drives the aileron, elevator, and rudder servos and their respective trim tabs. The Flight Director System takes air data, navigation, course/heading commands, radio altitude, and gyro data into a computer which conditions the various signals to provide a "fly-to" command on the ADI and HSI and steering signals to the AFCS. Conventional electro-mechanical instruments are being replaced with CRT and Active Matrix LCD electronic color displays that combine information on a composite display through the use of a symbol generator to provide the pilot with a greater degree of situational awareness.

Area Navigation Systems synthesize phantom VORTACs to allow a multiplicity of flight path vectors previously limited by the location of real VORTACs. Flight Management Systems accept inputs from all the navigational sensors and blends these inputs to provide a more accurate synergistic navigation solution than with any single sensor. The FMS performs flight path computations, displays the total picture on the EFIS, and provides signals to the autopilot and autothrottle.

What lies ahead for the future? Currently, in development is the next generation of avionics systems that promises to increase air travel safety and efficiency. Chief among these is cognitive decision-aiding systems that will assist the pilot in problem-solving and decision-aiding throughout the flight regime using artificial intelligence computer systems that emulate the human thinking process using a knowledge base of experts and a network of logic rules. Voice interactive systems will appear in commercial aircraft crew stations to provide voice recognition and response to control avionics systems rather than pushing buttons. This mode of control, along with head-up displays, will allow the pilot to interact more effectively with the aircraft systems while remaining "eyes-out" and "heads-up". Finally, differential GPS will perfect TCAS and ATC operations to allow Category II and III landings without the limitations of ILS or MLS.

The future is very bright for the Avionics Technician as is the present. The quantum leaps in electronics technology the last few decades have transformed today's aircraft into a very sophisticated and complex machine. As a result, the FAA is currently considering an Avionics Technician License on equal par with the FAA Airframe and Powerplant License. Such a certification is long overdue, but in the meantime we are depending on you, the highly-qualified Avionics Technician to maintain the systems described in this book to the utmost standard of flight safety. I truly expect the preceeding chapters will help you to achieve that goal.

REVIEW QUESTIONS

1. Describe what functions an autopilot should perform.
2. Define the sensors and loads that form the inputs and outputs of a typical autopilot system and flight director system.
3. What information is displayed on the ADI and HSI?
4. Describe the operation of the EFIS Symbol Generator.
5. What function does each of the seven pushbuttons serve on the EFIS Display Controller?
6. What is the purpose of the "CRS" and "HDG" knobs on the EFIS Display Controller?
7. What function does the EFIS Source Controller serve?
8. Define the purpose of the flight director command cues.
9. Describe the expanded localizer display available on the EADI.
10. What is displayed on the EHSI during partial compass mode?
11. Define the function of the components that make up the Collins EFIS-85 System.
12. What information does the MFD-85 display to the pilot?
13. What is a "waypoint" and how are waypoints used for Area Navigation?
14. What function does the RNAV Computer perform?
15. Define the benefits of a Flight Management System.
16. What is the purpose of the FMS lateral navigation function?
17. Describe the functions of the FMS Navigation Computer and its data base.
18. Explain the logic behind the RNAV mode hierarchy.
19. What causes VOR scalloping?
20. When is RNAV accuracy the greatest?

Appendix A

Glossary of Terms

Glossary of Terms

AC (Alternating Current)
Electric current that reverses direction periodically, usually many times per second.

ADI (Attitude Director Indicator)
A flight instrument that displays pitch and roll attitude information from the vertical gyro and command bars driven by the Flight Director System.

ADF (Automatic Direction Finder)
A direction finder that without manual manipulation indicates the direction of arrival of a radio signal. Also known as a radio compass.

AF (Audio-Frequency)
The frequency band from 20 Hz to 20,000 Hz in which sound can be heard.

AFC (Automatic Frequency Control)
A circuit used to maintain the frequency of an oscillator within specific limits.

AFCS (Automatic Flight Control System)
Autopilot; a system whereby a computer processes navigational signal inputs to electrical outputs to drive actuators which move the aircrafts' control surfaces.

AGC (Automatic Gain Control)
A control circuit that automatically changes the gain (amplification) of a receiver so that the desired output signal remains essentially constant despite variations in input signal strength. Also known as an Automatic Level Control (ALC).

AHRS (Attitude and Heading Reference System)
An inertial sensor system that provides aircraft altitude, heading, and flight dynamics information to cockpit displays, flight director, autopilot, weather radar antenna platform, and other avionic systems and instruments.

ALE (Automatic Link Establishment)
A procedure whereby a receiver will automatically scan a given set of frequencies to determine the optimum frequency to transmit to another receiver.

AM (Amplitude-Modulated)
Modulation in which the amplitude of a wave is the characteristic varied in accordance with the intelligence to be transmitted.

APU (Auxiliary Power Unit)
A device consisting of a gas turbine engine that mechanically drives a generator to supply electrical power to an aircraft while on the ground.

ATC (Air Traffic Control)

A service within a circular limit defined by a 5-statute-mile (approximately 8 km) horizontal radius from the geographical center of an airport at which an operative airport traffic control tower is located, and an upward extent to 2,000 feet (609.6m) above the surface.

ATCRBS (ATC Radio Beacon System)

A system for identifying aircraft flying within the terminal control area. Consists of a primary surveillance radar and a secondary surveillance radar.

ATE (Automatic Test Equipment)

Test equipment that makes two or more tests in sequence without manual intervention; it usually stops when the first out-of-tolerance value is detected.

BCD (Binary-Coded Decimal)

A system of number representation in which each digit of a decimal number is represented by a binary number.

BFO (Beat Frequency Oscillator)

An oscillator in which a desired signal frequency, such as an audio frequency, is obtained as the beat frequency produced by combining two different signal frequencies, such as two different radio frequencies. Also known as heterodyne oscillator.

BPF (Band-Pass Filter)

An electric filter which transmits more or less uniformly in a certain band, outside of which the frequency components are attenuated.

CADC (Central Air Data Computer)

An airborne computer that receives pneumatic inputs, in the form of pitot and static pressure, and converts these inputs to electrical signals to drive the flight instruments and other avionics equipment, such as the flight director system and/or autopilot.

CRT (Cathode-Ray Tube)

An electron tube in which a beam of electrons can be focused to a small area and varied in position and intensity on a surface.

CSD (Constant Speed Drive)

A mechanism transmitting motion from one shaft to another that does not allow the velocity ratio of the shafts to be varied, or allows it to be varied only in steps.

CW (Continuous-Wave)

A radio wave whose successive sinusoidal oscillations are identical under steady-state conditions.

dB (Decibel)

A unit for describing the ratio of two powers, or the ratio of a power to a reference power.

DC (Direct Current)

Electric current which flows in one direction only, as opposed to alternating current.

DG (Directional Gyro)

A 2°-of-freedom gyroscope with provisions for maintaining its spin axis in a horizontal orientation.

DH (Decision Height)

A height specified in MSL (mean sea level) above the highest runway elevation in the touchdown zone at which a missed approach shall be initiated if the required visual reference has not been established; this term is used only in procedures where an electronic glideslope provides a reference for descent in ILS (instrument landing systems) or PAR (precision approach radar).

DVM (Digital Voltmeter)

A voltmeter in which the unknown voltage is compared with an internally generated analog voltage, the result being indicated in digital form rather than by a pointer moving over a meter scale.

EFIS (Electronic Flight Instrumentation System)

A system consisting of CRT displays, Display/Source Controller, and a symbol generator. The symbol generator receives and processes all aircraft sensor inputs into graphical representations of flight information on the CRTs. Typical EFIS displays consist of electronic ADI, electronic HSI, and multi-function displays.

EHF (Extremely High-Frequency)

The frequency band from 30,000 MHz to 3,000,000 MHz in the radio spectrum.

ELT (Emergency Locator Transmitter)

A transmitter designed for use in locating downed aircraft in distress.

FDS (Flight Director System)

A system that processes inputs from the flight instruments and gyros and presents "How to Fly" information on the attitude director indicator command bars.

FET (Field Effect Transistor)

A transistor in which the resistance of the current path from source to drain is modulated by applying a transverse electric field between grid or gate electrodes.

FM (Frequency-Modulated)

Modulation in which the frequency of a wave is the characteristic varied in accordance with the intelligence to be transmitted.

FMS (Flight Management System)

An onboard computer-based system that blends inputs from both short-range and long-range navigation sensors, performs flight plan calculations, displays the total picture on the EFIS, and provides signals to the autopilot and autothrottle for automatic tracking.

Fr (Resonant Frequency)

A frequency in which a given antenna at a predetermined length will exhibit a sharp peak in power radiated or intercepted.

GNSS (Global Navigation Satellite System)

Composed of GPS, GLONASS and possibly other geostationary satellite systems. GNSS will provide navigation integrity and differential connections so that precision navigation can be achieved for all phases of flight.

GPS (Global Positioning System)

A network of 24 orbiting satellites developed by the U.S. Department of Defense used to provide precision navigation to aircraft, vessels, and ground vehicles.

GS (Glideslope)

An inclined electromagnetic surface which is generated by instrument-landing approach facilities and which includes a glide path supplying guidance in the vertical plane.

Gyroscope (Gyro)

A spinning wheel (rotor) mounted on moveable frames. When the rotor spins at high speed, the axle (spin axis) on which it turns continues to point in the same direction, no matter how the frames are moved. In a gyrocompass, the spin axis is automatically positioned parallel to the earth's axis.

HF (High-Frequency)

The frequency band from 3 to 30 MHz in the radio spectrum.

HSI (Horizontal Situation Indicator)

A flight instrument that displays aircraft heading information from the directional gyro and selected course and heading and course deviation from the Flight Director System.

I.C. (Integrated Circuit)

An interconnected array of active and passive elements integrated with a single semiconductor substrate or deposited on the substrate by a continuous series of compatible processes, and capable of performing at least one complete electronic circuit function.

IM (Inner Marker)

A 75-MHz marker beacon normally used with the Instrument Landing System (ILS) to indicate that the aircraft is over the boundary of the airport.

INS (Inertial Navigation System)

A device which determines position by automatic dead reckoning; this operation is performed by the double integration of the outputs of accelerometers stabilized with respect to inertial space.

kHz (kiloHertz)

A unit of frequency equal to 1,000 Hz.

KVA (KiloVolt Ampere)

A unit of apparent power in an alternating-current circuit, equal to 1,000 volt-amperes.

LASER (Light Amplification by Stimulated Emission of Radiation)

This device generates a very narrow intense beam of coherent light. It is used in Ring Laser Gyro Inertial Navigation Systems in place of mechanical gyros and accelerometers.

LCD (Liquid Crystal Display)

A display consisting of two sheets of glass with a liquid crystal material sandwiched between the layers of glass. The glass has a conductive coating on both sides, so that when a voltage is applied, the liquid crystal material darkens to form visible characters or images.

LED (Light Emitting Diode)

A semiconductor diode that converts electric energy efficiently into spontaneous and non-coherent electromagnetic radiation at visible and near-infrared wavelengths by electroluminescence at a forward-biased pn junction.

LO (Local Oscillator)

The oscillator in a superheterodyne receiver, whose output is mixed with the incoming modulated radio-frequency carrier signal in the mixer to give the frequency conversions needed to produce the intermediate-frequency signal.

LOC (Localizer)

A directional radio beacon to provide aircraft with signals for lateral guidance with respect to the runway centerline.

LORAN (Long-Range Radio Navigation)

A network of ground stations transmitting on a frequency of 100 kHz that provides the user with position location based on hyperbolic navigation.

LRN (Long-Range Navigation)

A term used to describe any one or more of navigation systems used for transoceanic flight, such as GPS, VLF/Omega, LORAN, and INS.

LSB (Lower Sideband)

The sideband containing all frequencies below the carrier-frequency value that are produced by an amplitude-modulation process.

LSI (Large-Scale Integration)

A very complex integrated circuit, which contains well over 100 interconnected individual devices, such as basic logic gates and transistors, placed on a single semiconductor chip.

MB (Marker Beacon)

A low-power radio beacon transmitting a vertical cone-like signal to designate a small area as an aid to navigation.

MHz (MegaHertz)

Unit of frequency equal to 1,000,000 hertz.

MLS (Microwave Landing System)

A system of ground equipment which generates guidance beams at microwave frequencies for guiding aircraft to landing; it is intended to replace the present lower-frequency instrument landing system.

NDB (Non-directional Beacon)

A beacon that provides navigational guidance over a 360° azimuth.

OBS (Omnibearing Selector)

A device capable of being set manually to any desired omnibearing, or its reciprocal, to control a course deviation indicator.

PEP (Peak Envelope Power)

The average power supplied to the antenna transmission line by a transmitter during one radio-frequency cycle at the highest crest of the modulation envelope.

PPI (Plan Position Indicator)

Returns from both the primary and secondary surveillance radar systems form a total air traffic control situation display on a single cathode-ray tube (CRT) radar scope.

PRF (Pulse Repetition Frequency)

The number of times per second that a pulse is transmitted.

PROM (Programmable Read-Only Memory)

A large-scale integrated-circuit chip for storing digital data.

PSR (Primary Surveillance Radar)

Radar in which the incident beam is reflected from the target to form the return signal.

RADBAR (RADio-BARometric)

An altimeter that provides both barometric and radar altitude information in a single instrument.

RAIM (Receiver Autonomous Integrity Monitor)

A fault detection technique whereby a combination of sensors can detect system failures so that safety of flight requirements can be met.

RAM (Random-Access Memory)

A data storage device having the property that the time required to access a randomly selected datum does not depend on the time of the last access or the location of the most recently accessed datum.

RA (Resolution Advisory)

A Terrain Collision Avoidance System mode whereby an optimum flight path strategy is computed and displayed in the form of a vertical maneuver for a pilot to avoid an airborne collision.

RCR (Reverse Current Relay)

Relay that operates whenever current flows in the reverse direction.

RMI (Radio Magnetic Indicator)

Consists of a rotating compass slaved by a directional gyro. The RMI has positioned over the face of the indicator, one or two pointers to indicate ADF and/or VOR magnetic and relative bearing.

RNAV (Area Navigation)

A method of point-to-point navigation along any desired course, within the service area of a VOR/DME station, without the need for flight over the station.

SATCOM (SATellite COMmunications)

A communication system that uses satellites as a relay system to transmit voice and data communications over very long distances.

SDC (Static Defect Correction)

A technique to correct the altimeter for static source error. A SDC module measures pitot and static pressures and, through the use of distinct calibration, provides the necessary corrections to the altimeter(s).

SELCAL (Selective Calling)

A radio communications system in which the central station transmits an assigned code to alert a specific aircraft that a message is about to be transmitted.

SLS (Side-Lobe Suppression)

Used to inhibit a transponder's reply in response to a side-lobe interrogation.

SPIP (Special Position Identification Pulse)

By momentarily pressing an IDENT button on the transponder control head, the SPIP displays a bright "pip" on the air traffic controller's plan position indicator.

SSB (Single Sideband)

One of the two sidebands used in amplitude modulation is suppressed.

SSR (Secondary Surveillance Radar)

The secondary radar that operates in conjunction with Air Traffic Control Radio Beacon System airborne transponder whereby the airborne transponder replies to SSR interrogations.

Stabilization

Maintenance of a desired orientation independent of the roll and pitch of an aircraft.

STC (Sensitivity Time Control)

In a radar receiver, a circuit which greatly reduces the gain at the time that the transmitter emits a pulse; following the pulse, the circuit increases the sensitivity; thus reflection from distance objects will be received and those from nearby objects will be prevented from saturating the receiver.

TA (Traffic Advisory)

The TCAS provides a TA, which is an indication of the relative position of an aircraft flying in close proximity to assist the pilot in visual acquisition.

TACAN (TACtical Air Navigation)

Short-range ultra high-frequency air navigation system that provides accurate slant-range distance and bearing information.

TCA (Terminal Control Areas)

A control area or a portion thereof normally situated at the confluence of air-traffic service routes in the vicinity of one or more major airfields.

TCAS (Terminal Collision Avoidance System)

A system that uses a Mode-S transponder to detect potential collisions of nearby aircraft and computes an optimum flight path strategy, or resolution advisory, for the pilot to avoid a collision.

TRU (Transformer-Rectifier Unit)

A device consisting of a transformer to reduce 115 VAC to 28 VAC and rectifier to convert the 28 VAC to a 28-VDC output to the DC electrical bus.

VCO (Voltage-Controlled Oscillator)

An oscillator whose frequency of oscillation can be varied by changing an applied voltage.

VG (Vertical Gyro)

A 2°-of-freedom gyroscope with provisions for maintaining its spin axis in a vertical orientation.

VHF (Very-High Frequency)

The band of frequencies from 20 MHz to 300 MHz in the radio spectrum, corresponding to wavelengths of 1 meter to 10 meters.

VLF (Very-Low Frequency)

The band of frequencies from 3 kHz to 30 kHz in the radio spectrum.

VLF/Omega Navigation

A long-range navigation system operating in the frequency range of 10.2 kHz to 13.6 kHz that is based on measurement of the phase angles between the received pulses.

VNAV (Vertical Navigation)

A technique in which the vertical axis of the aircraft is guided or controlled from computed outputs from the CADC, RNAV or FMS.

VOR (VHF Omnirange Navigation)

A means of enroute navigation whereby an airborne receiver detects magnetic bearings (radials) transmitted from a ground station to determine direction from magnetic north.

Appendix B

DME Channel/Frequency/Spacing Correlation

DME Channel/Frequency/Spacing Correlation				
	Airborne DME		VORTAC	
VORTAC Freq (MHz)	Freq (MHz)	Spacing (μsec)	Freq (MHz)	Spacing (μsec)
108.00	1041	12	978	12
108.05	1041	36	1104	30
108.10	1042	12	979	12
108.15	1042	36	1105	30
108.20	1043	12	980	12
108.25	1043	36	1106	30
108.30	1044	12	981	12
108.35	1044	36	1107	30
108.40	1045	12	982	12
108.45	1045	36	1108	30
108.50	1046	12	983	12
108.55	1046	36	1109	30
108.60	1047	12	984	12
108.65	1047	36	1110	30
108.70	1048	12	985	12
108.75	1048	36	1111	30
108.80	1049	12	986	12
108.85	1049	36	1112	30
108.90	1050	12	987	12
108.95	1050	36	1113	30
109.00	1051	12	988	12
109.05	1051	36	1114	30
109.10	1052	12	989	12
109.15	1052	36	1115	30
109.20	1053	12	990	12
109.25	1053	36	1116	30
109.30	1054	12	991	12
109.35	1054	36	1117	30
109.40	1055	12	992	12
109.45	1055	36	1118	30
109.50	1056	12	993	12
109.55	1056	36	1119	30
109.60	1057	12	994	12
109.65	1057	36	1120	30

DME Channel/Frequency/Spacing Correlation (Continued)				
	Airborne DME		VORTAC	
VORTAC Freq (MHz)	Freq (MHz)	Spacing (μsec)	Freq (MHz)	Spacing (μsec)
109.70	1058	12	995	12
109.75	1058	36	1121	30
109.80	1059	12	996	12
109.85	1059	36	1122	30
109.90	1060	12	997	12
109.95	1060	36	1123	30
110.00	1061	12	998	12
110.05	1061	36	1124	30
110.10	1062	12	999	12
110.15	1062	36	1125	30
110.20	1063	12	1000	12
110.25	1063	36	1126	30
110.30	1064	12	1001	12
110.35	1064	36	1127	30
110.40	1065	12	1002	12
110.45	1065	36	1128	30
110.50	1066	12	1003	12
110.55	1066	36	1129	30
110.60	1067	12	1004	12
110.65	1067	36	1130	30
110.70	1068	12	1005	12
110.75	1068	36	1131	30
110.80	1069	12	1006	12
110.85	1069	36	1132	30
110.90	1070	12	1007	12
110.95	1070	36	1133	30
111.00	1071	12	1008	12
111.05	1071	36	1134	30
111.10	1072	12	1009	12
111.15	1072	36	1135	30
111.20	1073	12	1010	12
111.25	1073	36	1136	30
111.30	1074	12	1011	12
111.35	1074	36	1137	30

DME Channel/Frequency/Spacing Correlation (Continued)				
	Airborne DME		**VORTAC**	
VORTAC Freq (MHz)	**Freq (MHz)**	**Spacing (μsec)**	**Freq (MHz)**	**Spacing (μsec)**
111.40	1075	12	1012	12
111.45	1075	36	1138	30
111.50	1076	12	1013	12
111.55	1076	36	1139	30
111.60	1077	12	1014	12
111.65	1077	36	1140	30
111.70	1078	12	1015	12
111.75	1078	36	1141	30
111.80	1079	12	1016	12
111.85	1079	36	1142	30
111.90	1080	12	1017	12
111.95	1080	36	1143	30
112.00	1081	12	1018	12
112.05	1081	36	1144	30
112.10	1082	12	1019	12
112.15	1082	36	1145	30
112.20	1083	12	1020	12
112.25	1083	36	1146	30
112.30	1094	12	1157	12
112.35	1094	36	1031	30
112.40	1095	12	1158	12
112.45	1095	36	1032	30
112.50	1096	12	1159	12
112.55	1096	36	1033	30
112.60	1097	12	1160	12
112.65	1097	36	1034	30
112.70	1098	12	1161	12
112.75	1098	36	1035	30
112.80	1099	12	1162	12
112.85	1099	36	1036	30
112.90	1100	12	1163	12
112.95	1100	36	1037	30
113.00	1101	12	1164	12
113.05	1101	36	1038	30

DME Channel/Frequency/Spacing Correlation (Continued)				
	Airborne DME		VORTAC	
VORTAC Freq (MHz)	Freq (MHz)	Spacing (μsec)	Freq (MHz)	Spacing (μsec)
113.10	1102	12	1165	12
113.15	1102	36	1039	30
113.20	1103	12	1166	12
113.25	1103	36	1040	30
113.30	1104	12	1167	12
113.35	1104	36	1041	30
113.40	1105	12	1168	12
113.45	1105	36	1042	30
113.50	1106	12	1169	12
113.55	1106	36	1043	30
113.60	1107	12	1170	12
113.65	1107	36	1044	30
113.70	1108	12	1171	12
113.75	1108	36	1045	30
113.80	1109	12	1172	12
113.85	1109	36	1046	30
113.90	1110	12	1173	12
113.95	1110	36	1047	30
114.00	1111	12	1174	12
114.05	1111	36	1048	30
114.10	1112	12	1175	12
114.15	1112	36	1049	30
114.20	1113	12	1176	12
114.25	1113	36	1050	30
114.30	1114	12	1177	12
35114.	1114	36	1051	30
114.40	1115	12	1178	12
114.45	1115	36	1052	30
114.50	1116	12	1179	12
114.55	1116	36	1053	30
114.60	1117	12	1180	12
114.65	1117	36	1054	30
114.70	1118	12	1181	12
114.75	1118	36	1055	30

DME Channel/Frequency/Spacing Correlation (Continued)				
	Airborne DME		VORTAC	
VORTAC Freq (MHz)	Freq (MHz)	Spacing (μsec)	Freq (MHz)	Spacing (μsec)
114.80	1119	12	1182	12
114.85	1119	36	1056	30
114.90	1120	12	1183	12
114.95	1120	36	1057	30
115.00	1121	12	1184	12
115.05	1121	36	1058	30
115.10	1122	12	1185	12
115.15	1122	36	1059	30
115.20	1123	12	1186	12
115.25	1123	36	1060	30
115.30	1124	12	1187	12
115.35	1124	36	1061	30
115.40	1125	12	1188	12
115.45	1125	36	1062	30
115.50	1126	12	1189	12
115.55	1126	36	1063	30
115.60	1127	12	1190	12
115.65	1127	36	1064	30
115.70	1128	12	1191	12
115.75	1128	36	1065	30
115.80	1129	12	1192	12
115.85	1129	36	1066	30
115.90	1130	12	1193	12
115.95	1130	36	1067	30
116.00	1131	12	1194	12
116.05	1131	36	1068	30
116.10	1132	12	1195	12
116.15	1132	36	1069	30
116.20	1133	12	1196	12
116.25	1133	36	1070	30
116.30	1134	12	1197	12
116.35	1134	36	1071	30
116.40	1135	12	1198	12
116.45	1135	36	1072	30

DME Channel/Frequency/Spacing Correlation (Continued)				
	Airborne DME		VORTAC	
VORTAC Freq (MHz)	Freq (MHz)	Spacing (μsec)	Freq (MHz)	Spacing (μsec)
116.50	1136	12	1199	12
116.55	1136	36	1073	30
116.60	1137	12	1200	12
116.65	1137	36	1074	30
116.70	1138	12	1201	12
116.75	1138	36	1075	30
116.80	1139	12	1202	12
116.85	1139	36	1076	30
116.90	1140	12	1203	12
116.95	1140	36	1077	30
117.00	1141	12	1204	12
117.05	1141	36	1078	30
117.10	1142	12	1205	12
117.15	1142	36	1079	30
117.20	1143	12	1206	12
117.25	1143	36	1080	30
117.30	1144	12	1207	12
117.35	1144	36	1081	30
117.40	1145	12	1208	12
117.45	1145	36	1082	30
117.50	1146	12	1209	12
117.55	1146	36	1083	30
117.60	1147	12	1210	12
117.65	1147	36	1084	30
117.70	1148	12	1211	12
117.75	1148	36	1085	30
117.80	1149	12	1212	12
117.85	1149	36	1086	30
117.90	1150	12	1213	12
117.95	1150	36	1087	30